Basic Ship Theory

76123

76123

Basic Ship Theory

K.J. Rawson
MSc, DEng, FEng, RCNC, FRINA, WhSch

E.C. Tupper
BSc, CEng, RCNC, FRINA, WhSch

Fifth edition
Volume 2
Chapters 10 to 16

Ship Dynamics and Design

BUTTERWORTH
HEINEMANN

OXFORD   AUCKLAND   BOSTON   JOHANNESBURG   MELBOURNE   NEW DELHI

Butterworth-Heinemann
Linacre House, Jordan Hill, Oxford OX2 8DP
225 Wildwood Avenue, Woburn, MA 01801-2041
A division of Reed Educational and Professional Publishing Ltd

A member of the Reed Elsevier plc group

First published by Longman Group Limited 1968
Second edition 1976 (in two volumes)
Third edition 1983
Fourth edition 1994
Fifth edition 2001

British Library Cataloguing in Publication Data
Rawson, K. J. (Kenneth John), 1926–
  Basic ship theory. – 5th ed.
  Vol. 2, ch. 10–16: Ship dynamics and design K. J. Rawson,
  E. C. Tupper
  1. Naval architecture   2. Shipbuilding
  I. Title   II. Tupper, E. C. (Eric Charles), 1928–
  623.8'1

Library of Congress Cataloguing in Publication Data
A catalogue copy of this book is available from the Library of Congress

ISBN 0 7506 5397 3

For information on all Butterworth-Heinemann
publications visit our website at www.bh.com

Typeset in India by Integra Software Services Pvt Ltd,
Pondicherry, India 605005; www.integra-india.com
Printed and bound in Great Britain by Biddles Ltd, *www.biddles.co.uk*

Contents

Foreword to the fifth edition

Over the last quarter of the last century there were many changes in the maritime scene. Ships may be now much larger; their speeds are generally higher; the crews have become drastically reduced; there are many different types (including hovercraft, multi-hull designs and so on); much quicker and more accurate assessments of stability, strength, manoeuvring, motions and powering are possible using complex computer programs; on-board computer systems help the operators; ferries carry many more vehicles and passengers; and so the list goes on. However, the fundamental concepts of naval architecture, which the authors set out when *Basic Ship Theory* was first published, remain as valid as ever.

As with many other branches of engineering, quite rapid advances have been made in ship design, production and operation. Many advances relate to the effectiveness (in terms of money, manpower and time) with which older procedures or methods can be accomplished. This is largely due to the greater efficiency and lower cost of modern computers and the proliferation of information available. Other advances are related to our fundamental understanding of naval architecture and the environment in which ships operate. These tend to be associated with the more advanced aspects of the subject; more complex programs for analysing structures, for example, which are not appropriate to a basic text book.

The naval architect is affected not only by changes in technology but also by changes in society itself. Fashions change as do the concerns of the public, often stimulated by the press. Some tragic losses in the last few years of the twentieth century brought increased public concern for the safety of ships and those sailing in them, both passengers and crew. It must be recognized, of course, that increased safety usually means more cost so that a conflict between money and safety is to be expected. In spite of steps taken as a result of these experiences, there are, sadly, still many losses of ships, some quite large and some involving significant loss of life. It remains important, therefore, to strive to improve still further the safety of ships and protection of the environment. Steady, if somewhat slow, progress is being made by the national and international bodies concerned. Public concern for the environment impacts upon ship design and operation. Thus, tankers must be designed to reduce the risk of oil spillage and more dangerous cargoes must receive special attention to protect the public and nature. Respect for the environment including discharges into the sea is an important aspect of defining risk through accident or irresponsible usage.

A lot of information is now available on the Internet, including results of much research. Taking the Royal Institution of Naval Architects as an example of a learned society, its website makes available summaries of technical papers and enables members to join in the discussions of its technical groups. Other data is available in a compact form on CD-rom. Clearly anything that improves the amount and/or quality of information available to the naval architect is to be welcomed. However, it is considered that, for the present at any rate, there remains a need for basic text books. The two are complementary. A basic under-standing of the subject is needed before information from the Internet can be used intelligently. In this edition, we have maintained the objective of convey-ing principles and understanding to help student and practitioner in their work.

The authors have again been in a slight dilemma in deciding just how far to go in the subjects of each chapter. It is tempting to load the books with theories which have become more and more advanced. What has been done is to provide a glimpse into developments and advanced work with which students and practitioners must become familiar. Towards the end of each chapter, a section giving an outline of how matters are developing has been included which will help to lead students, with the aid of the Internet, to all relevant references. Some web site addresses have also been given.

It must be appreciated that standards change continually, as do the titles of organizations. Every attempt has been made to include the latest at the time of writing but the reader should always check source documents to see whether they still apply in detail at the time they are to be used. What the reader can rely on is that the principles underlying such standards will still be relevant.

2001   K J R   E C T

Acknowledgements

The authors have deliberately refrained from quoting a large number of references. However, we wish to acknowledge the contributions of many practitioners and research workers to our understanding of naval architecture, upon whose work we have drawn. Many will be well known to any student of engineering. Those early engineers in the field who set the fundamentals of the subject, such as Bernoulli, Reynolds, the Froudes, Taylor, Timoshenko, Southwell and Simpson, are mentioned in the text because their names are synonymous with sections of naval architecture.

Others have developed our understanding, with more precise and comprehensive methods and theories as technology advanced and the ability to carry out complex computations improved. Some notable workers are not quoted as their work has been too advanced for a book of this nature.

We are indebted to a number of organizations which have allowed us to draw upon their publications, transactions, journals and conference proceedings. This has enabled us to illustrate and quantify some of the phenomena discussed. These include the learned societies, such as the Royal Institution of Naval Architects and the Society of Naval Architects and Marine Engineers; research establishments, such as the Defence Evaluation and Research Agency, the Taylor Model Basin, British Maritime Technology and MARIN; the classification societies; and Government departments such as the Ministry of Defence and the Department of the Environment, Transport and the Regions; publications such as those of the International Maritime Organisation and the International Towing Tank Conferences.

Introduction

Volume 1 of *Basic Ship Theory* has presented fundamental work on ship shape, static behaviour, hazards and protection and upon ship strength. It has also described in detail the environment in which marine vehicles have to work and the properties of the sea and the air. Now we are in a position to discuss the dynamic behaviour of ships and other vehicles in the complex environment in which they operate and how those surroundings can be controlled to the maximum comfort of vehicle and crew. We can also enter upon the creative activity of ship design.

Familiarity with Volume 1 has been assumed throughout but for convenience, certain conversion factors, preferred values and symbols and nomenclature are repeated here.

Special names have been adopted for some of the derived SI units and these are listed below together with their unit symbols:

Physical quantity	SI unit	Unit symbol
Force	newton	$N = kg\,m/s^2$
Work, energy	joule	$J = N\,m$
Power	watt	$W = J/s$
Electric charge	coulomb	$C = A\,s$
Electric potential	volt	$V = W/A$
Electric capacitance	farad	$F = A\,s/V$
Electric resistance	ohm	$\Omega = V/A$
Frequency	hertz	$Hz = s^{-1}$
Illuminance	lux	$lx = lm/m^2$
Self inductance	henry	$H = V\,s/A$
Luminous flux	lumen	$lm = cd\,sr$
Pressure, stress	pascal	$Pa = N/m^2$
	megapascal	$MPa = N/mm^2$
Electrical conductance	siemens	$S = 1/\Omega$
Magnetic flux	weber	$Wb = V\,s$
Magnetic flux density	tesla	$T = Wb/m^2$

In the following two tables are listed other derived units and the equivalent values of some UK units respectively:

Physical quantity	SI unit	Unit symbol
Area	square metre	m^2
Volume	cubic metre	m^3
Density	kilogramme per cubic metre	kg/m^3
Velocity	metre per second	m/s
Angular velocity	radian per second	rad/s

Acceleration	metre per second squared	m/s^2
Angular acceleration	radian per second squared	rad/s^2
Pressure, Stress	newton per square metre	N/m^2
Surface tension	newton per metre	N/m
Dynamic viscosity	newton second per metre squared	$N\,s/m^2$
Kinematic viscosity	metre squared per second	m^2/s
Thermal conductivity	watt per metre degree kelvin	$W/(m\,°K)$

Quantity	UK unit	Equivalent SI units
Length	1 yd	0.9144 m
	1 ft	0.3048 m
	1 in	0.0254 m
	1 mile	1609.344 m
	1 nautical mile (UK)	1853.18 m
	1 nautical mile (International)	1852 m
Area	$1\,in^2$	$645.16 \times 10^{-6}\,m^2$
	$1\,ft^2$	$0.092903\,m^2$
	$1\,yd^2$	$0.836127\,m^2$
	$1\,mile^2$	$2.58999 \times 10^6\,m^2$
Volume	$1\,in^3$	$16.3871 \times 10^{-6}\,m^3$
	$1\,ft^3$	$0.0283168\,m^3$
	1 UK gal	$0.004546092\,m^3 = 4.546092$ litres
Velocity	1 ft/s	0.3048 m/s
	1 mile/hr	0.44704 m/s; 1.60934 km/hr
	1 knot (UK)	0.51477 m/s; 1.85318 km/hr
	1 knot (International)	0.51444 m/s; 1.852 km/hr
Standard acceleration, g	$32.174\,ft/s^2$	$9.80665\,m/s^2$
Mass	1 lb	0.45359237 kg
	1 ton	1016.05 kg = 1.01605 tonnes
Mass density	$1\,lb/in^3$	$27.6799 \times 10^3\,kg/m^3$
	$1\,lb/ft^3$	$16.0185\,kg/m^3$
Force	1 pdl	0.138255 N
	1 lbf	4.44822 N
Pressure	$1\,lbf/in^2$	$6894.76\,N/m^2$; 0.0689476 bars
Stress	$1\,tonf/in^2$	$15.4443 \times 10^6\,N/m^2$
		$15.4443\,MPa$ or N/mm^2
Energy	1 ft pdl	0.0421401 J
	1 ft lbf	1.35582 J
	1 cal	4.1868 J
	1 Btu	1055.06 J
Power	1 hp	745.700 W
Temperature	1 Rankine unit	5/9 Kelvin unit
	1 Fahrenheit unit	5/9 Celsius unit

Prefixes to denote multiples and sub-multiples to be affixed to the names of units are:

Factor by which the unit is multiplied	Prefix	Symbol
$1\,000\,000\,000\,000 = 10^{12}$	tera	T
$1\,000\,000\,000 = 10^{9}$	giga	G
$1\,000\,000 = 10^{6}$	mega	M
$1\,000 = 10^{3}$	kilo	k
$100 = 10^{2}$	hecto	h
$10 = 10^{1}$	deca	da
$0.1 = 10^{-1}$	deci	d
$0.01 = 10^{-2}$	centi	c
$0.001 = 10^{-3}$	milli	m
$0.000\,001 = 10^{-6}$	micro	μ
$0.000\,000\,001 = 10^{-9}$	nano	n
$0.000\,000\,000\,001 = 10^{-12}$	pico	p
$0.000\,000\,000\,000\,001 = 10^{-15}$	femto	f
$0.000\,000\,000\,000\,000\,001 = 10^{-18}$	atto	a

We list, finally, some proposed metric values (values proposed for density of fresh and salt water are based on a temperature of 15 °C (59 °F).)

Item	Accepted Imperial figure	Direct metric equivalent	Preferred SI value
Gravity, g	$32.17\,\text{ft/s}^2$	$9.80665\,\text{m/s}^2$	$9.807\,\text{m/s}^2$
Mass density salt water	$64\,\text{lb/ft}^3$ $35\,\text{ft}^3/\text{ton}$	$1.0252\,\text{tonne/m}^3$ $0.9754\,\text{m}^3/\text{tonne}$	$1.025\,\text{tonne/m}^3$ $0.975\,\text{m}^3/\text{tonne}$
Mass density fresh water	$62.2\,\text{lb/ft}^3$ $36\,\text{ft}^3/\text{ton}$	$0.9964\,\text{tonne/m}^3$ $1.0033\,\text{m}^3/\text{tonne}$	$1.0\,\text{tonne/m}^3$ $1.0\,\text{m}^3/\text{tonne}$
Young's modulus, E (Steel)	$13,500\,\text{tonf/in}^2$	$2.0855 \times 10^7\,\text{N/cm}^2$	$209\,\text{GN/m}^2$ or GPa
Atmospheric pressure	$14.7\,\text{lbf/in}^2$	$101,353\,\text{N/m}^2$ $10.1353\,\text{N/cm}^2$	$10^5\,\text{N/m}^2$ or Pa or 1.0 bar
TPI (salt water)	$\dfrac{A_\text{w}}{420}\,\text{tonf/in}$ $A_\text{w}\,(\text{ft}^2)$	$1.025\,A_\text{w}\,\text{tonnef/m}$ $A_\text{w}\,(\text{m}^2)$	$1.025\,A_\text{w}\,\text{tonnef/m}$
NPC	$A_\text{w}\,(\text{m}^2)$	$100.52\,A_\text{w}\,(\text{N/cm})$	
NPM		$10,052\,A_\text{w}\,(\text{N/m})$	$10^4\,A_\text{w}\,(\text{N/m})$
MCT $1''$ (salt water) (Units of tonf and feet)	$\dfrac{\Delta \overline{\text{GM}}_\text{L}}{12L}\,\dfrac{\text{tonf ft}}{\text{in}}$		
One metre trim moment (Δ in MN or $\dfrac{\text{tonnef/m}}{\text{m}}$, Δ in tonnef)		$\dfrac{\Delta \overline{\text{GM}}_\text{L}}{L} \left(\dfrac{\text{MN m}}{\text{m}}\right)$	$\dfrac{\Delta \overline{\text{GM}}_\text{L}}{L} \left(\dfrac{\text{MN m}}{\text{m}}\right)$
Force displacement Δ	1 tonf	$1.01605\,\text{tonnef}$ $9964.02\,\text{N}$	$1.016\,\text{tonnef}$ $9964\,\text{N}$
Mass displacement Σ	1 ton	$1.01605\,\text{tonne}$	$1.016\,\text{tonne}$
Weight density: Salt water Fresh water			$0.01\,\text{MN/m}^3$ $0.0098\,\text{MN/m}^3$
Specific volume: Salt water Fresh water			$99.5\,\text{m}^3/\text{MN}$ $102.0\,\text{m}^3/\text{MN}$

Of particular significance to the naval architect are the units used for displacement, density and stress. The force displacement Δ, under the SI scheme must be expressed in terms of newtons. In practice the meganewton (MN) is a more convenient unit and 1 MN is approximately equivalent to 100 tonf (100.44 more exactly). The authors have additionally introduced the tonnef (and, correspondingly, the tonne for mass measurement) as explained more fully in Chapter 3.

REFERENCES AND THE INTERNET

References for each chapter are given in a Bibliography at the end of each volume with a list of works for general reading. Because a lot of useful information is to be found these days on the Internet, some relevant web sites are quoted at the end of the Bibliography.

Symbols and nomenclature

GENERAL

a	linear acceleration
A	area in general
B	breadth in general
D, d	diameter in general
E	energy in general
F	force in general
g	acceleration due to gravity
h	depth or pressure head in general
h_w, ζ_w	height of wave, crest to trough
H	total head, Bernoulli
L	length in general
L_w, λ	wave-length
m	mass
n	rate of revolution
p	pressure intensity
p_v	vapour pressure of water
p_∞	ambient pressure at infinity
P	power in general
q	stagnation pressure
Q	rate of flow
r, R	radius in general
s	length along path
t	time in general
t°	temperature in general
T	period of time for a complete cycle
u	reciprocal weight density, specific volume
u, v, w	velocity components in direction of x-, y-, z-axes
U, V	linear velocity
w	weight density
W	weight in general
x, y, z	body axes and Cartesian co-ordinates
	Right-hand system fixed in the body, z-axis vertically down, x-axis forward. Origin at c.g.
x_0, y_0, z_0	fixed axes
	Right-hand orthogonal system nominally fixed in space, z_0-axis vertically down, x_0-axis in the general direction of the initial motion.
α	angular acceleration
γ	specific gravity
Γ	circulation
δ	thickness of boundary layer in general
θ	angle of pitch
μ	coefficient of dynamic viscosity
ν	coefficient of kinematic viscosity
ρ	mass density
ϕ	angle of roll, heel or list
χ	angle of yaw
ω	angular velocity or circular frequency
∇	volume in general

GEOMETRY OF SHIP

A_M	midship section area
A_W	waterplane area
A_x	maximum transverse section area
B	beam or moulded breadth
$\overline{BM}$	metacentre above centre of buoyancy
C_B	block coefficient
C_M	midship section coefficient
C_P	longitudinal prismatic coefficient
C_{VP}	vertical prismatic coefficient
C_{WP}	coefficient of fineness of waterplane
D	depth of ship
F	freeboard
$\overline{GM}$	transverse metacentric height
$\overline{GM}_L$	longitudinal metacentric height
I_L	longitudinal moment of inertia of waterplane about CF
I_P	polar moment of inertia
I_T	transverse moment of inertia
L	length of ship—generally between perps
L_{OA}	length overall
L_{PP}	length between perps
L_{WL}	length of waterline in general
S	wetted surface
T	draught
Δ	displacement force
λ	scale ratio—ship/model dimension
∇	displacement volume
Σ	displacement mass

PROPELLER GEOMETRY

A_D	developed blade area
A_E	expanded area
A_O	disc area
A_P	projected blade area
b	span of aerofoil or hydrofoil
c	chord length
d	boss or hub diameter
D	diameter of propeller
f_M	camber
P	propeller pitch in general
R	propeller radius
t	thickness of aerofoil
Z	number of blades of propeller
α	angle of attack
ϕ	pitch angle of screw propeller

RESISTANCE AND PROPULSION

a	resistance augment fraction
C_D	drag coeff.
C_L	lift coeff.
C_T	specific total resistance coeff.
C_W	specific wave-making resistance coeff.
D	drag force
F_n	Froude number
I	idle resistance
J	advance number of propeller
K_Q	torque coeff.
K_T	thrust coeff.
L	lift force

P_D	delivered power at propeller
P_E	effective power
P_I	indicated power
P_S	shaft power
P_T	thrust power
Q	torque
R	resistance in general
R_n	Reynolds' number
R_F	frictional resistance
R_R	residuary resistance
R_T	total resistance
R_W	wave-making resistance
s_A	apparent slip ratio
t	thrust deduction fraction
T	thrust
U	velocity of a fluid
U_∞	velocity of an undisturbed flow
V	speed of ship
V_A	speed of advance of propeller
w	Taylor wake fraction in general
w_F	Froude wake fraction
W_n	Weber number
β	appendage scale effect factor
β	advance angle of a propeller blade section
δ	Taylor's advance coeff.
η	efficiency in general
η_B	propeller efficiency behind ship
η_D	quasi propulsive coefficient
η_H	hull eff.
η_O	propeller eff. in open water
η_R	relative rotative efficiency
σ	cavitation number

SEAKEEPING

c	wave velocity
f	frequency
f_E	frequency of encounter
I_{xx}, I_{yy}, I_{zz}	real moments of inertia
I_{xy}, I_{xz}, I_{yz}	real products of inertia
k	radius of gyration
m_n	spectrum moment where n is an integer
M_L	horizontal wave bending moment
M_T	torsional wave bending moment
M_v	vertical wave bending moment
s	relative vertical motion of bow with respect to wave surface
$S_\zeta(\omega), S_\theta(\omega)$, etc.	one-dimensional spectral density
$S_\zeta(\omega, \mu), S_\theta(\omega, \mu)$, etc.	two-dimensional spectral density
T	wave period
T_E	period of encounter
T_z	natural period in smooth water for heaving
T_θ	natural period in smooth water for pitching
T_ϕ	natural period in smooth water for rolling
$Y_{\theta\zeta}(\omega)$	response amplitude operator—pitch
$Y_{\phi\zeta}(\omega)$	response amplitude operator—roll
$Y_{\chi\zeta}(\omega)$	response amplitude operator—yaw
β	leeway or drift angle
δ_R	rudder angle
ε	phase angle between any two harmonic motions
ζ	instantaneous wave elevation
ζ_A	wave amplitude

ζ_w	wave height, crest to trough
θ	pitch angle
θ_A	pitch amplitude
κ	wave number
ω_E	frequency of encounter
Λ	tuning factor

MANOEUVRABILITY

A_C	area under cut-up
A_R	area of rudder
b	span of hydrofoil
c	chord of hydrofoil
K, M, N	moment components on body relative to body axes
O	origin of body axes
p, q, r	components of angular velocity relative to body axes
X, Y, Z	force components on body
α	angle of attack
β	drift angle
δ_R	rudder angle
χ	heading angle
ω_C	steady rate of turn

STRENGTH

a	length of plate
b	breadth of plate
C	modulus of rigidity
ε	linear strain
E	modulus of elasticity, Young's modulus
σ	direct stress
σ_y	yield stress
g	acceleration due to gravity
I	planar second moment of area
J	polar second moment of area
j	stress concentration factor
k	radius of gyration
K	bulk modulus
l	length of member
L	length
M	bending moment
M_p	plastic moment
M_{AB}	bending moment at A in member AB
m	mass
P	direct load, externally applied
P_E	Euler collapse load
p	distributed direct load (area distribution), pressure
p'	distributed direct load (line distribution)
τ	shear stress
r	radius
S	internal shear force
s	distance along a curve
T	applied torque
t	thickness, time
U	strain energy
W	weight, external load
y	lever in bending
δ	deflection, permanent set, elemental (when associated with element of breadth, e.g. δb)
ρ	mass density
ν	Poisson's ratio
θ	slope

NOTES

(*a*) A distance between two points is represented by a bar over the letters defining the two points, e.g. $\overline{\text{GM}}$ is the distance between G and M.

(*b*) When a quantity is to be expressed in non-dimensional form it is denoted by the use of the prime $'$. Unless otherwise specified, the non-dimensionalizing factor is a function of ρ, L and V, e.g. $m' = m/\frac{1}{2}\rho L^3$, $x' = x/\frac{1}{2}\rho L^2 V^2$, $L' = L/\frac{1}{2}\rho L^3 V^2$.

(*c*) A lower case subscript is used to denote the denominator of a partial derivative, e.g. $Y_u = \partial Y/\partial u$.

(*d*) For derivatives with respect to time the dot notation is used, e.g. $\dot{x} = dx/dt$.

10 Powering of ships: general principles

The power required to drive a ship through the water depends upon the resistance offered by the water and air, the efficiency of the propulsion device adopted and the interactions among them. Because there is interaction it is vital to consider the design of the hull and the propulsion device as an integrated system. When the water surface is rough, the problem is complicated by increased resistance and by the propulsion device working in less favourable conditions. Powering in waves is considered in Chapter 12. This chapter is devoted to the powering of ships in calm water and concentrates on displacement monohulls. In multihull displacement forms there will be effects on both viscous and wavemaking resistance due to interference between the separate hulls. In planing, surface effect or hydrofoil craft special considerations apply. These can only be touched upon briefly in this book (Chapter 16).

For the merchant ship, the speed required is dictated by the conditions of service. It may have to work on a fixed schedule, e.g. the cruise ship, or as one of a fleet of ships maintaining a steady supply of material. Therefore a designer must be able to predict accurately the speed a new design will attain. The fuel bill is a significant feature in the operating costs of any ship, so the designer will be anxious to keep the power needed for the operating speed to a minimum. Oil is also a dwindling natural resource.

The speed of a warship is dictated by the operational requirements. An anti-submarine frigate must be sufficiently fast to close with an enemy submarine and destroy it. At the same time, excessive speed and fuel consumption can only be met at the expense of the amount of armament the ship carries.

In all ships, the power needed should be reduced to a minimum consistent with other design requirements to minimize the weight, cost and volume of the machinery and fuel. It follows, that an accurate knowledge of a design's powering characteristics is of considerable importance and that a fair expenditure of effort is justified in achieving it. For predicting full-scale resistance, the designer can use full-scale data from ships built over a considerable period of years, theoretical analysis or models.

Generally speaking, full-scale data is limited in usefulness because of the process of evolution to which ships are subject. To mention two factors, the introduction of welding led to a smoother hull, and ships have tended over the years to become larger. Again, the new ship is often required to go faster so that data from her predecessors cannot be used directly for assessing her maximum power. Clearly, this method is not valid when a new ship form is introduced such as the SWATH (Small Waterplane Twin Hull) ship or the trimaran.

Theory has been used as an aid to more practical methods and continues to develop. Computational fluid dynamics is a very powerful tool which is increasingly used by researchers to study problems of fluid flow, including those involving cavitation but the main contribution of theory is still generally to guide the model experimenter, providing a more rational and scientific background to his work, suggesting profitable lines of investigation and indicating the relative importance of various design parameters.

Where a methodical series of tests has been carried out on a form embracing the new design, the details should be obtained from the literature. Even without a methodical series, systematic plotting of previous data can provide a first estimate of power needs.

The main tool of the designer has been, and remains, the model with theory acting as a guide and full-scale data providing the all-essential check on the model prediction. The model is relatively cheap and results can be obtained fairly rapidly for a variety of changes to enable the designer to achieve an optimum design.

An example of the results obtainable by a judicious blend of theory and model data is provided by what is known as regression analysis. Basically, a mathematical expression is produced for the resistance of the ship, in terms of various ship parameters such as L/B ratio, C_P, etc. This expression is then used to deduce the required trend on these parameters to minimize resistance and produces a form superior to those currently in use.

These various considerations are developed more fully later but first it is necessary to consider some of the properties of the fluids in which the ship moves. These are fundamental to the prediction of full-scale performance from the model and for any theoretical investigation.

Fluid dynamics

There are two fluids with which the naval architect is concerned, air and water. Unless stated otherwise water is the fluid considered in the following sections. Air resistance is treated as a separate drag force. Models are used extensively and it is necessary to ensure that the flow around the model is 'similar' to that around the ship in order that results may be scaled correctly. Similarity in this sense requires that the model and ship forms be geometrically similar (at least that portion over which the flow occurs), that the streamlines of the fluid flow be geometrically similar in the two cases and that the fluid velocities at corresponding points around the bodies are in a constant ratio.

Water possesses certain physical properties which are of the same order of magnitude for the water in which a model is tested and for that in which the ship moves. These are:

the density, ρ
the surface tension, σ
the viscosity, μ
the vapour pressure, p_v

the ambient pressure, p_∞
the velocity of sound in water, a

The quantitative values of some of these properties are discussed in Chapter 9. Other factors involved are:

a typical length, usually taken as the wetted length L for resistance work, and as the propeller diameter D for propeller design;

velocity,	V
propeller revolutions,	n
resistance,	R
thrust,	T
torque,	Q
gravitational acceleration,	g.

Dimensional analysis provides a guide to the form in which the above quantities may be significant. The pi theorem states that the physical relationship between these quantities can be represented as one between a set of non-dimensional products of the quantities concerned. It also asserts that the functionally related quantities are independent and that the number of related quantities will be three less (i.e. the number of fundamental units—mass, length, time) than the number of basic quantities.

Applying non-dimensional analysis to the ship powering problem, it can be shown that:

$$\frac{R}{\rho V^2 L^2} = F\left\{ \frac{VL\rho}{\mu}, \frac{V}{\sqrt{(gL)}}, \frac{V}{a}, \frac{\sigma}{g\rho L^2}, \frac{p_\infty - p_v}{\rho V^2} \right\}$$

$$\frac{T}{\rho n^2 D^4} \quad \text{and} \quad \frac{Q}{\rho n^2 D^5} = F\left\{ \frac{V}{nD}, \frac{VD\rho}{\mu}, \frac{V^2}{gD}, \frac{\sigma}{\rho g L^2}, \frac{p_\infty - p_v}{\rho V^2} \right\}$$

Expressed in another way, it is physically reasonable to suggest that if data can be expressed in terms of parameters that are independent of scale, i.e. non-dimensional parameters, the same values of these data will probably be obtained from experiments at different scales if the parameters are constant. Where the governing parameters cannot be kept constant, data will change in going from the model to full scale. The above are not the only non-dimensional parameters that can be formed but they are those in general use. Each has been given a name as follows:

$\dfrac{R}{\rho V^2 L^2}$ is termed the resistance coefficient

$\dfrac{VD\rho}{\mu}$ or $\dfrac{VL\rho}{\mu}$ is termed the Reynolds' number (the ratio μ/ρ is called the kinematic viscosity and is represented by v)

$\dfrac{V}{\sqrt{(gD)}}$ or $\dfrac{V}{\sqrt{(gL)}}$ is termed the Froude number

$\dfrac{V}{a}$ is termed the Mach number ⎫ These two quantities are not
                                          ⎪ significant in the context of
                                          ⎬ the present book and are not
$\dfrac{\sigma}{g\rho L^2}$ is termed the Weber number ⎭ considered further

$\dfrac{p_\infty - p_v}{\rho V^2}$ is termed the cavitation number

$\dfrac{T}{\rho n^2 D^4} = K_T = $ thrust coefficient

$\dfrac{Q}{\rho n^2 D^5} = K_Q = $ torque coefficient

$\dfrac{V}{nD} = J = $ advance coefficient

Unfortunately, it is not possible to set up model scale experiments in which all the above parameters have the same values as in the full-scale. This is readily seen by considering the Reynolds' and Froude numbers. Since ρ, μ and g are substantially the same for model and ship, it would be necessary for both VL and $V/\sqrt{L}$ to be kept constant. This is physically impossible. By using special liquids instead of water in which to test models, two parameters could be satisfied but not all of them.

Fortunately, certain valid results can be obtained by keeping one parameter constant in the model tests and limiting those tests to certain measurements. For example, model resistance tests are conducted at corresponding Froude number and model propeller cavitation tests at corresponding cavitation number. This means that resistance forces which depend on Reynolds' number will have to be modified in going from model to ship. It will be shown that, had this difficulty of achieving physical similarity not been present, the early experimenters would not have experienced so much difficulty in predicting the resistance of the full-scale ship.

Components of resistance and propulsion

It is necessary to provide a propulsive device to drive the ship through the water. It has been explained, that since the propulsion device interacts on the resistance of the ship the two cannot be treated in isolation. However, as a matter of convenience, the overall problem is considered as the amalgamation of a number of smaller problems. The actual divisions are largely arbitrary but are well established. In the following, it is assumed that the propulsive device is a propeller.

If the naked hull of the ship could be driven through the water by some device which in no way interacted with the hull or water, it would experience a total resistance R_T which would be the summation of several types of resistance as is explained later. The differentiation between types of resistance is necessary because they scale differently in going from model to full-scale. The product of

R_T and the ship's speed V defines a horsepower which is known as the *effective power* (P_E). This e.h.p. can be regarded as the useful work done in propelling the ship.

The power actually delivered to the shafts for propelling a ship is the *shaft power* (P_S). The ratio between the shaft and effective powers is a measure of the overall propulsive efficiency achieved and is termed the *propulsive coefficient* (PC). It should be noted that some authorities take the PC as the ratio of P_{EA} to P_S, P_{EA} defined as below. The propulsive coefficient arises partly from the efficiency of the propeller, and partly from the interaction of propeller and hull. In addition, it has to be modified to make model and full-scale data compatible.

Following on from the above, four basic components of the powering problem suggest themselves:

(*a*) P_E or the hull resistance
(*b*) the propeller
(*c*) hull/propeller interaction
(*d*) ship/model correlation.

EFFECTIVE POWER

One watt is the rate of performing 1 joule of work per second. As far as propelling the ship through the water is concerned, the 'useful' or 'effective' work is that done in overcoming the resistance of the ship by its speed of advance. The resistance concerned is conventionally taken to be that of the 'naked' hull, i.e. without any appendages. This leads to the following definition:

The *effective power* of a ship is the product of the resistance of the naked hull and the speed of the hull. Therefore,

$$P_E = R_T \times V$$

A corresponding definition can be evolved using the resistance of the hull including that of appendages and this is conventionally denoted by P_{EA}.

The ratio of P_{EA} to P_E is known as the *Appendage coefficient*, i.e.

$$\text{Appendage coefficient} = P_{EA}/P_E$$

EXAMPLE 1. At 251 m/min the tow rope pull of a naked hull is 35.6 kN. Find the effective power of the hull at this speed.

Solution:

$$P_E = 251\,\text{m/min} \times 35.6\,\text{kN} \times \frac{1\,\text{min}}{60\,\text{s}} = 149\,\text{kW}$$

EXAMPLE 2. A 6000 tonnef destroyer develops a total power of 44.74 MW at 30 knots. Assuming that the effective power is 50 per cent of this total power, calculate the resistance of its naked hull.

Solution:

$$P_E = \frac{1}{2} \times 44.74 = 22.37\,\text{MW}$$

Therefore $22.37 \times 10^6\,\text{W} = (\text{Resistance})\ \text{newtons} \times \left(\dfrac{30 \times 1852}{3600}\right)\ \text{m/s}$

i.e.

Resistance $= 1.449 \times 10^6\ \text{newtons}$

TYPES OF RESISTANCE

The classical theory of hydrodynamics has shown that a body deeply immersed in fluid of zero viscosity experiences no resistance. No matter how the streamlines may be deflected as they pass the body, they return to their undisturbed state a long way downstream of the body (see Fig. 10.1) and the resultant force on the body is zero. There are pressure variations in the fluid as the streamlines are deflected and particle velocities change. In this respect, Bernoulli's theorem is obeyed, i.e. increased velocities are associated with pressure reductions. Thus, the body can be acted upon by forces of considerable magnitude but they all act so as to cancel each other out.

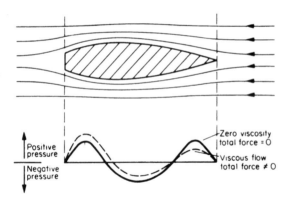

Fig. 10.1

In a practical case the fluid is viscous and a deeply immersed body would suffer a frictional drag. In addition, when the body approaches a free surface, the pressure variations around the body can manifest themselves as elevations or depressions of the water surface. That is to say, waves are formed on the surface. This process upsets the balance of pressures acting on the body which results in a drag force. The magnitude of the drag force is related to the energy of the wave system created.

The total resistance of a ship moving on a calm water surface has several components. They are: wave-making resistance; skin frictional resistance; viscous pressure resistance; air resistance; appendage resistance.

Each component can now be studied separately provided it is remembered that each will have some interaction with the others.

WAVE-MAKING RESISTANCE

It is common experience, that a body moving across an otherwise undisturbed water surface produces a wave system. This system arises from the pressure field around the body and the energy possessed by it must be derived from the body. As far as the body is concerned the transfer of energy will manifest itself as a force opposing the forward motion. This force is termed the *wave-making resistance*.

A submerged body also experiences a drag due to the formation of waves on the free surface, the magnitude of this drag reducing with increasing depth of submergence until it becomes negligible at deep submergence. This typically occurs at depths equal to approximately half the length of the body. An exception to this general rule can occur with submarines at sea if they are moving close to the interface between two layers of water of different density. In this case, a wave system is produced at the interface resulting in a drag on the submarine.

A gravity wave, length λ, in deep water moves with a velocity C defined by

$$C^2 = \frac{g\lambda}{2\pi}$$

Because the wave pattern moves with the ship, C must be equal to the ship velocity V and λ being a length measurement can, for dimensional analysis, be represented as proportional to the ship length L for a given speed.

Thus it is seen that of the non-dimensional parameters deduced earlier it is V^2/gL or $V/\sqrt{(gL)}$ which is significant in the study of wave-making resistance. As stated in the section on fluid dynamics, the quantity $V/\sqrt{(gL)}$ is usually designated the *Froude number*. In many cases, the simpler parameter $V/\sqrt{L}$ is used for plotting results but the plot is no longer non-dimensional.

Hydrodynamically, the ship can be regarded as a moving pressure field. Kelvin considered mathematically the simplified case of a moving pressure point and showed that the resulting wave pattern is built up of two systems. One system is a divergent wave system and the other a system of waves with crests more or less normal to the path of the pressure point. Both systems travel forward with the speed of the pressure point (Fig. 10.2).

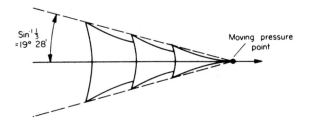

Fig. 10.2 Wave system associated with moving pressure point

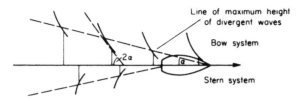

Fig. 10.3 Ship wave pattern

The wave system associated with a ship is more complicated. To a first approximation, however, the ship can be considered as composed of a moving pressure field sited near the bow and a moving suction field near the stern. The bow produces a wave pattern similar to that produced by Kelvin's pressure point with a crest at the bow. The stern on the other hand produces a wave system with a trough at the stern.

If the line of maximum height of crests of the divergent system is at α, then the wave crests at these positions subtend an angle of approximately 2α to the ship middle line as in Fig. 10.3.

The two transverse wave systems, i.e. at bow and stern, have a wave-length of $2\pi V^2/g$. The transverse waves increase in width as the divergent waves spread out. The total energy content per wave is constant, so that their height falls progressively with increasing distance from the ship.

In general, both divergent systems will be detectable although the stern system is usually much weaker than that from the bow. Normally, the stern transverse system cannot be detected as only the resultant of the two systems is visible astern of the ship.

In some ships, the wave pattern may be made even more complex by the generation of other wave systems by local discontinuities in the ship's form.

Since at most speeds both the bow and stern systems are present aft of the ship, there is an interaction between the two transverse wave systems. If the systems are so phased that the crests are coincident, the resulting system will have increased wave height, and consequently greater energy content. If the crest of one system coincides with the trough of the other the resulting wave height and energy content will be less. The wave-making resistance, depending as it does on the energy content of the overall wave system, varies therefore with speed and also effective length between the bow and stern pressure systems. Again, the parameters V and L are important.

Froude studied the effect on resistance of the length of the ship by towing models with the same endings but with varying lengths of parallel middle body. The results are in line with what could be expected from the above general reasoning.

The distance between bow and stern pressure systems is typically $0.9L$. The condition that crests or troughs of the bow system should coincide with the first trough of the stern system is therefore

$$\frac{V^2}{0.9L} = \frac{g}{N\pi}$$

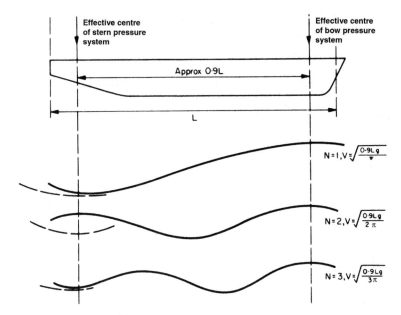

Fig. 10.4 *Interaction of bow and stern wave systems*

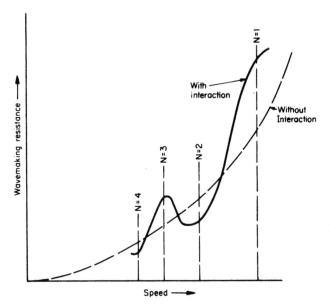

Fig. 10.5 *'Humps' and 'hollows' in wave-making resistance curves*

For $N = 1, 3, 5, 7$, etc., the troughs will coincide and for $N = 2, 4, 6$, etc., the crests from the bow system coincides with the trough from the after system as in Fig. 10.4.

If there were no interaction between the bow and stern wave systems, the resistance would increase steadily with speed as shown in Fig. 10.5 ('Without interaction' curve). Because interaction occurs at speeds discussed above, the actual resistance curve will oscillate about the curve as indicated.

A 'hump' occurs when N is an odd integer and a 'hollow' when N is an even integer. It is to be expected that the most pronounced hump will be at $N = 1$, because the speed is highest for this condition and this hump is usually referred to as the *main hump*. The hump associated with $N = 3$ is often called the *prismatic hump* as its influence is greatly affected by the prismatic coefficient of the form considered.

Since the Froude number $F_n = V/\sqrt{(gL)}$, the values of F_n corresponding to the humps and hollows are shown in Table 10.1.

Table 10.1

N	F_n
1	$\sqrt{\left(\dfrac{0.9}{\pi}\right)} = 0.54$
2	$\sqrt{\left(\dfrac{0.9}{2\pi}\right)} = 0.38$
3	$\sqrt{\left(\dfrac{0.9}{3\pi}\right)} = 0.31$
4	$\sqrt{\left(\dfrac{0.9}{4\pi}\right)} = 0.27$

Clearly, a designer would not deliberately produce a ship whose normal service speed was at a 'hump' position. Rather, the aim would be to operate in a 'hollow', although other considerations may be overriding in deciding on the length of the ship.

FRICTIONAL RESISTANCE

The water through which a ship moves has viscosity which is a property of all practical fluids. It was shown earlier, that when viscosity is involved the conditions for dynamic similarity are geometrically similar boundaries and constancy of Reynolds' number.

When a body moves through a fluid which is otherwise at rest, a thin layer of fluid adheres to the surface of the body and has no velocity relative to the body. At some distance from the body the fluid remains at rest.

The variation of velocity of the fluid is rapid close to the body (Fig. 10.6) but reduces with increasing distance from the body. The region in which there is a rapid change in velocity is termed the *boundary layer*.

The definition of boundary layer thickness is to some extent arbitrary since in theory it extends to infinity. It is common practice to define the thickness as the distance from the surface of the body at which the velocity of the fluid is 1 per cent of the body velocity.

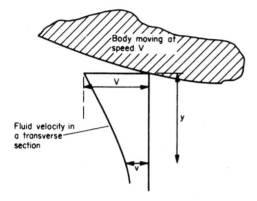

Fig. 10.6 The boundary layer

Due to the velocity gradient across the boundary layer, the fluid is in shear and the body experiences a resistance which is termed the *frictional resistance*. If the fluid velocity is v at distance y from the body the shear stress in the fluid is given by

$$\tau = \mu \frac{dv}{dy}$$

This applies to the case of *laminar flow* in which each fluid particle follows its own streamline path with no mass transfer between adjacent fluid layers. The shear in this case is due solely to molecular action. Laminar flow conditions are only likely to apply at relatively low Reynolds' numbers. At higher Reynolds' numbers the steady flow pattern breaks down and is replaced by a more confused pattern which is termed *turbulent flow*. The value of R_n at which this breakdown in flow occurs is termed the critical Reynolds' number, and its actual value depends upon the smoothness of the surface and the initial turbulence present in the fluid. For a smooth flat plate, breakdown occurs at a Reynolds' number between 3×10^5 and 10^6. In turbulent flow, the concept of a boundary layer still applies but in this case, besides the molecular friction force, there is an interaction due to the momentum transfer of fluid masses between adjacent layers. The exact mechanism of the turbulent boundary layer is incompletely understood, but it follows that the velocity distribution curve at Fig. 10.6 can represent only a mean velocity curve.

The *transition* from laminar to turbulent flow is essentially one of stability. At low Reynolds' numbers, disturbances die out and the flow is stable. At a certain *critical* value of Reynolds' number, the laminar flow becomes unstable and the slightest disturbance will cause turbulence. The critical R_n for a flat surface is a function of l the distance from the leading edge. Ahead of a point defined by l as follows:

$$(R_n) \text{ critical} = \frac{Vl}{v}$$

the flow is laminar. At distance *l* transition begins and after a certain *transition region*, turbulence is fully established.

For a flat surface, the critical Reynolds' number is approximately 10^6. For a curved surface, the pressure gradient along the surface has a marked influence on transition. Transition is delayed in regions of decreasing pressure, i.e. regions of increasing velocity. Use is made of this fact in certain aerodynamic low drag forms such as the 'laminar flow' wing. The gain arising from retaining laminar flow is shown by the fact that a flat plate suffers seven times the resistance in all turbulent as opposed to all laminar flow.

The thickness of the turbulent boundary layer is given approximately by

$$\frac{\delta x}{L} = 0.37(R_L)^{-\frac{1}{5}}$$

where *L* is the distance from the leading edge and R_L is the corresponding Reynolds' number. For example, at $15\,\text{m/s}$, with $L = 150\,\text{m}$, $\delta\tau$ is about $0.75\,\text{m}$.

Even in turbulent flow, the fluid particles adjacent to the body's surface are at rest relative to the body. It follows that there exists a *laminar sub-layer* although in practice this is extremely thin. It is nevertheless of importance in that a body appears smooth if surface roughness does not protrude through this sub-layer. Such a body is said to be *hydraulically smooth* and a plot of drag against Reynolds' number would be as shown by the basic curve in Fig. 10.7.

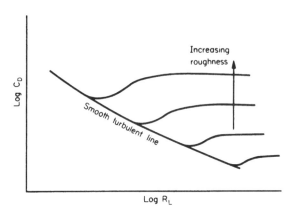

Fig. 10.7 Effect of roughness

For a rough surface, resistance follows the smooth curve as Reynolds' number is increased until a certain value and it then breaks away and eventually becomes horizontal, i.e. the drag coefficient becomes independent of R_n and drag varies as the square of the velocity. The rougher the surface the smaller the value of R_n at which the breakaway occurs.

Owing to the increase in boundary layer thickness, the ratio of roughness (i.e. effective granularity of surface) to the boundary layer thickness decreases along the length of a surface. For this reason, protrusions from a hull of a given size have less effect on resistance at the after end of a ship than they do forward.

For all practical purposes, the complete boundary layer of a ship at sea can be regarded as turbulent. In a model in a towing tank, a portion may be laminar but the extent of this is sensitive to external conditions and it can vary considerably in a given model. Because of the difference in resistance associated with the two types of boundary layer, this phenomenon has led to inconsistent model test data in the past and this has caused most ship tanks to artificially stimulate turbulent flow conditions to ensure reproducible conditions. A number of devices are used to stimulate turbulence but that now most commonly used is a row of studs a short distance from the bow of the model.

For convenience, the frictional resistance of a ship is usually divided into two components. The first component is that resistance which would be experienced by a 'flat plate' of equivalent surface area. The second component is the increased frictional resistance occasioned by the actual form of the ship and this component is known as the *frictional form resistance*.

Hull roughness is a complex subject. It depends on a wide range of features, each varying with time in a different way. The plating, as built, will have an inherent roughness, welding deposits and distortion due to fabrication. Distortions will increase in service due to water loading and damage. The paint films will gradually break down and corrosion will occur. Marine fouling can occur quite rapidly in some areas of the world. There is also the difficulty of defining the rough surface. Amplitude variations can be measured using a hull roughness gauge (that developed by the British Ship Research Association is commonly used) to give the mean apparent amplitude. But what may be termed the general 'texture' of the surface is also important.

VISCOUS PRESSURE RESISTANCE

Total ship resistance comprises the fore-and-aft component of all pressures normal to the hull. That part of the pressure resistance which manifests itself as waves has already been discussed; the remainder of the pressure resistance is due to viscous effects which inhibit that build-up of pressure around the after end of the ship predicted for a perfect fluid. Part of this resistance will be due to the generation of vortices from form discontinuities such as the turn of the bilge. Another part arises from the thickening of the boundary layer and this may be increased by flow separation. Because these last two elements are affected by the form of the ship they are together known as form drag or form resistance. Form drag is likely to be most significant in full bodied ships. Pressure energy lost to the sea is thus seen as waves and as eddies or vortices. Examination of the energy dissipated in the wake and in the waves may enable some of the resistance due to form to be calculated. That due to the transfer of energy between wave and wake is sometimes isolated for examination and is called wave breaking resistance.

AIR RESISTANCE

Air is a fluid, as is water, and as such will resist the passage of the upper portions of the ship through it. This resistance will comprise both frictional and eddy-making components.

In an artist's impression of a ship it is possible to depict a very smooth streamlined above water form. In practice, the weight penalty associated with such fairing and the difficulties of fabrication are not justified by the reduction in air resistance or by the relatively small gain in usable internal volume. In practice, therefore, air flowing over the superstructure meets a series of discontinuities which cause separation, i.e. streamlines break down and eddies are formed. As expected, air resistance like water eddy resistance will vary as V^2.

At full speed in conditions of no wind, it is probable that the air resistance will be some 2–4 per cent of the total water resistance. Should the ship be moving into a head wind of the same speed as the ship, the relative wind speed will be doubled and the air resistance quadrupled. Thus, clearly, in severe weather conditions such as in a full gale the air resistance can contribute materially to slowing down the ship.

APPENDAGE RESISTANCE

The discussion up to this point has been concerned mainly with the resistance of the naked hull, i.e. without appendages. Typical appendages are rudders, shaft brackets or bossings, stabilizers, bilge keels, docking keels. Each appendage has its own typical length, which is much smaller than the ship length, and accordingly is running at its own Reynolds' number. Each appendage, therefore, has a resistance which would scale differently to full-size if run at model size, although obeying the same scaling laws.

To include appendages in a normal resistance model would, therefore, upset the scaling of the hull resistance. It is for this reason that models are run naked, and the resulting total ship resistance must be modified by adding in estimates of the resistance due to each full-scale appendage.

The resistance of the appendages may be estimated from formulae based on previous experience or by running models both with and without appendages and scaling the difference to full-scale using different scaling laws from those used for the hull proper. Fortunately, appendage resistances are usually small (of the order of 10 per cent of that of the hull) so that errors in their assessment are not likely to be critical. It is usual to assume that the appendage resistance varies as V^2, so that the contribution to the non-dimensional resistance coefficient is constant.

In addition to the above resistances, the ship in service generally has her resistance to ahead motion increased by the presence of waves and spray generated by the wind. In rough weather, this effect can be of considerable magnitude and often causes a significant fall off in speed. This is discussed in Chapter 12.

RESIDUARY RESISTANCE

For the practical evaluation of ship resistance for normal ship forms, it is usual to group wave-making resistance, form resistance, eddy resistance and frictional form resistance into one force termed *residuary resistance*. This concept is not theoretically correct, but, in practice, provides a sufficiently accurate answer.

Thus the total resistance is given by

$$R_T = R_R + R_F$$

where R_R = residuary resistance, and R_F = frictional resistance of an equivalent flat plate.

Having examined how the resistance of a ship arises, it is necessary to examine the effects of the propulsion device and how consideration of the two cannot be separated. In returning to the evaluation of ship resistance in the next chapter, the resistance will then be considered as the summation of the frictional and residuary resistances.

THE PROPULSION DEVICE

The force needed to propel the ship must be obtained from a reaction against the air, water or land, e.g. by causing a stream of air or water to move in the opposite direction. The sailing ship uses air reaction. Devices acting on water are the paddle wheel, oar and screw propeller. Reaction on land is used by the punt pole or the horse towing a barge.

For general applications, the land reaction is not available and the naval architect must make use of water or air. The force acting on the ship arises from the rate of change of momentum induced in the fluid.

Consider a stream of fluid, density ρ, caused to move with velocity V in a 'tube' of cross-sectional area A. Then the mass of fluid passing any section per second $= \rho A V$ and the momentum of this fluid $= mV = \rho A V^2$. Since fluid is initially at rest the rate of change of momentum $= \rho A V^2$.

In a specific application, the force required is governed by the speed desired and the resistance of the ship. Since the force produced is directly proportional to the mass density of the fluid used, it is reasonable to use the more massive of the two fluids available, i.e. water. If air were used, then either the cross-sectional area of the jet must be large or the velocity must be high.

This explains why most ships employ a system by which water is caused to move aft relative to the ship. A variety of means is available for producing this stream of water aft, but by far the most commonly used is the screw propeller and this is dealt with first.

THE SCREW PROPELLER

Basically, the screw propeller may be regarded as part of a helicoidal surface which, on being rotated, 'screws' its way through the water driving water aft and the ship forward. Some propellers have adjustable blades—they are called *controllable pitch propellers*—but by far the greater majority of propellers have fixed blades. The ones we are concerned with first are *fixed pitch propellers*.

Propellers can be designed to turn in either direction in producing an ahead thrust. If they turn clockwise when viewed from aft, they are said to be *right-handed*; if anti-clockwise, they are said to be *left-handed*. In a twin screw ship, the starboard propeller is normally right-handed and the port propeller left-handed, i.e. they turn as in Fig. 10.8. They are said to be outward turning and this reduces cavitation which is discussed later.

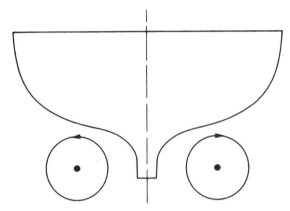

Fig. 10.8 Usual handing of propellers in a twin screw ship. Ship view from aft

Considering each blade of the propeller, the *face* is the surface seen when viewed from aft, i.e. it is the driving surface when producing an ahead thrust. The other surface of the blade is called the *back*. The *leading edge* of the blade is that edge which thrusts through the water when producing ahead thrust and the other edge is termed the *trailing edge*.

Other things being equal, the thrust developed by a propeller varies directly with the surface area, ignoring the boss itself. This area can be described in a number of ways. The *developed blade area* of the propeller is the sum of the face area of all the blades. The *projected area* is the projection of the blades on to a plane normal to the propeller axis, i.e. the shaft axis. The *disc area* is the area of a circle passing through the tips of the blades and normal to the propeller axis.

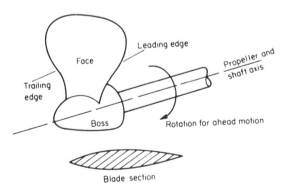

Fig. 10.9 The propeller blade

In non-dimensional work, the *blade area ratio* (BAR) is now generally used. This is the ratio of the developed blade area to the disc area, i.e.

$$\text{BAR} = \frac{A_D}{A_0} = \frac{4A_D}{\pi D^2}, \ A_D \text{ obtained by drawing}$$

If the variation of helical chord length with radius is known, then the true blade area can be obtained analytically by integration. This is known as the *expanded area* and the *expanded area ratio* (EAR) is defined by

$$\text{EAR} = \frac{4A_E}{\pi D^2}$$

In some earlier work, the concept of a *disc area ratio* (DAR) was employed in which the developed area was increased to allow for the boss. Froude proposed a boss allowance of 25 per cent of the developed area but Gawn used 12.5 per cent.

A true helicoidal surface is generated by a line rotated about an axis normal to itself and advancing in the direction of this axis at constant speed. The distance the line advances in making one complete revolution is termed the *pitch*. For simple propellers, the pitch is the same at all points on the face of the blade. This is the *face pitch* of the propeller and the ratio of this to the propeller diameter is the *face pitch ratio*

$$\text{i.e. face pitch ratio} = \frac{P}{D}$$

The distance advanced by a propeller during one revolution when delivering no thrust is termed the *analysis pitch*. In practice, this is rather greater than the geometrical pitch of the propeller. When developing thrust, the propeller advance per revolution is less than the analysis pitch. The difference is termed the *slip*. That is,

slip = analysis pitch – advance per revolution

The ratio of the slip to the analysis pitch is correctly called the *slip ratio s*, but by common usage is often referred to simply as slip.

Most modern propellers have pitch varying with radius and to define the geometry of the propeller the variation must be specified. For convenience, a nominal pitch is often quoted which is the pitch at a radius of 0.7 times maximum radius.

The projected shape of a propeller blade is generally symmetrical about a radial line called the median. Some propellers have what is known as *skew back* and this is when the median is curved back, relative to the direction of rotation

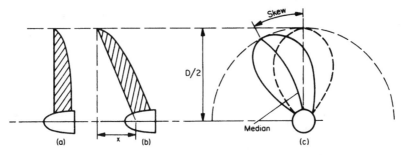

Fig. 10.10

of the propeller, as shown in Fig. 10.10(*c*). Skew back is defined by the circumferential displacement of the tip of the blade. It is of some advantage when the propeller is working in a flow with a pronounced circumferential variation, as not all the blade is affected at the same time and variations in thrust and torque are smoothed out.

For some applications the blade face is not normal to the propeller axis. In such a case, e.g. Fig. 10.10(*b*), the blade is said to be *raked*. It may be raked either forward or aft, but generally the latter to increase the clearance between the blade tip and the hull. Referring to the figure

Rake ratio $= x/D$

The blade section shape at any radius is the shape of the intersection between the blade and a co-axial cylinder when the cylindrical surface has been rolled out flat. The *median* or *camber line* is the line through the mid-thickness of the blade. The *camber* is the maximum distance separating the median line and the straight line, the chord *c*, joining the leading and trailing edges.

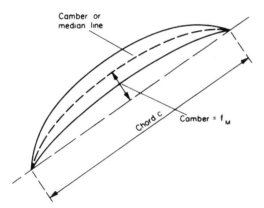

Fig. 10.11

Camber is normally expressed as the *camber ratio*, where

camber ratio $= f_M/c$

Similarly the *thickness ratio* of the section is t/c where t is the maximum thickness of the section. In most modern propellers, the thickness varies non-linearly with radius. The *thickness distribution* must be specified for complete definition of the propeller geometry.

SPECIAL TYPES OF PROPELLER

Most of this section deals with the fixed-pitch propeller which is the most common propulsor. Other types are:

(a)   Controllable pitch propeller

In such propellers the blades can be rotated about axes normal to the driving shaft so that thrust and torque can be varied at constant shaft revolutions. This can help match the propeller and machinery characteristics. If the blade rotation is great enough the propeller can produce astern thrust removing the need for a reversing gearbox. Manoeuvring can be faster as the blade angle can be varied more rapidly than can the shaft revolutions but there will be an optimum rate of change to produce maximum acceleration or deceleration.

Another suitable application of the CP propeller is to ships which must operate efficiently at two quite different loading conditions, e.g. the tug when towing or running free, the trawler when trawling or when on passage to or from the fishing grounds.

Limitations of the CP propeller include the power that can be satisfactorily transmitted (installations for more than 20 MW are uncommon), the complication of the mechanisms controlling the blade angle and the limitation of BAR to about 0.8 which affects the cavitation performance. The control mechanism must pass down the shaft and into the boss. The boss is enlarged to take this gear and to house the bearings for the blades. This increased boss size slightly reduces the maximum efficiency obtainable. The blade sections at the root are governed by the rotation on the boss and are poor for cavitation.

(b)   Contra-rotating propeller (CRP)

In this case there are two propellers in line on a double shaft rotating in opposite directions, the forward screw on the outer shaft and the after screw on the inner shaft. Generally the two propellers will be of different diameters and rpm. The design of CRPs is more complicated because of the interactions between the screws and the need for the contra-rotating gear system, but they have higher efficiencies than conventional propellers. This is because the after screw recovers the rotational energy imparted to the wake by the forward screw. Compared with conventional propellers the CRP achieves its optimum performance at a smaller diameter for a given rpm, or at lower rpm for the same diameter. The CRP also generally has a superior cavitation performance and reduced noise emission due to the lighter blade loading.

(c)   The self pitching propeller

In this type of propeller the blades are free to rotate through 360° about a bearing axis substantially normal to the shaft axis. The blade pitch is determined solely by the action of hydrodynamic and centrifugal forces. Particular applications are for auxiliary yachts and motorsailers.

(d)   The vertical axis propeller

This propeller consists basically of a horizontal disc rotating about a vertical axis. Projecting vertically down from this disc are a number of spade-like blades and these feather as they and the disc rotate. By varying the sequence in which the blades feather a thrust can be produced in any desired direction.

An obvious advantage of such a propeller is that it confers good manoeuvrability on any ship so fitted. This is touched upon in Chapter 13. With most conventional machinery units, the drive shaft is horizontal and to drive the

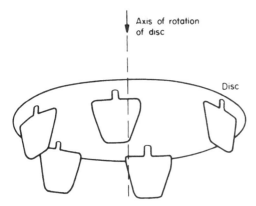

Fig. 10.12 Vertical axis type propeller

horizontal disc it is necessary to introduce a bevel gear with consequential limitations on the maximum power that can be transmitted.

(e)   Ducted propellers

A typical arrangement is sketched in Fig. 10.13. Improvements over the conventional propeller performance arise from the enlargement of the tail race and the thrust that can be produced by suitable shaping of the duct to offset the drag of the shroud and its supports. Most applications have been made in ships with heavily loaded propellers, e.g. tugs, but the range of use is increasing.

Other advantages of the shroud are that it protects the propeller from physical damage and acts as a cloak masking the propeller noise.

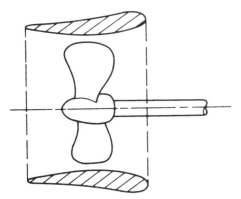

Fig. 10.13 Shrouded propeller

Van Manen gave data for various types of propeller based on open water tests as shown in Fig. 10.14. It indicates the type of propeller which will give the best efficiency for a given type of ship. Efficiency is not always the only factor to be considered, of course, in choosing the propulsion device.

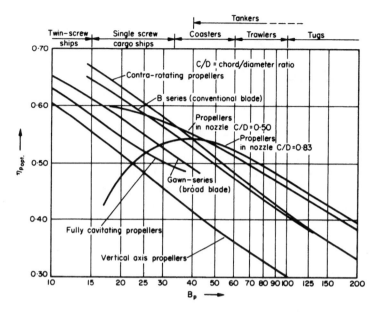

Fig. 10.14 Comparison of optimum efficiency values for different types of propulsion

(f) Pump jets

In this propulsor, a rotor with a relatively large number of blades operates between sets of stator blades, the whole being surrounded by a suitably shaped duct. When correctly designed there is no resultant heeling moment on the body being propelled, no rotational losses in the wake and no cavitation. For these reasons pump jets have been used to propel underwater vehicles.

(g) Propulsion pods

Electrically driven propellers in pods have been developed since high hysteresis motors enabled significant increases in the power that can be transmitted. The pods are on a vertical axis which enables the direction of thrust to be changed by 'azimuthing'.

ALTERNATIVE MEANS OF PROPULSION

These can only be touched upon very briefly, and the following list is by no means exhaustive:

(a) Hydraulic or jet propulsion

If water is drawn into the ship and then thrust out at the stern by means of a pump then the ship can be regarded as jet propelled.

Since the pump or impeller is basically a propeller, the overall efficiency of such a system is lower than the corresponding screw propeller, i.e. of diameter equal to the jet orifice diameter, because of the resistance to flow of water through the duct in the ship. It is attractive, however, where it is desirable to

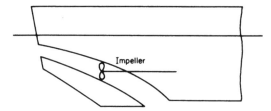

Fig. 10.15 Jet propulsion

have no moving parts outside the envelope of the main hull. This is the case of craft operating in very shallow water and a very successful class of boat has been designed using this principle for operating on shallow rivers. Many fast craft use water jet propulsion with water discharging into air and some high powered units are now available.

(b)   Paddle wheels

In essence, the paddle wheel is a ring of paddles rotating about an athwartship horizontal axis.

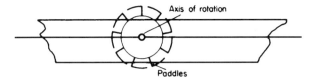

Fig. 10.16 Paddle wheel

In the simplest form, the paddles are fixed but greater efficiency is obtained by feathering them as the blades enter and leave the water. They can confer good manoeuvrability on a ship when fitted on either side amidships and for this reason many tugs have been designed using this principle. For operation in narrow waters, the large beam of this arrangement may be unacceptable and this was the consideration that led to the development of the 'stern wheeler' on the rivers of the USA.

(c)   Wind/air reaction

In a sailing ship the sail, when stretched under the action of the wind, can be regarded as an aerofoil section developing lift and drag as would a solid body (see Chapter 16). The early 1980s saw a revival of interest in sails in merchant ships as an economy measure. A Japanese 1600 dwt tanker achieved 12 per cent fuel economy with a sail area of 194 m^2.

The Flettner Rotor concept gave the *Buckau* a speed of 5 knots or so in a 10 knot wind in 1925. Blown circulation control can be used to avoid the need for a rotating 'thruster' by employing a circulation controlled aerofoil device based on the phenomenon that an accelerated stream of fluid from a tangential jet tends to remain attached to a curved surface.

MOMENTUM THEORY APPLIED TO THE SCREW PROPELLER

It was shown above, that the force available for propelling a ship could be related to the momentum in the screw race. Let us now develop this idea in a little more detail. The propeller will cause water to accelerate from some distance ahead of the propeller disc and, because water is virtually incompressible, the flow of water through the disc will be as in Fig. 10.17.

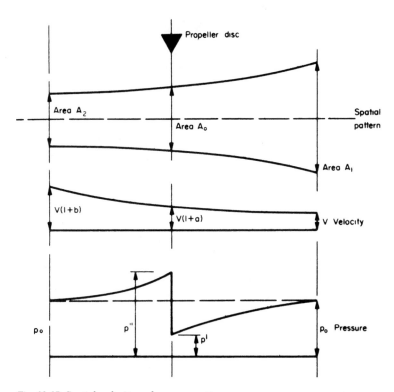

Fig. 10.17 *Spatial, velocity and pressure patterns*

Let A_1 and A_2 be points sufficiently ahead of and abaft the actual propeller disc so that the pressure at these points is effectively that in the free field. Due to the contraction of the screw race, the velocity will increase as shown and pressure will decrease between A_1 and the disc, suffer a jump at the disc and then decrease again between the disc and A_2. Now

thrust on propeller $= T = A_0(p'' - p')$

and applying Bernoulli's principle to both sides of the disc,

$$p_0 + \rho\frac{V^2}{2} = p' + \rho\frac{V^2}{2}(1 + a)^2$$

$$p_0 + \rho\frac{V^2}{2}(1 + b)^2 = p'' + \rho\frac{V^2}{2}(1 + a)^2$$

Subtracting

$$p'' - p' = \rho \frac{V^2}{2}(1+b)^2 - \rho \frac{V^2}{2}$$

Hence

$$T = \frac{\rho}{2} A_0 V^2 (2b + b^2)$$

But, also, thrust = rate of increase of axial momentum

$$\therefore \ T = \rho A_0 V(1+a)bV = \rho A_0 V^2(1+a)b$$

Comparing these two expressions for thrust, it is seen that

$$a = b/2$$

That is to say, half the velocity increase experienced in the screw race is caused by the suction created by the propeller and takes place before the water enters the propeller disc. This factor of increase, a, is known as the *axial inflow factor*. This factor controls the propeller efficiency that can be obtained since

$$\text{Propeller efficiency} = \frac{\text{useful work done by propeller}}{\text{power absorbed by the propeller}}$$

$$= \frac{\text{thrust} \times \text{propeller speed}}{\text{overall change in kinetic energy}}$$

$$= \frac{\rho A_0 V^2 2a(1+a)V}{\frac{1}{2}\rho A_0 V(1+a)V^2[(1+2a)^2 - 1]}$$

$$= \frac{2\rho A_0 V^3 a(1+a)}{\frac{1}{2}\rho A_0 V^3(1+a)(4a^2 + 4a)}$$

$$= \frac{1}{1+a}$$

This shows that even in the ideal case, high propeller efficiency is only possible with a small inflow factor, i.e. with a large diameter propeller.

In actual propellers the efficiencies will be less than this ideal due to a wide range of effects including the finite number of blades, the propeller hub, the thickness of blades, wake variations, cavitation and viscous losses.

THE BLADE ELEMENT APPROACH

The momentum theory is useful in indicating the influence of the propeller on the water ahead of its own disc, and in demonstrating that even theoretically there is a limit to the efficiency which can be achieved. It is not, however, of direct value in assessing the torque and thrust developed in a propeller of a particular geometry.

One approach is to consider each blade of the propeller as made up of a series of annular elements such as the shaded portion in Fig. 10.18 which represents

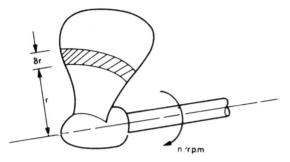

Fig. 10.18 Annular element of propeller blade

that portion of the blade between radii r and $r + \delta r$. If the propeller is turning at n r.p.m., then the element will have a tangential velocity of $2\pi rn$ besides a velocity of advance V_1 relative to the water. The element, which can be regarded as a short length of aerofoil section, will experience a relative water velocity as shown in Fig. 10.19.

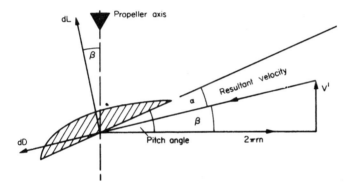

Fig. 10.19 Water flow relative to blade element

It will be seen that the blade element is at an angle α, to the resultant velocity. This angle is known as the *angle of attack*. To explain what happens now, it is necessary to introduce the concept of a *vortex*. In a potential vortex, fluid circulates about an axis, the circumferential velocity of any fluid particle being inversely proportional to its distance from the axis. The strength of the vortex is defined by the *circulation* Γ. If the blade element were in an inviscid fluid, the

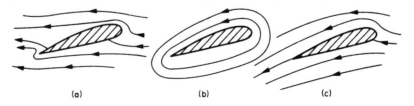

(a)   (b)   (c)

Fig. 10.20

potential flow pattern around it would be as in Fig. 10.20(*a*). In a real fluid, the very high velocities at the sharp trailing edge produce an unstable situation in the viscous fluid due to shear stresses. The potential flow pattern breaks down and a stable flow pattern is established as in Fig. 10.20(*c*). This consists of the original uniform flow with a superimposed vortex (Fig. 10.20(*b*)) having the foil as a core. The strength of the circulation depends upon the shape of the section and its angle of attack.

If, in an inviscid fluid, a circulation of this same strength could somehow be established, the blade element would be acted on by a lift force normal to the resultant velocity and the force would be proportional to the circulation. In a real fluid, the viscosity which gives rise to the circulation also introduces a small drag force whilst having little influence on the lift.

In Fig. 10.19, the lift and drag acting on the blade element are shown as dL and dD respectively. As already stated, the circulation depends upon the shape of the section and the angle of attack. A number of so-called aerofoil sections are available which produce high lift for small drag. If one of these sections is being used, its characteristics will be available from standard tests. Hence, the lift and drag on each element of the blade can be calculated. By resolving parallel and normal to the propeller axis, the contributions of the element to the overall thrust and torque of the propeller are

$$\mathrm{d}T = \mathrm{d}L\cos\beta - \mathrm{d}D\sin\beta$$

and

$$\mathrm{d}Q = (\mathrm{d}L\sin\beta + \mathrm{d}D\cos\beta) \times (\text{radius of element})$$

By repeating this process for each element and integrating over the blade, the thrust and torque on each blade and hence of the propeller can be obtained. Account can be taken of propellers in which the pitch angle varies with radius, but a really comprehensive theory of propellers must also take into account the interference between blades, and the tendency for pressures on the face and back of the blade to be equalized by flow around the tip of the blade.

In more advanced theories the lifting surfaces of the propeller are represented by lines, surfaces or panels of vortices and source–sink distributions to derive lift and drag by mathematical analysis. The theories can be used to design individual propellers or to indicate broad lines of development for methodical series. They enable the design of the blades to be optimized for the variations in water velocity at the propeller disc. This in turn has increased interest in determining these velocity distributions accurately by means of wake surveys and prediction. This is not an easy matter because the propulsor itself affects the distribution. Laser measuring techniques can be used to avoid introducing physical probes into the area which themselves would affect the flow pattern. Some averaging is necessary of the changing pattern experienced by individual blade elements as the propeller rotates.

A full treatment of these matters is not possible in a book of this nature but two major factors—cavitation and the generalized interaction between hull and propulsor—are now discussed.

CAVITATION

The thrust and torque of the propeller depend upon the lift and drag character-istics of the blade sections. The lift on the section is produced partly from the suction on the back of the blade and partly from positive pressure on the face. In this context, suction and positive pressure are relative to the free field pressure at the blade. A typical pressure distribution is shown in Fig. 10.21.

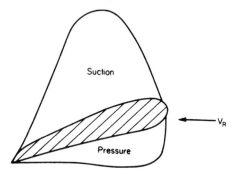

Fig. 10.21 Pressure distribution over an aerofoil section

As the pressure on the back falls lower and lower, with increasing propeller r.p.m., say, the absolute pressure will eventually become low enough for the water to vaporize and local cavities form. This phenomenon is known as *cavitation*. Since the water cannot accept lower pressures, the lift cannot increase as rapidly and the presence of cavitation manifests itself, therefore, as a fall off in thrust, torque and efficiency. There is also a marked increase in underwater noise and pressure variations in the vicinity of the propeller. Radial variation of pitch can improve cavitation performance and need have little effect on propeller efficiency. The pressure at which cavitation occurs depends upon temperature, the amount of dissolved air or other gases present and the surface tension. Without gasses in solution, the pressure might typically be of the order $3,500 \, \text{N/m}^2$ for fresh water and the presence of air can increase this to about $10,500 \, \text{N/m}^2$. As a corollary to the circulation around the blade, which produces the lift forces, the blade sheds trailing vortices associated with the blade tip, propeller hub and the face. The core of a vortex will be at a relatively low pressure and the first cavitation usually occurs in the tip vortex—the *tip vortex cavitation*. As σ, the cavitation number, is reduced, the cavitation spreads across the back of the blade giving the appearance of a sheet or surface of separation. This is known as *sheet cavitation* and can occur on both the face and back of the blade.

Other forms of cavitation are *bubble cavitation* which usually occurs at the thick root sections of the blades and is particularly susceptible to local irregu-larities in the blade surface. Unfortunately, the bubbles formed in a region of low pressure can be swept away into regions of high pressure where they collapse. This can lead to intense local pressures which may cause pitting or erosion of the propeller blades. Such pitting not only produces a weaker propeller but also increases the surface irregularities of the blades. *Face cavitation* can occur when the blade sections are working at very small or negative angles of incidence which can arise when a propeller operates in a varying velocity field. An undesirable aspect of cavitation in warships is the noise associated with the cavitation bubbles and their collapse. This can betray the ship's presence to an enemy and can attract homing weapons.

It follows from the simple discussion above of the nature of the phenomenon, that cavitation is likely to be delayed and less severe after onset if:

1. blades are lightly loaded and the pressure distribution is kept as smooth as possible. That is peaks should be avoided;
2. the pressure loading is reduced towards the blade tips;
3. the wake is kept as uniform as possible avoiding large changes in velocity and angle of inflow.

Providing high blade skew is found to be favourable for cavitation performance. In one single screw cargo ship, a tip skew of 60 degrees increased the cavitation inception speed by 3 knots compared with a straight propeller. High skew also reduces cavitation-induced vibration.

A special situation arises in propellers running at high r.p.m., as might be the case in a high speed motor boat. In these circumstances, it is impossible to eliminate cavitation but reasonable efficiency can be obtained by using a propeller designed to have the back completely covered by cavitation; although the lift generated by the back is limited by the vapour pressure, the torque component due to skin friction on the back of the blade is eliminated. Such a propeller is known as a *super-cavitating* propeller. It is very inefficient at speeds lower than 40–50 knots.

SINGING

Before the onset of cavitation, the blades of a propeller may emit a high-pitched note. This singing, as it is termed, is due to the elastic vibration of the material excited by the resonant shedding of non-cavitating eddies from the trailing edge of the blades. Heavy camber appears to be conducive to singing. Cures can be effected by changing the shape of the trailing edge of increased damping of the blade.

INTERACTION BETWEEN THE SHIP AND PROPELLER

The interaction manifests itself in the following ways:

(a) the hull carries with it a certain mass of water as was pointed out in considering the boundary layer. This means that the average velocity of water relative to the propeller disc is no longer equal to the velocity of advance of the propeller relative to still water;
(b) the water velocity will vary in both magnitude and direction across the propeller disc and the performance of the propeller will differ from that in open water even allowing for the difference in average velocity;

(*c*) the propeller causes variation in local pressures in the water and these will react upon the hull, leading to an effective increase in resistance.

Let us proceed to consider each of these effects in more detail.

The difference between the ship speed and the speed of the water relative to the ship is termed the *wake*. The wake is the combination of the boundary layer associated with skin friction, the flow velocities occasioned by the streamlined form of the ship and the orbital velocities of the waves created by the ship. If the water is moving in the same direction as the ship, the wake is said to be positive. If the ship speed is V and the average velocity of the water relative to the hull at the propeller position is V_1, then:

$$\text{Wake} = V - V_1$$

To non-dimensionalize this relation, the wake can be divided by either V_1 or V. The former was proposed by Froude and the latter by Taylor leading to two wake factors as follows:

> *Froude wake factor* $= w_F = (V - V_1)/V_1$
> *Taylor wake factor* $= w = (V - V_1)/V$

Clearly, these are merely different ways of expressing the same phenomenon.

Apart from this *average* flow of water relative to the hull there will be variations in velocity over the propeller disc. As the hull is approached more closely, the water moves less fast relative to the ship. Apart from this general effect of the hull there will be local perturbations due to the shaft, shaft bossings or shaft brackets and other appendages. Due to the fact that the water must 'close-in' around the stern the flow through the propeller disc will not be everywhere the same and will not, in general, be parallel to the shaft line. These effects are combined and expressed as a *relative rotative efficiency* (RRE) which is defined as

$$\text{RRE} = \eta_R = \frac{\text{efficiency of propeller behind the ship}}{\text{efficiency of propeller in open water at speed } V_1}$$

Finally, there is the influence on the hull of pressure variations induced by the propeller action. As far as the propeller is concerned it has to produce a thrust T which is greater than the resistance R of the hull without propeller.

As with the wake, there are two ways of expressing this physical phenomenon. It can be considered as an *augment of resistance*, *a*, where

$$a = \frac{T - R}{R}$$

or, it can be regarded as a *thrust deduction factor*, *t*, where

$$t = \frac{T - R}{T}$$

HULL EFFICIENCY

The thrust power (P_T), developed by the propeller is given by the product of T and V_1. On the other hand, the effective power is given by the product RV.

Now

$$P_T = TV_1 = R(1+a)\frac{V}{1+w_F} = \frac{RV(1+a)}{1+w_F} = RV(1+a)(1-w)$$

$$= RV\frac{(1-w)}{(1-t)}$$

therefore

$$\frac{P_E}{P_T} = \frac{1+w_F}{1+a} \quad \text{or} \quad \frac{1-t}{1-w}$$

This ratio is known as the *hull efficiency* and seldom differs very greatly from unity.

To complete the picture of the propeller acting behind the ship, the concept of relative rotative efficiency must be added in. The three factors, augment, wake and RRE are referred to collectively as the *hull efficiency elements*. Augment and wake are functions of Reynolds' number but variation between ship and model is ignored and the error so introduced is taken account of by the trials factors.

OVERALL PROPULSIVE EFFICIENCY

The shaft power (P_S) is the power needed to propel the complete ship. The ratio between the P_E and P_S is a measure of the overall propulsive efficiency achieved and is termed the *propulsive coefficient* (PC)

$$PC = \frac{P_E}{P_S}$$

The overall efficiency can be regarded as the cumulative effect of a number of factors. Consider the following in addition to P_E and P_S

P_{EA} = power to tow hull complete with appendages,

P_T = thrust power developed by propellers = TV_1,

P_D = power delivered to propellers when propelling the ship,

P'_D = power delivered to propellers when developing a thrust T in open water at a speed V_1.

Now the propulsive coefficient can be defined as:

$$PC = \frac{P_E}{P_S} = \frac{P_E}{P_{EA}} \times \frac{P_{EA}}{P_T} \times \frac{P_T}{P'_D} \times \frac{P'_D}{P_D} \times \frac{P_D}{P_S}$$

where

$$\frac{P_E}{P_{EA}} = \frac{1}{\text{appendage coefficient}}$$

$$\frac{P_{EA}}{P_T} = \text{hull efficiency, } \eta_H$$

$$\frac{P_T}{P'_D} = \text{propeller efficiency } \eta_0 \text{ in open water at speed } V_1$$

$$\frac{P'_D}{P_D} = \text{relative rotative efficiency, } \eta_R$$

$$\frac{P_D}{P_S} = \text{shaft transmission efficiency, } \eta_S$$

That is

$$\text{PC} = \left[\frac{\eta_H \times \eta_0 \times \eta_R}{\text{appendage coefficient}} \right] \times \text{transmission efficiency}$$

It is recommended that the transmission efficiency be taken as 0.97 for ships with machinery amidships and 0.98 for ships with machinery aft. For modern warships appendage coefficients vary from about 1.05 to 1.10. In using PC it is necessary to check the definition. Some authorities use $\text{PC} = P_{EA}/P_S$.

The quantity in the brackets is known as the *quasi-propulsive coefficient* (QPC), η_D, and can be obtained from model results. There is some error in applying this to the full-scale ship and to allow for this and transmission efficiency and any differences between the ship and model test conditions, e.g. wind, waves, cavitation, use is made of a *QPC factor* which is defined as

$$\text{QPC factor} = \frac{\text{PC from ship trial}}{\text{QPC from model}}$$

The value to be assigned to the QPC factor when estimating power requirements for a new design is usually determined from results of a similar ship.

The National Physical Laboratory (now part of British Maritime Technology) used a *load factor* instead of the QPC factor, where

$$\text{load factor} = 1 + x = \frac{\text{transmission efficiency}}{(\text{QPC factor})(\text{appendage coefficient})}$$

In the NPL analysis, the *overload fraction* x is intended to allow for the basic shell roughness, fouling, weather conditions and depends on ship length and type. It is recommended that whatever value of x is used in estimates a standard power estimate should also be made with a load factor of unity, i.e. with $x = $ zero, and an appendage scale-effect factor $\beta = 1$, i.e. assuming appendage resistance scales directly from the model to the ship.

SHIP–MODEL CORRELATION

The conduct of a ship speed trial is dealt with later together with the analysis by which the ship's actual speed is deduced. This demonstrates whether the ship meets its specification but does not tell the designer much about the soundness of his prediction method. If the specified speed is not reached it may be that he wrongly estimated the ship's resistance or hull efficiency elements, the propeller design may have been incorrect or the machinery may not have developed the intended power. A much more comprehensive analysis of the trials data is required by the designer to assist him with later designs. Even if the speed prediction was acceptable, it is still possible that several errors in assessing various factors cancelled each other out.

The analysis method used must depend upon the design methods to be checked. Froude developed the following method using 'circular' functions defined as below:

$$\textcircled{E} = \frac{1000(P_E)}{\Delta^{\frac{2}{3}}V^3}, \quad \text{using naked model } P_E$$

$$\textcircled{E}_A = \frac{1000(P_A)}{\Delta^{\frac{2}{3}}V^3}, \quad \text{where } P_A = \text{power to tow the appendages}$$

$$\textcircled{E}_{WP} = \frac{1000(P_{WP})}{\Delta^{\frac{2}{3}}V^3}, \quad \text{where } P_{WP} = \text{power to overcome windage and}$$
$$\text{fouling under trials conditions}$$

$$\textcircled{E}_T = \textcircled{E} + \textcircled{E}_A + \textcircled{E}_{WP}$$

N.B. If the speed trial is carried out under good conditions, $\textcircled{E}_{WP}$ should be negligible.

$$\textcircled{T}_M = \frac{1000(P_T)}{\Delta^{\frac{2}{3}}V^3} = \textcircled{E}_T(1 + a)(1 - w)$$

$$\textcircled{T}_R = \frac{H \times 1000}{\Delta^{\frac{2}{3}}V^3} = H\textcircled{T}_M/P_T$$

where

H = thrust power from open water propeller data using the trial
r.p.m. and speed and the model wake

$$\textcircled{D} = \frac{1000(P_D)}{\Delta^{\frac{2}{3}}V^3}$$

$$\textcircled{I} = \frac{1000(P_S)}{\Delta^{\frac{2}{3}}V^3}$$

Each of the above parameters is calculated for each run and plotted to a base of speed. The propulsive coefficient, equal to $\textcircled{E}/\textcircled{I}$ or $\left[\textcircled{E} + \textcircled{E}_A\right]/\textcircled{I}$

with alternative definition of PC is also plotted with the QPC factor which is the ratio of the PC from the ship trial to the QPC from model tests.

If the predictions from model experiments were exact the QPC factor would equal the shaft transmission efficiency, and:

$$\textcircled{T}_\mathrm{R} = \textcircled{T}_\mathrm{M}$$

In general, this relationship is not precise as $\textcircled{T}_\mathrm{R}$ includes some scale effects including those due to cavitation on the ship propeller. The ratio $\textcircled{T}_\mathrm{R}/\textcircled{T}_\mathrm{M}$ is known as the propeller thrust correlation factor

Thus, both the QPCF and the ratio $\textcircled{T}_\mathrm{R}/\textcircled{T}_\mathrm{M}$ show the essential differences between the model and full-scale data. In a new design the designer uses these quantities, deduced from previous trials, to assist in scaling from the model to the ship. He would make allowance for any differences in the two designs such as different appendage coefficients.

Model testing

RESISTANCE TESTS

Many great men attempted to use models or to show how they could be used to predict full-scale behaviour, including Bouguer, Tiedemann, Newton, Chapman, Euler and Beaufoy, but it was not until the time of William Froude that full-scale prediction became a practical proposition in the late 19th century.

It was William Froude who postulated the idea of splitting the total resistance into the residuary resistance and the frictional resistance of the equivalent flat plate. He also argued that air resistance and the effects of rough water could be treated separately. By studying the wave patterns created by geometrically similar forms at different speeds, Froude found that the patterns appeared identical, geometrically, when the models were moving at speeds proportional to the square root of their lengths. This speed is termed the *corresponding speed*, and this is merely another way of expressing constancy of Froude number. He also noted that the curves of resistance against speed were generally similar if the resistance per unit displacement was plotted for corresponding speeds. Proceeding further, he found that by subtracting from the total resistance an allowance for the frictional resistance, determined from flat plates, the agreement was very good indeed.

This led to *Froude's law of comparison* which may be stated as:

> If two geometrically similar forms are run at corresponding speeds (i.e. speeds proportional to the square root of their linear dimensions), then their residuary resistances per unit of displacement are the same.

Thus the essentials are available for predicting the resistance of the full-scale ship from a model. The steps as used by Froude are still used today, refinements being restricted to detail rather than principle. For each particular value of the ship speed:

(*a*) measure the resistance of a geometrically similar model at its corresponding speed,

(*b*) estimate the skin friction resistance from data derived from experiments on flat plates,

(*c*) subtract the skin friction resistance from the total resistance to obtain the residuary resistance,

(*d*) multiply the model residuary resistance by the ratio of the ship to model displacements to obtain the ship residuary resistance,

(*e*) add the skin friction resistance estimated for the ship to obtain the total ship resistance.

It should be noted that any error in estimating frictional resistance applies both to the model and ship. Thus, only the effect on the difference of the two is significant.

It is now possible to see why earlier attempts to correlate the total resistance of ship and model failed. Two models with identical resistances could only represent ships with identical resistances if the ratios of their residuary and skin friction resistance were the same. In general, this could not be true unless the forms were themselves the same. Indeed, if model A had less total resistance than model B it did not even follow that ship A would be less resistive than ship B. Thus, even the qualitative comparisons made between models, used so frequently even today in many branches of naval architecture, may be invalid.

RESISTANCE TEST FACILITIES AND TECHNIQUES

With the aid of a grant from the Admiralty, Froude constructed the world's first model tank at Torquay in 1871 where R. E. Froude continued his father's work on the latter's death in 1879. The work of the Froudes proved so useful that, when the lease on the Torquay site expired in 1885, a grant was made to erect another at Haslar in 1887. This was the beginning of the Admiralty Experiment Works (now used by the Defence Evaluation and Research Agency

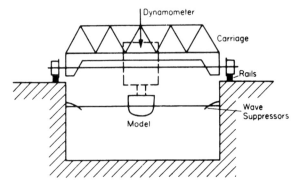

Fig. 10.22 Typical ship tank section

and British Maritime Technology), which has grown over the years and has remained one of the world's leading establishments in this field.

Modern ship tanks for measuring model resistance are fundamentally the same as the first tank made by Froude. Such a facility is essentially a long tank, of approximately rectangular cross-section, spanned by a carriage which tows the model along the tank. Improvements have been made over the years in respect of the methods of propelling the carriage, in the constancy of speed holding, in the instrumentation and analysis of data. Digital recording and computers on carriages have reduced data reduction times significantly.

In a typical run, the carriage is accelerated up to the required speed, resistance records and measurement of hull sinkage and trim are taken during a period of constant speed and then the carriage is decelerated. With increasing ship lengths and service speeds, there has arisen a demand for longer and longer tanks to cope with the longer acceleration and deceleration runs.

Model test procedures today rely much on the methodical and painstaking approaches of W. and R. E. Froude. As early as 1880, R. E. Froude was aware of unexplained variations in the resistance measured in repeat experiments on a given model. He suspected currents set up in the tank by the passage of the model and variations in skin friction resistance due to temperature changes. Methodical investigation into the first of these two features led to the adoption at AEW of small propeller type logs to record the speed of the model relative to the water. Investigation of the temperature effect led Froude to postulate that a 3 per cent decrease in skin resistance for every 10 °F rise in temperature could be adopted as a fair working allowance and linked this with a standard temperature of 55 °F.

In the temperature experiments, R. E. Froude used the model of HMS *Iris*, a 91 m, 3760 tonnef despatch vessel, as a 'standard' model to be tested at various times throughout the year. Final proof that, even after correcting for tank currents and temperature, significant variations in resistance were occurring, came in tests on the *Iris* model in the tank at Haslar to correlate with those previously run at Torquay. This led to the application of a so-called *Iris correction* obtained by running the standard model at frequent intervals and applying a correcting factor to the resistance of a new model depending on the variation of the *Iris* resistance from its standard value. Generally, the *Iris* correction varies between 1 and 6 per cent, but during abnormal periods, commonly referred to as 'storms', the correction can be more than 10 per cent. The cause of the storms is now known to be due to the presence in the water of substances having long chain molecules. The concept of a standard model has since been adopted by other ship tanks.

The concept of deliberately introducing additives to the boundary layer in order to reduce frictional resistance has since been tried. Although substantial reductions have been achieved it is not at present an economic proposition.

MODEL DETERMINATION OF HULL EFFICIENCY ELEMENTS

Experiments must be carried out with the hull and propeller correctly combined as illustrated in Fig. 10.23.

With the model at the correct speed, corresponding to that of the ship under study, a series of runs is made over a range of propeller r.p.m. straddling the

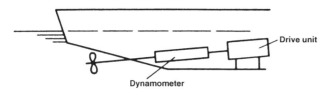

Fig. 10.23 Experimental technique

self-propulsion point of the model. Model speed and resistance are recorded together with the thrust, torque and r.p.m. of the propeller. Results are plotted to a base of propeller r.p.m., as shown for thrust in Fig. 10.24, to find the model self-propulsion point.

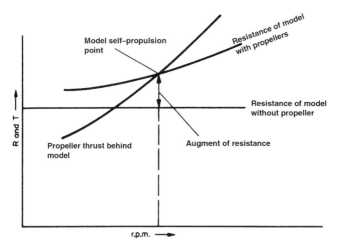

Fig. 10.24 Determination of model self-propulsion point

The model propeller then has its thrust and torque measured in open water at a speed of advance estimated to be that of the flow through the propeller when behind the hull, i.e. making allowance for the wake. By comparing this curve with that obtained in the combined experiment, the correct speed for the propeller in open water can be calculated. The difference between the model speed in the combined experiment and the corrected open water speed is the *wake*. The relative rotative efficiency follows as the ratio between the torques measured in the open water and combined experiments at self-propulsion r.p.m. The *augment of resistance* is obtained as illustrated in Fig. 10.24.

It should be noted that, although the propeller used in these experiments is made as closely representative of the ship propeller as possible, at least the first estimate of its geometry, the scale is too small to enable the thrust and torque figures to be used directly. Instead, the hull efficiency elements calculated as above are used with either methodical series data or specific cavitation tunnel measurements in order to produce the propeller design.

PROPELLER TESTS IN OPEN WATER

It is important that the designer has data available on which to base selection of the geometric properties of a propeller and to determine likely propeller efficiency. Such data is obtained from methodical series testing of model propellers in open water. Such testing eliminates the effects of cavitation and the actual flow of water into a propeller behind a particular ship form, and makes comparisons of different propellers possible on a consistent basis.

The tests are carried out in a ship tank with the propeller mounted forward of a streamlined casing containing the drive shaft. The propeller is driven by an electric motor on the carriage. Thrust, torque, propeller r.p.m. and carriage speed are recorded and from these K_T, K_Q, J and η can be calculated. Usually runs are carried out at constant r.p.m. with different speeds of advance for each run.

It will be appreciated that towing tanks can be used for a wide range of hydrodynamic tests other than those associated with resistance and propulsion. These are discussed in later chapters.

CAVITATION TUNNEL TESTS

It is impossible to run a model propeller in open water so that all the non-dimensional factors are kept at the same values as in the ship. In particular, it is difficult to scale pressure because the atmospheric pressure is the same for ship and model and scaling the depth of the propeller below the surface does not provide an adequate answer. If cavitation is important, the pressure of air above the water must be reduced artificially and this is the reason for using *cavitation tunnels* to study propeller performance. Such a tunnel is shown diagrammatically in Fig. 10.25, and is usually provided with means for reducing the air content of the water to improve viewing.

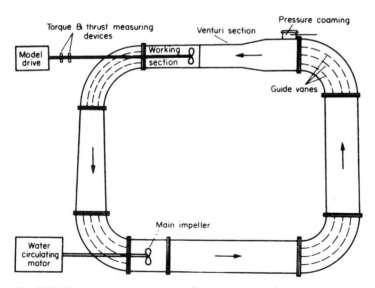

Fig. 10.25 Diagrammatic arrangement of a cavitation tunnel

In practice, experiments are usually run under the following conditions:

(*a*) the water speed is made as high as possible to keep Reynolds' number high to avoid serious scaling of skin friction;
(*b*) the model propeller is selected to have as large a diameter as is compatible with the tunnel size (tunnel wall effects must be avoided);
(*c*) model is run at the correct *J* value. This fixes the rate of propeller revolutions;
(*d*) the pressure in the tunnel is lowered to produce the correct cavitation number at the propeller axis.

Since the propeller revolutions are the most easily adjusted variable, it is usual to set the tunnel water speed, adjust the tunnel pressure to give the correct cavitation number and then vary the propeller r.p.m. systematically to cause a variation in the advance coefficient. The whole series can then be repeated for other σ values.

The tunnel shown in Fig. 10.25 is a fairly simple one and suffers from the fact that it is difficult to simulate the actual flow conditions at the after end of the ship. In some cases, attempts to reproduce this have been made using specially designed grids to control the local flow conditions. Also, the flow is from right to left in the working section so that the drive shaft on the model propeller is aft of the disc rather than forward of it as is the case for the ship. In big tunnels, both objections can be overcome by modelling the after end of the hull complete inside the tunnel and driving the propeller from inside this model hull.

A large tunnel at Hamburg, completed in 1989, has a test section 11 m long with a cross section 2.8 m × 1.6 m, water velocity up to 12 m/s and a pressure range 0.15 to 2.5 bar absolute. It can test integrated hull/propulsor arrangements for surface ships, submarines or underwater weapons. Low noise levels within the facility permit the conduct of acoustical studies.

DEPRESSURIZED TOWING TANK

In the 1970s the Netherlands Ship Model Basin brought into service a towing tank in which the air pressure can be reduced above the whole water surface. It is 240 m long and 18 m wide, with a water depth of 8 m. The pressure in the tank can be lowered to 0.03 bar. The main advantage is an ability to carry out propulsion tests under more representative conditions and to study cavitation of the hull or appendages.

CIRCULATING WATER CHANNELS

It is only the relative movement of model and water that is important. Thus an alternative to towing the model through the water is to hold the model steady, whilst allowing freedom of vertical movement, and cause the water to flow past it. This is achieved in a circulating water channel (CWC) which may also be able to modify the air pressure above the water free surface. A major advantage of the CWC is the possibility of measurements being made over a much longer time span—recording time is no longer limited by the tank length. Uniform flow is difficult to achieve and for this reason smaller CWCs are often restricted to flow visualization studies (again easier with a stationary model) and approximate force measurements.

Ship trials

SPEED TRIALS

When a ship has been completed, speed trials are carried out to confirm that the ship has met its specification as regards design speed. Such trials also provide useful data to help the designer in producing subsequent designs.

The trials are carried out over a known distance. The distance may be defined by precisely located land markers (Fig. 10.26) or in more open water by use of an accurate positional satellite navigation system. The following description relates to a land-based measured distance but the underlying principles are the same.

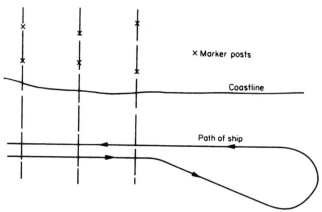

Fig. 10.26 Measured mile trials
(Note: Drawing not to scale. A straight approach run of about 3 miles is used.)

The ship approaches on a course normal to the lines joining corresponding pairs of 'mile' posts and sufficiently far off shore to ensure adequate depth of water to eliminate the effect of depth of water on resistance. The time to traverse the measured distance is accurately noted together with shaft thrust, torque and revolutions. A fine day with little wind and calm seas is chosen. To reduce its effect upon resistance the use of the rudder is kept to a minimum during the run. At the end of the run the rudder is put over to a moderate angle and the ship is taken round in a large sweep, as illustrated, to provide adequate run-up for the next pass to ensure that the ship has stopped accelerating by the time it passes the first pair of posts.

The trial is carried out for a range of powers up to the maximum the machinery can generate. At each power, several runs are made in each direction to enable the effect of any tide to be eliminated. If the runs are made at regular time intervals, it is adequate to take a mean of means, i.e. by meaning each consecutive pair of speeds, taking means of consecutive pairs of results so obtained and so on. The process is illustrated by the following example in which the mean speed is 15 knots.

In order that the ship's condition may be assessed as accurately as possible, the state of the hull should be recorded at the undocking preceding the trial and the time out of dock noted. Before the ship leaves harbour, draughts and water properties should be noted to provide a measure of its displacement. The use of fuel before and during the trials will help assess the variations in the displacement and trim during the trials. During the trials the properties of the surrounding water should be recorded. Speed measurements are made of shaft rpm, torque and thrust. The meters used must be accurately calibrated.

CAVITATION VIEWING TRIALS

In warships, where noise signature is important, the first ship of a new class undertakes cavitation viewing trials. A suitable port is arranged in the hull to enable the propeller to be viewed with the ship underway. Stroboscopic lighting is used to 'freeze' the propeller so that cavitation patterns can be seen clearly and photographed. Such trials can be combined with the ship's speed trial. Barometric pressure and water properties must be recorded.

EXAMPLE 3. A ship on a measured mile course records the speeds of 14.82, 15.22, 14.80, 15.20, 14.78 and 15.18 knots for six consecutive runs at regular time intervals. Calculate the mean speed.

If runs are not carried out at regular time intervals, it is necessary to assume that the tide varies with time according to a mathematical equation such as

$$\text{Speed of tide} = v = a + a_1 t + a_2 t^2$$

where t is the time measured from the initial run made.

It is then assumed that the speed without tide would be V, say, and that the readings obtained represent $V + v$ where v is the value appropriate to the time the run was made.

Solution:

Measured Speeds (knots)	Means				
	First	Second	Third	Fourth	Fifth
14.82					
	15.02				
15.22		15.015			
	15.01		15.010		
14.80		15.005		15.005	
	15.00		15.000		15.00
15.20		14.995		14.995	
	14.99		14.990		
14.78		14.985			
	14.98				
15.18					

Mean ship speed = 15.00 knots.

EXAMPLE 4. A ship on a measured mile course records the speeds of 15.22, 14.82, 15.20 and 14.80 at times of 1200, 1300, 1430 and 1530 hours. Calculate the speed of the ship and the equation governing the variation of the tidal current with time.

Solution: It is convenient to take the times at $t = 0$, $t = 1$, $t = 2.5$, $t = 3.5$, i.e. measuring in hours from the time of the initial run. Then, assuming that the ship speed is V and the tide is given by $v = a_0 + a_1 t + a_2 t^2$, we can write

at 1200 hrs; $15.22 = V + a_0$

at 1300 hrs; $14.82 = V - a_0 - a_1 - a_2$

at 1430 hrs; $15.20 = V + a_0 + 2.5a_1 + 6.25a_2$

at 1530 hrs; $14.80 = V - a_0 - 3.5a_1 - 12.25a_2$

Solving these equations,

$$V = 15.01 \, \text{knots}$$

$$v = (0.21 - 0.028t + 0.008t^2) \, \text{knots}, \ t \text{ in hours}$$

The difference in sign in alternate equations merely denotes that the tide is with or against the ship. Clearly, the tide is with the ship when it records its higher speeds but this is not significant to the mathematics since, if the wrong assumption is made, the tidal equation will lead to a negative tide.

SERVICE TRIALS

The above trials are carried out under calm conditions as a means of confirming that a ship meets its contractual requirement as regards speed. What is of greater interest to the owner, and designer, is the ship's performance in service under average or typical service conditions. Average figures for speed over long distances and associated fuel consumption have been extracted for many years from ship's logs. These could be related to the sea conditions as estimated by the crew. With the advent of satellites and satellite navigation systems it is possible to measure a ship's speed over shorter periods of time and link that performance with the sea system in which it was operating. Strictly the ship's speed is relative to land but this can be corrected for estimated tide and current effects. For special trials the satellite could track a buoy to provide these corrections. The specific sea conditions at any time can be obtained by retrospective analysis of wave data measured by satellite.

EXPERIMENTS AT FULL SCALE

Ship trials over a measured distance in calm water can confirm, or otherwise, the accuracy of the prediction of ship speed for a given power. They cannot, however, prove that the fundamental arguments underlying these estimates are valid. In particular, they cannot prove that the estimation of P_E was accurate because the influence of the ship propulsion system is always present.

William Froude realized this and with Admiralty assistance carried out full-scale resistance measurements on HMS *Greyhound* in 1874. More recently, full-scale resistance trials were carried out using the *Lucy Ashton* and HMS *Penelope*.

In the earlier trials, the screw sloop *Greyhound* was towed from an outrigger fitted to HMS *Active*, a vessel of about 3100 tonnef displacement. This method (Fig. 10.27) was adopted to avoid, as far as possible, any interference between the towing and the towed ship. Trials were carried out with the *Greyhound* at three displacements and covered a speed range of $3-12\frac{1}{2}$ knots. Some trials were with and some without bilge keels. For some runs the tow rope was slipped and the deceleration of the ship noted.

William Froude concluded that the experiments:

> ... substantially verify the law of comparison which has been propounded by me as governing the relation between the resistance of ships and their models.

In the *Lucy Ashton* trials some problems of towing a vessel were over-come by fitting the ship with four jet engines mounted high on the ship and out-board of the main hull to avoid the jet efflux impinging either on the hull or on the water in the immediate vicinity of the hull. Accurate measurement of thrust, totalling just over 6 tonnef from the four engines, was achieved by using hydraulic load measuring capsules. Speeds were measured over measured mile distances and special measures were taken to ensure accurate results and also to measure the surface roughness of the hull.

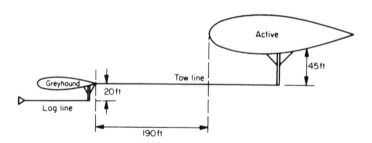

Fig. 10.27 Greyhound experiments

Resistance tests were made over a speed range of 5–15 knots with a clean naked hull with first a red oxide paint surface and then a bituminous aluminium paint. Each trial was repeated for sharp seams of plating and with the seams faired off with a plastic composition. Additional trials were run to study the effect of dummy twin-screw bossings, with twin-screw 'A' brackets and shafts and with a hull surface which had been allowed to foul for about a month.

The main purpose of the trial was to compare the various methods available for scaling model resistance to full-scale. The results indicated that Froude's law of comparison is valid for the scaling-up of wave-making resistance, but that the usual assumption that the skin friction of models and ships is the same as that of the corresponding plane surface of the same length and wetted surface is not strictly correct. Fortunately, the error is not very important in practical calculations. The results also indicate that over the range of models tested, the interference between the skin-frictional and wave-making resistance is not significant.

The results of the trials proved that full-scale ship resistance is sensitive to small roughnesses. For instance, the bituminous aluminium paint, which was the smoother of the two surface finishes, gave about $3\frac{1}{2}$ per cent less total resistance which was estimated to be equivalent to about 5 per cent of the skin frictional resistance. Fairing the seams gave about a 3 per cent reduction in total resistance. The effect of 40 days fouling on the bituminous aluminium painted hull was to increase the skin frictional resistance by about 5 per cent, i.e. about $\frac{1}{8}$th of one per cent per day.

Trials in HMS *Penelope* were conducted by the Admiralty Experiment Works while the ship was operating as a special trials ship. *Penelope* was towed by another frigate using a mile-long nylon rope. Although the main purpose of the trial was to measure radiated noise from, and vibration in, a dead ship, the opportunity was taken to measure resistance and wake pattern of *Penelope* in calm water and in waves. For this purpose both propellers were removed and a pitot rake fitted to one shaft. Propulsion data were recorded in the towing ship also. Propulsion data for *Penelope* were obtained from separate measured mile trials with three different sets of propellers fitted.

Correlation of ship and model data showed the resistance of *Penelope* to be some 14 per cent higher than predicted over the range 12–13 knots but indicated no significant wake scale effects. The hull roughness, using a wall roughness gauge was found to be about 0.3 mm mean apparent amplitude per 50 mm. The mean apparent amplitude per 50 mm is the standard parameter used in the UK to represent the average hull roughness. The propulsion results showed that thrust, torque and efficiency of the ship's propellers were higher than predicted by model tests.

Summary

In studying the powering of ships, it is essential that the hull and propulsion device be considered together. The shaft power required to drive a ship at a given speed can be derived from a series of model tests and calculations. The basic elements in the assessment of the shaft power have been established and are summarized in Fig. 10.28.

It remains to show how model data is presented and the necessary calculations carried out. This is done in the next chapter.

Problems

1. A 50 MN displacement ship, length 120 m, is to be represented by a model 3 m long. What is the displacement of the model? At what speed must it be run to represent a speed of 20 knots in the ship and what is the ratio of the ship to model effective power at this speed?

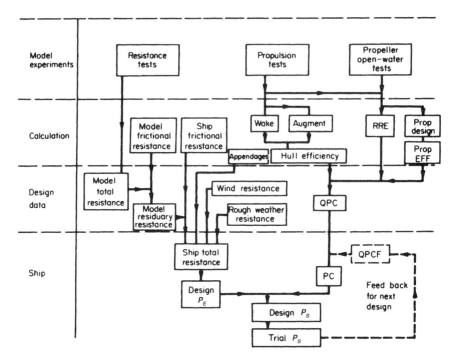

Fig. 10.28 Assessment of ship shaft power

2. Show how wave-making resistance at a given speed is affected by varying the length of parallel middle body, entrance angle and run remaining unchanged.

   On the resistance curve for a ship length 70 m with 20 m of parallel middle body, one hump occurs at a speed of 14 knots and an adjacent one at $\sqrt{\frac{3}{5}}$ times this. At what speed would the main hump occur? If the length of parallel middle body is increased 15 m, at what sequence of speeds will humps occur on the resistance curve?

3. How are waves created when a typical warship form passes through the water? With the aid of sketches describe a typical ship wave pattern and explain what effect 'shoulders' on the curve of areas would have on this pattern. Draw a typical resistance curve for a ship indicating the main features of its characteristic shape and explaining why humps and hollows occur. Derive the expressions from which the position of these humps and hollows may be determined and, hence, determine the speeds at which the two most prominent humps occur for a ship of effective wave-length 99 m.

4. Sketch and describe a typical ship wave pattern. What is meant by interference between wave systems? Show, with the use of diagrams, why humps and hollows occur in the curve of wave-making resistance against $V/\sqrt{L}$, giving approximate values of $V/\sqrt{L}$ at which the humps and hollows occur.

   A destroyer, length 122 m, is observed to be steaming at high speed. The first trough of the bow wave system is seen to coincide with the stern trough.

Estimate the speed of the destroyer assuming that the wave system distance is $0.9L$.

5. A ship at full speed has an effective power naked of 7×10^6 W. The appendage coefficient is 1.15, the hull efficiency 0.98, the propeller efficiency 0.69, the RRE 0.99 and the QPCF 0.90. Calculate the propulsive coefficient, the quasipropulsive coefficient and the shaft power required.

6. A 50 MN displacement ship 100 m long is towed in the naked condition on a long tow rope at a speed of 20 knots. The force in the tow rope is 1 MN. Find the effective power for the ship. Deduce the shaft power for a geometrically similar ship 120 m long at 20 knots assuming that shaft power in this speed region is proportional to the cube of the speed and that:

appendage coefficient	1.20
hull efficiency	0.97
propeller efficiency	0.72
RRE	1.00
QPCF	0.95

7. Describe how speed trials are conducted, listing the items recorded. What factors would you consider important when choosing a site for a new measured mile course?

Full-power trials of a new frigate involved five passes over the measured mile, each pass being followed by one in the opposite direction. The times of the start of each run and the speeds attained are:

Time of start	1045	1103	1127	1227	1245
Speed of run (knots)	27.59	28.66	27.64	28.60	27.69

Making suitable adjustments to the time intervals and assuming the tide speed is given by $v = a + bt + ct^2 + dt^3$, determine the true speed of the ship.

8. List the measurements which are made, during sea trials, on each run over the measured mile, explaining briefly how each measurement is made.

The following data were obtained during progressive speed trials on a merchant ship. Assuming that the tidal velocity may be expressed in the form $v = a + bt + ct^2$, calculate the true speed at each power.

Run no.	Direction	Time of day	Recorded speed (knots)	r.p.m	P_S (MW)
1	N	0830	10.35	83	0.86
2	S	0900	9.60		
3	N	0930	12.52	102	1.52
4	S	1000	11.70		
5	N	1130	14.30	126	2.42
6	S	1200	13.96		

9. A vessel on successive runs on the measured mile obtains the following speeds in knots:

27.592, 28.841, 27.965, 28.943, 27.777, 28.426

Calculate (i) ordinary average speed, (ii) mean of means of six runs, (iii) mean of means of first four runs, (iv) mean of means of second four runs, (v) mean of means of last four runs.

10. Assuming that the speed runs reported in the last question were obtained as a result of runs at intervals of one hour, deduce the true speed of the ship assuming that the tide is governed by an equation

$$v = a + bt + ct^2 + dt^3 + et^4$$

Determine the values of the coefficients in this equation.

11. A propeller 3 m in diameter moves ahead at 15 knots in 'open' sea water. If the propeller race has a 3 knots increase in speed, approximate by the axial momentum theory to the thrust developed.

12. The propellers of a twin-screwed ship operate in a wake of 2 knots, the ship moving ahead at 21 knots. The P_E naked is 4.47 MW, the appendage coefficient is 1.12. If the thrust developed by each propeller is 0.264 MN, calculate (a) the P_T of each propeller, (b) the hull efficiency, (c) the augment of resistance factor.

11 Powering of ships: application

Presentation of data

Any method of data presentation should bring out clearly the effect of the parameters concerned on the resistance of the ship. A non-dimensional form of plotting is desirable but further than this it is difficult to generalize. The best plot for a designer may not be the best for research. The best form of plotting may depend upon how data is to be processed and then the type of calculation in which it is to be used.

It was shown, in the previous chapter, how dimensional analysis can be used to derive a suitable form for non-dimensional presentation of the data involved in ship hydrodynamics.

RESISTANCE DATA

The Froude approach

It was William Froude who first postulated that a ship's resistance is made up of two main components, one due to friction and one to wavemaking. He assumed that these components followed different scaling laws and that they did not interact with each other. He further assumed that the skin friction component of resistance was the same as the resistance of a thin flat plate of the same length (model or ship) with the same wetted surface area at the same speed. He realized there was a smaller component due to eddy making but assumed this could be treated in the same way as the wavemaking resistance. The elements of resistance not due to friction he called the *residuary resistance*.

To present his work, Froude developed what is known as the *circular notation*. Although a truly non-dimensional form of presentation it looks strange to the modern eye. Because it was so important in developing knowledge in this field, and because a lot of data exists in this form, the *Froude notation* and its use are dealt with in the Annex at the end of the book.

ITTC presentation

The International Towing Tank Conference use the following notation

$$C = \text{Resistance coefficient} = \frac{\text{Resistance}}{\frac{1}{2}\rho S V^2}, \quad S = \text{wetted surface area}$$

Subscripts, T, V, R, F and AA are used to denote total, viscous, residuary, frictional and air resistance respectively. Further qualification is by means of subscripts S and M denoting the ship and model respectively.

The following relationships have been adopted:

$$C_{VM} = (1 + k)C_{FM}$$

where k is a form factor from low-speed resistance tests

$$C_{RS} = C_{RM} = C_{TM} - C_{VM}$$
$$C_{VS} = (1 + k)C_{FS} + \Delta C_F$$

where ΔC_F is a roughness allowance

$$C_{TS} = C_{VS} + C_R + C_{AAS}$$

The use of this notational method in arriving at the ship performance is discussed later. Although the ITTC presentation is now that generally used much useful data still exists within the framework of two old presentational methods described in the Annex and below.

Taylor's method

Taylor (1943) expressed resistance, both frictional and residuary, in lbf per tonf of displacement (i.e. R/Δ). For similar models at corresponding speeds such quantities are constant for resistances following Froude's law of comparison. They are compared on the basis of the following parameters:

$$\text{Speed coefficient} = \frac{V}{\sqrt{L}}, \quad V \text{ in knots}, \quad L \text{ in ft}$$

$$\text{Displacement/length ratio} = \frac{\Delta}{(L/100)^3}, \quad \Delta \text{ in tonf}$$

$$\text{Prismatic coefficient}$$

$$\text{Beam/draught ratio}$$

Taylor chose the displacement/length ratio as a quantity which is independent of displacement for similar ships. Length is used in the denominator as being the linear dimension having most influence on resistance. In this expression, Δ is in tonf of salt water whether ship or model is under consideration.

Unfortunately, this type of presentation is not truly non-dimensional and care must be taken with units in applying Taylor's data. Some typical curves are reproduced as Figs 11.1 and 11.2.

Taylor studied the influence of bow shape on resistance by considering the slope of the curve of sectional areas at the bow. This slope is expressed by a quantity t obtained as follows. Draw the tangent at the bow to the curve of sectional areas. This will cut the vertical at the centre of length, intercepting it on a certain ordinate. Then t is the ratio between this ordinate and the ordinate of the sectional area curve at the centre of length (Fig. 11.3).

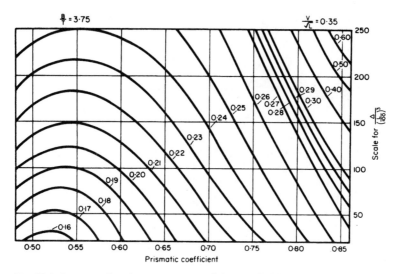

Fig. 11.1 Contours of residuary resistance in lbf per tonf of displacement

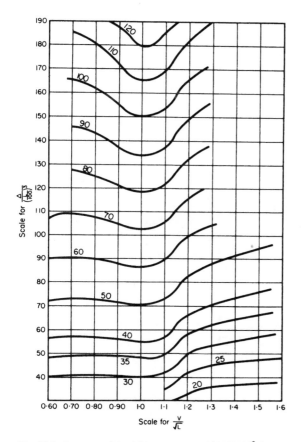

Fig. 11.2 Contours of (midship section area)/(L/100)² for minimum residuary resistance

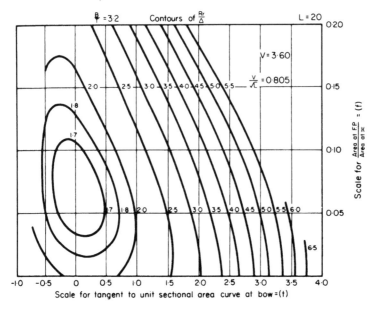

Fig. 11.3 *Variation of resistance with tangent value*

Taylor suggested that the wetted surface area of vessel could be obtained from the formula:

$$S = C(\Delta L)^{\frac{1}{2}}$$

where C is defined by Fig. 11.4.

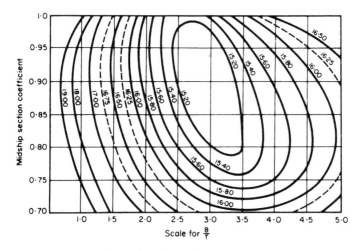

Fig. 11.4 *Contours of wetted surface coefficients*

Statistical analysis

A statistical analysis of resistance data for destroyers and frigates used regression equations for $\textcircled{C}$ for a 4.9 m model in the form:

$$\textcircled{C} = b_0 + b_1 x_1 + b_2 x_2 + \cdots + b_n x_n$$

The first nine terms of the right-hand side of these equations represented form parameters as follows:

$x_1 = \overline{F_{WL}G}/L_{WL}$, where $\overline{F_{WL}G}$ is the horizontal distance of the c.g. from fore end of the static waterline.

$x_2 = $ half angle of entrance.

$x_3 = B/T$

$x_4 = $ (max. transverse sectional area below WL)$/BT$

$x_5 = $ transom area as a percentage of the max. section area

$x_6 = $ (distance of ACU from FE of static waterline)$/L_{WL}$

$x_7 = \textcircled{M}$

$x_8 = $ prismatic coefficient

The remaining terms represented various combinations of these basic variates, i.e. they represented non-linear terms. The best fit was obtained with the more complex equations containing the non-linear terms. This is to be expected but it does not follow that the complex equations give better predictions for a new design. As an example, Table 11.1 gives the b coefficients for the simpler equations for four $\textcircled{K}$ values.

This method of analysis can be very useful but large extrapolations from the generality of forms on which the data are derived must be treated with care. The parameters used are those which are thought to be most significant.

Table 11.1*

Coefficient	$\textcircled{K}$			
	1.8	3.2	4.6	6.0
b_0	0.1032/01	0.3269/01	0.7393/01	0.4589/01
b_1	0.2798/00	−0.02364/01	−0.5395/01	−0.2332/01
b_2	0.6697/−02	0.1531/−01	−0.1972/−02	−0.6813/−02
b_3	0.3620/−01	0.8618/−01	0.1225/00	0.7545/−01
b_4	−0.4145/00	−0.1035/01	−0.5765/00	−0.1309/00
b_5	0.5635/−02	0.7369/−03	0.1501/−01	0.1200/−01
b_6	−0.1618/00	0.1293/00	0.6622/00	0.1280/00
b_7	0.5690/−01	−0.1536/00	−0.3577/00	−0.1669/00
b_8	−0.3060/00	0.1384/01	−0.5878/00	−0.6869/00

Note: The 'b' coefficients are given in exponential form, i.e. 0.1032/01 represents 1.032; −0.4145/00 represents −0.4145; 0.1200/−01 represents 0.01200, etc.

Experience may show that other parameters are more suitable. So far $\textcircled{M}$ has proved the most significant and a useful relationship is:

$$\textcircled{C} = a + b/\textcircled{M} + c\,\textcircled{M}^{2}$$

In general it is not possible to vary one parameter without some consequential change in others. This may explain why the simple equations above indicate that an increase in transom area is always bad. This is not in accord with common experience and illustrates the danger of attributing any particular physical sig-nificance to the sign and magnitude of the regression coefficients. The interaction between various parameters also influences the extent to which optimization can be achieved. Values obtained from a mathematical optimization process may not be achievable in one form and in any case do not uniquely define the ship form. The regression equations should only be applied to forms of the same general type as those used to derive the equations.

PROPELLER DATA

In Chapter 10, dimensional analysis led to the derivation of three basic coeffi-cients, viz.:

$$K_{\mathrm{T}} = \frac{T}{\rho n^{2} D^{4}} = \text{thrust coefficient}$$

$$K_{\mathrm{Q}} = \frac{Q}{\rho n^{2} D^{5}} = \text{torque coefficient}$$

$$J = \frac{V}{nD} = \text{advance coefficient}$$

In these coefficients, the product nD is a measure of the rotative speed of the propeller.

The other basic parameter is the propeller efficiency η which is given by

$$\eta = \frac{\text{useful output}}{\text{input}}$$

$$= \frac{TV}{Q \times 2\pi n} = \frac{K_{\mathrm{T}}}{K_{\mathrm{Q}}}\frac{J}{2\pi}$$

For a given advance coefficient, it is only necessary to define two of the factors K_{T}, K_{Q} and η as the third follows from the above relationship. The two usually quoted are K_{T} and η. It was pointed out in Chapter 10 that a propeller designer makes considerable use of the results of methodical model series representing the propeller in open water. Such series are reported in various technical papers and a typical plot for a given blade area ratio is shown in Fig. 11.5. Similar plots are available other BARs.

In most design problems, the speed of advance and the power P_{D} to be absorbed are known. In addition, the propeller r.p.m. are also often fixed by considerations of gear ratios and vibration. Diagrams such as that in Fig. 11.5 can be used to obtain, by interpolation, the propeller diameter for maximum

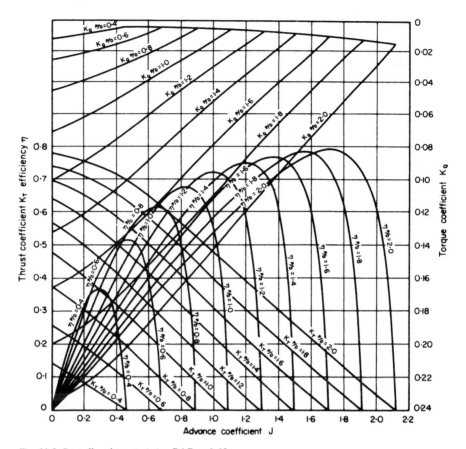

Fig. 11.5 Propeller characteristics, BAR = 0.65

efficiency. This process is described later. Another type of presentation can, however, be adopted to simplify this common type of problem.

This is a plot of B_p against δ where

$$B_p = \frac{nP_D^{\frac{1}{2}}}{V^{2.5}} = 33.08 \left(\frac{K_Q}{J^5} \right)^{\frac{1}{2}}$$

$$\delta = 3.2808 \frac{nD}{V}$$

where n is in r.p.m., P_D is in the power, V is in knots, and D is in metres.

Such a plot is presented in Fig. 11.6.

For given values of n, P_D and V, B_p is fixed, and by drawing a vertical ordinate at this value on the figure the maximum obtainable η and corresponding propeller diameter can be determined. In fact, the curve for the optimum efficiency for the most favourable diameter can be plotted on the figure. It connects the points on the $\eta = $ constant curves at which these curves are vertical, i.e. $B_p = $ constant. This line is shown dotted on Fig. 11.6. If the diameter is limited in some ways, the optimum within this limitation is readily deduced.

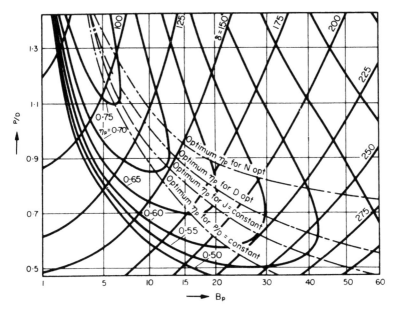

Fig. 11.6 $B_p - \delta$ diagrams

When cavitation occurs, as may be the case of a ship at sea or with a propeller in a cavitation tunnel, the η and K_T values are modified as shown in Fig. 11.7.

The more advanced propeller theories which were touched upon in the previous chapter give the designer the ability to design propellers suited to the specific ship wakes in which they are to operate, or to give reduced levels of noise and propeller-induced vibrations. They call for better knowledge of the flow into the propeller so that wake measurements assume greater importance.

Power estimation

RESISTANCE PREDICTION

Froude's law of comparison is the key to most forms of extrapolation from model to ship. By this law, the residuary resistance per unit of displacement is the same for model and ship at corresponding speeds. It remains then necessary to know how the frictional resistance varies with Reynolds' number to enable a plot such as Fig. 11.8 to be produced. Let AA′ represent the variation of total resistance of the model with Reynolds' number. Then, provided the skin friction line is a correct one, $\overline{A_1 A_2}$ and $\overline{A_2 A_3}$ are the residuary and skin friction components at a Reynolds' number $(R_n)_m$. By Froude's law of comparison, if $(R_n)_S$ is the corresponding ship R_n, $\overline{A_1 A_2}$ will be equal to $\overline{B_1 B_2}$. Thus, the total ship resistance curve can be obtained by drawing curves through points on the model curve parallel to the skin friction line to intersect vertical lines through the R_n values appropriate to the corresponding speeds.

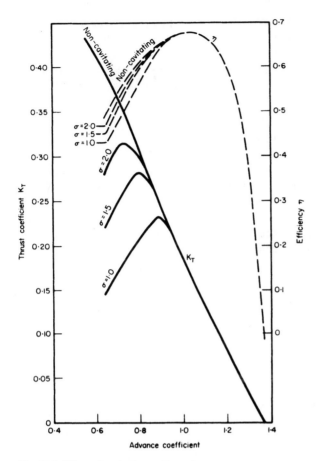

Fig. 11.7 Effect of cavitation on K_T and η

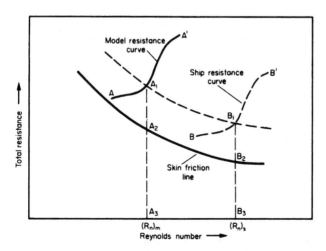

Fig. 11.8 Extrapolation from model to full-scale resistance

Clearly, the accuracy of prediction of such a method is dependent on the accuracy of the curve defining the variation of skin friction resistance with Reynolds' number. Not unnaturally, several curves have been proposed over the years each having its particular advocates. The ship resistance prediction resulting from the different skin friction curves are but marginally different.

The ITTC 1957 model–ship correlation line

Obviously it is desirable to have a single line for correlating ship and model results which is accepted by all practitioners and can be used as a 'standard'. Much effort has been devoted over the years to trying to reach agreement on such a standard line as the volumes of the Royal Institution of Naval Architects bear eloquent witness. One way of studying the relative merits of different formulations is to carry out tests on a series of models of various sizes so that a range of Reynolds' numbers is covered. By presenting all the results in a single plotting such as Fig. 11.8, the shape of the skin friction line is determined by passing curves through points on each model curve at corresponding Reynolds' numbers.

Successive International Towing Tank Conferences studied this problem and in 1957, in Madrid, agreed to a standard line. This ITTC line is defined by

$$C_F = \frac{0.075}{(\log R_n - 2)^2} = \frac{R_F}{\frac{1}{2}\rho S V^2}$$

Values of C_F for various values of R_n are given in Table 11.2.

The term 'correlation line' was used quite deliberately in recognition of the fact that the extrapolation from model to ship is not governed only by variation in skin friction.

APPENDAGE RESISTANCE

A ship has a number of appendages each of which will have associated Froude and Reynolds' numbers depending upon its characteristic length. Thus in going from the model to full scale the resistance of each appendage will scale differently from that of the main hull. Whilst an appended model can be (and is by some authorities) run and scaled as a whole this introduces an approximation which must be allowed for in some overall correlation factor. Many authorities prefer to calculate the resistance of appendages such as bilge keels, stabilizers, rudders, shaft brackets and so on. Provided a consistent method of calculation is adopted for a series of designs any errors in absolute values will be taken care of by correlation factors deduced from comparison of ship trial data with model tests.

(a)   Bilge keels

Because care is taken to align the bilge keels with the flow around the hull the resistance of the keels may be taken as the skin frictional resistance of the total wetted surface, based on the characteristic length of the keels. Allowance must

Table 11.2
Coefficients for ITTC 1957 model–ship correlation
line. Coefficients must be multiplied by 10^{-3}

Reynolds' number	$10^5 \times$	$10^6 \times$	$10^7 \times$	$10^8 \times$	$10^9 \times$	$10^{10} \times$
1.0	8.333	4.688	3.000	2.083	1.531	1.172
1.5	7.435	4.301	2.799	1.966	1.456	1.122
2.0	6.883	4.054	2.669	1.889	1.407	1.088
2.5	6.496	3.878	2.574	1.832	1.370	1.063
3.0	6.203	3.742	2.500	1.788	1.342	1.044
3.5	5.971	3.632	2.440	1.751	1.318	1.027
4.0	5.780	3.541	2.390	1.721	1.298	1.014
4.5	5.620	3.464	2.347	1.694	1.280	1.002
5.0	5.482	3.397	2.309	1.671	1.265	0.991
5.5	5.361	3.338	2.276	1.651	1.252	0.982
6.0	5.254	3.285	2.246	1.632	1.240	0.973
6.5	5.159	3.238	2.220	1.616	1.229	0.966
7.0	5.073	3.195	2.195	1.601	1.219	0.959
7.5	4.995	3.156	2.173	1.587	1.209	0.952
8.0	4.923	3.120	2.152	1.574	1.201	0.946
8.5	4.857	3.087	2.133	1.562	1.193	0.941
9.0	4.797	3.056	2.115	1.551	1.185	0.935
9.5	4.740	3.027	2.099	1.540	1.178	0.931

be made for the fact that part of the hull is shielded from the water flow. Possible refinements are to allow for the actual mean flow velocity over the keels and for an interference drag arising from the junction between the hull and keel. An allowance for fouling can be made in line with that for the hull itself.

(b) Rudders, stabilizer fins and shaft bracket arms

These are all aerofoil type sections and drag can be deduced from the characteristics of the aerofoil section adopted. The velocity assumed can be taken as that from model flow experiments although some augmentation of the velocity used for rudders is usual (typically 10 per cent increase) to allow for the propulsor influence. An allowance for fouling can be made based on the surface area of the appendage. Shaft bracket arms will experience interference effects where they enter the hull and where they join the barrel. The resistance of the barrel itself will depend upon the projected area presented to local flow. Usually there is a significant cross-flow velocity at the shaft brackets because of the local hull shape and the need to provide good propeller/hull clearance.

(c) Large inlets

Typical of these are main condenser inlets and the resistance of these will depend upon whether a circulating pump is used. Essentially the resistance due to flow through the system is deduced from the momentum changes as the water enters, transits and then leaves the system.

1978 ITTC PERFORMANCE PREDICTION METHOD

The prediction method proposed by the ITTC in 1978 for single-screw ships follows closely the general analysis presented in Chapter 10. The steps are:

(a) the viscous resistance is taken as $(1 + k)$ times the frictional resistance where k is determined from the model test and assumed independent of speed and scale;

(b) a roughness allowance is calculated from

$$\Delta C_F = \left[105 \left(\frac{k_s}{L} \right)^{\frac{1}{3}} - 0.64 \right] \times 10^{-3}$$

where k_s = roughness of hull = 150×10^{-6} m

L = length of waterline

(c) air resistance is calculated from

$$C_{AAS} = 0.001 \frac{A_T}{S}$$

A_T = transverse projected area of ship above the waterline

(d) Taylor wake factor, w_T, and thrust deduction factor, t, are used, the latter being assumed the same in the ship as model, i.e. $t_S = t_M$;

(e) the thrust deduction factor, $t(= t_M = t_S)$ is obtained from the difference between the self propulsion thrust and the hull resistance without propeller, corrected if necessary for temperature differences at the time of the separate tests;

(f) the wake fraction is calculated from the self-propulsion tests and model propeller characteristics. From the thrust, T, and torque, Q, measured in the former

$$K_{TM} = \frac{T}{\rho D^4 n^2}, \quad K_{QM} = \frac{Q}{\rho D^5 n^2}$$

The model propeller characteristics give J_{TM} and K_{QTM} for the K_{TM} value. Hence

$$w_{TM} = \frac{V - V_1}{V} = 1 - \frac{J_{TM} D n}{V}$$

(g) the full-scale wake is taken as

$$w_{TS} = (t + 0.04) + (w_{TM} - t - 0.04) \frac{C_{VS}}{C_{VM}}$$

where the value 0.04 is introduced to take account of rudder effects;

(h) the relative rotative efficiency is assumed the same for the ship as model

$$\eta_{RS} = \eta_{RM} = \eta_R = \frac{K_{QTM}}{K_{QM}}$$

(*i*) the total ship resistance coefficient without bilge keels is given by

$$C_{TS} = (1+k)C_{FS} + C_R + \Delta C_F + C_{AAS}$$

where

C_{FS} = frictional coefficient of ship according to the ITTC 1957 ship model correlation line

C_R = residual resistance calculated from the total and viscous resistance of the model

$$= C_{TM} - (1+k)C_{FM}$$

(*j*) bilge keels can be allowed for by multiplying the C_{FS} and ΔC_F terms by the ratio

$$\frac{S + S_{BK}}{S}, \quad S_{BK} = \text{surface area of the bilge keels}$$

(*k*) scale effect corrections are applied to the propeller characteristics as follows:

$$K_{TS} = K_{TM} - \Delta K_T$$
$$K_{QS} = K_{QM} - \Delta K_Q$$

where $\Delta K_T = -\Delta C_D \left(0.3 \dfrac{P}{D}\right) \dfrac{cZ}{D}$

$$\Delta K_Q = \Delta C_D (0.25) \frac{cZ}{D}$$

ΔC_D = difference in drag coefficient = $C_{DM} - C_{DS}$

$$C_{DM} = 2\left(1 + \frac{2t}{c}\right)\left[\frac{0.044}{(R_{nco})^{1/6}} - \frac{5}{(R_{nco})^{2/3}}\right]$$

$$C_{DS} = 2\left(1 + \frac{2t}{c}\right)\left[1.89 + 1.62 \log \frac{c}{k_p}\right]^{-2.5}$$

R_{nco} = local Reynolds' number at radius = 0.75 maximum
  (this value not to be less than 2×10^5 in open water test)

k_p = blade roughness = 30×10^{-6} m

c = chord length

t = maximum blade thickness

$\dfrac{P}{D}$ = pitch ratio;

(*l*) the load of the full-scale propeller is obtained from

$$\frac{K_{TS}}{J^2} = \frac{S}{2D^2} \times \frac{C_{TS}}{(1-t)(1-w_{TS})^2}$$

From this value J_{TS} and K_{QS} follow from the full-scale propeller character-istics. From these it follows that

$$\text{full-scale revs} \quad = n_S = \frac{(l - w_{TS})V_S}{J_{TS}D} \text{ (r.p.s.)}$$

$$\text{delivered power} = P_{DS} = 2n\rho D^5 n_S^3 \frac{K_{QTS}}{\eta_R} \times 10^{-3} \text{ (kW)}$$

$$\text{propeller thrust} = T_s = \frac{K_T}{J^2} \times J_{TS}^2 \rho D^4 n_S^2 \text{ (N)}$$

$$\text{propeller torque} = Q_s = \frac{K_{QTS}}{\eta_R} \rho D^5 n_S^2 \text{ (Nm)}$$

$$\text{effective power} \quad = P_E = C_{TS} \times \tfrac{1}{2}\rho V_S^3 \times S \times 10^{-3} \text{ (kW)}$$

$$\text{total efficiency} \quad = \eta_D = \frac{P_{DS}}{P_E}$$

$$\text{hull efficiency} \quad = \eta_H = \frac{1 - t}{1 - w_{TS}}$$

EFFECT OF SMALL CHANGES OF DIMENSIONS

Froude's formula for frictional resistance may be written

$$R_F = f\Delta^{\frac{2}{3}}V^{1.825}$$

For geometrically similar ships at corresponding speeds

$$V \propto L^{\frac{1}{2}}; \ \Delta \propto L^3$$

Hence

$$R_F \propto fL^{2.9125}$$

By Froude's law of comparison the residuary resistance varies as L^3. Hence, for small changes in dimensions no large error is introduced if it is assumed that the total resistance varies in the same way, i.e.

$$R_T \propto L^3$$

Variation in residuary resistance with size and speed

At a given speed for any condition, the residuary resistance will vary with displacement and speed as follows:

$$R_R = K\Delta^m V^{n-1}$$

The power then varies as $\Delta^m V^n$. For a geometrically similar form at corres-ponding speed

$$R'_R = K(\Delta')^m (V')^{n-1}$$

But by Froude's law of comparison

$$\frac{R_R}{R'_R} = \frac{\Delta}{\Delta'}$$

and

$$\frac{V}{V'} = \left(\frac{L}{L'}\right)^{\frac{1}{2}} = \left(\frac{\Delta}{\Delta'}\right)^{\frac{1}{6}}$$

Hence

$$\frac{\Delta}{\Delta'} = \left(\frac{\Delta}{\Delta'}\right)^{m} \left(\frac{\Delta}{\Delta'}\right)^{(n-1)/6}$$

i.e.

$$m + \frac{n-1}{6} = 1$$

$$6m + n = 7$$

The value of n can be deduced from the slope of the resistance/speed curve as follows:

$$C_R = \text{Const.} \; \frac{R}{V^2}$$

Hence

$$C_R = \text{Const.} \; \frac{V^{n-1}}{V^2} = \text{Const.} \; V^{n-3}$$

Differentiating,

$$\frac{\partial C_R}{\partial V} = (n-3) \, \text{Const.} \; V^{n-4}$$

Eliminating the constant in these equations

$$\frac{\partial C_R}{\partial V} = (n-3)\frac{C_R}{V}$$

Hence

$$n = 3 + \frac{V}{C_R} \cdot \frac{\partial C_R}{\partial V}$$

The value of m follows from

$$m = \frac{7-n}{6}$$

It may be necessary to depart from the form in the early design stages in a way that destroys geometric similarity. Beam may have to be increased to improve initial stability, length may have to be increased to provide an acceptable weather deck layout, and so on. Mathematically the variation of power with length, beam and draught can be expressed in terms of performance coefficients α, β and γ as

$$\frac{\partial P_E}{P_E} = \alpha\frac{\partial L}{L} + \beta\frac{\partial B}{B} + \gamma\frac{\partial T}{T}$$

The values of the performance coefficients are obtained from tests with three models—known as *triplets*. One is the parent form, one has beam changed by a small percentage and one has modified length. The change in dimension is typically 10 per cent and is a simple linear stretch in the relevant direction.

VARIATION OF SKIN FRICTIONAL RESISTANCE WITH TIME OUT OF DOCK

When a ship enters the water having been freshly cleaned and painted its resistance is a minimum. With the passage of time, seaweed and barnacles attach themselves to the surface so presenting a rougher surface to the passage of water. This roughening of the surface leads to an increase in the skin frictional resistance. It is to be expected that the amount of fouling as it is called will depend upon the area in which the ship is operating and the time spent at sea compared with the time at rest in harbour.

The MOD used to assume an increase in skin friction resistance of a quarter of one per cent per day and took as standard a 'deep and dirty' condition with the ship at deep displacement and six months out of dock. Then:

$$\text{Increase in } \textcircled{C}_F = \delta\textcircled{C}_F = \frac{365}{200} \times 0.25\textcircled{C}_F = 0.456\textcircled{C}_F$$

Other authorities assume a standard percentage of the available power, e.g. 20 per cent, is used up in overcoming the increased resistance due to fouling and also in overcoming the extra resistance due to running through waves. Allowances for fouling can be less with modern protection systems and more recent practice is to allow a 15 per cent margin on endurance power. The corresponding figure for the USN is 10 per cent but they use full load displacement in calculating endurance whereas the RN uses an average displacement which reflects the fact that the ship gets lighter as fuel is consumed.

EXAMPLE 1. Data for a model 5 m long of a ship 178 m long and 11,700 tonnef displacement, corrected to standard temperature, is defined by the following table.

$R_n \times 10^6$	6.600	6.894	7.188	7.522	7.817	8.111	8.445
$C_T \times 10^3$	3.944	3.944	3.951	3.972	3.979	3.982	4.014

$R_n \times 10^{-6}$	8.740	8.995	9.295	9.624	9.918	10.213
$C_T \times 10^3$	4.024	4.034	4.090	4.164	4.310	4.478

Deduce a plot, power against speed for the clean and dirty conditions assuming that the wetted surface area is $3650\,\text{m}^2$.

Solution: The data having been presented in the form of C_T against R_n, the 1957 ITTC analysis is applied

Reynolds' number, $R_n = \dfrac{VL}{v}$

The standard values (i.e. at 15 °C) for v are

$$v = 1.139 \times 10^{-6}\,\text{m}^2/\text{s for fresh water}$$

$$v = 1.188 \times 10^{-6}\,\text{m}^2/\text{s for sea water}$$

If V is the speed of the ship, the corresponding speed for the 5 m model is $V\sqrt{(5/178)}$. Hence for the ship

$$R_n = \frac{1.139}{1.188}\left(\frac{178}{5}\right)^{\frac{3}{2}}(R_n)_{\text{model}}$$

$$= 203.7(R_n)_{\text{model}}$$

The ship speed

$$1.188 \times 10^{-6}R_n/L = 6.674 \times 10^{-9}(R_n)_{\text{ship}}\, m/s.$$

The C_T values for the model have to be corrected for the skin friction difference in going from model to ship. This correction is the difference between the ordinates of the ITTC line at the Reynolds' numbers appropriate to the model and the ship. Hence, the total resistance C_T can be deduced for the ship. Since

$$C_T = \frac{\text{resistance}}{\frac{1}{2}\rho S V^2}$$

power $= R \times V = \frac{1}{2}\rho S V^3 C_T = 1.872 V^3 C_T$ MW, V in m/s.

The increase in resistance in the dirty condition is given by

$$\delta C_F = 0.456 C_F \text{ using the ship value}$$

The calculation can now be completed in tabular form as shown in Table 11.3. The results are plotted in Fig. 11.9.

RESISTANCE IN SHALLOW WATER

Chapter 9 concentrated on waves in deep water. As can be found in any standard textbook on hydrodynamics, in water of depth, h, the speed of propagation of a wave is given by:

$$c^2/gh = \lambda/2\pi h[\tanh(2\pi h/\lambda)]$$

In deep water $[\tanh(2\pi h/\lambda)]$ tends to unity giving $c^2 = g\lambda/2\pi$. When h/λ is small $[\tanh(2\pi h/\lambda)]$ tends to $2\pi h/\lambda$ and $c^2 = gh$. This is constant for any given depth, that is it does not depend upon the wavelength. $(gh)^{\frac{1}{2}}$ is the maximum speed of a wave in shallow water and is known as the *critical speed*.

Table 11.3
1957 ITTC analysis

$(R_n)_{model}$ $\times 10^6$	$(R_n)_{ship}$ $\times 10^9$	V Ship (m/s)	$C_F \times 10^{-3}$ Model	$C_F \times 10^{-3}$ Ship	SFC $\times 10^{-3}$	$C_T \times 10^{-3}$ Model	Clean ship $C_T \times 10^{-3}$	MW	$\delta C_F \times 10^{-3}$	Dirty ship $C_T \times 10^{-3}$	MW
6.600	1.344	8.97	3.229	1.476	1.753	3.944	2.191	1.534	0.673	2.864	2.005
6.894	1.404	9.37	3.204	1.468	1.736	3.944	2.208	1.817	0.669	2.877	2.368
7.188	1.464	9.77	3.180	1.461	1.719	3.951	2.232	2.082	0.666	2.898	2.704
7.522	1.532	10.22	3.154	1.453	1.701	3.972	2.271	2.423	0.663	2.934	3.131
7.817	1.592	10.63	3.132	1.446	1.686	3.979	2.293	2.736	0.659	2.952	3.557
8.111	1.652	11.03	3.112	1.440	1.672	3.982	2.310	3.100	0.657	2.967	3.982
8.445	1.720	11.48	3.090	1.433	1.657	4.014	2.357	3.566	0.653	3.010	4.554
8.740	1.780	11.88	3.071	1.427	1.644	4.024	2.380	3.991	0.651	3.031	5.082
8.995	1.832	12.23	3.056	1.422	1.634	4.034	2.400	4.390	0.648	3.048	5.575
9.295	1.893	12.63	3.038	1.417	1.621	4.090	2.469	4.975	0.646	3.115	6.277
9.624	1.960	13.08	3.020	1.410	1.610	4.164	2.554	5.716	0.643	3.197	7.155
9.918	2.020	13.46	3.004	1.405	1.599	4.310	2.711	6.639	0.641	3.352	8.209
10.213	2.080	13.88	2.990	1.400	1.590	4.478	2.888	7.723	0.638	3.526	9.429

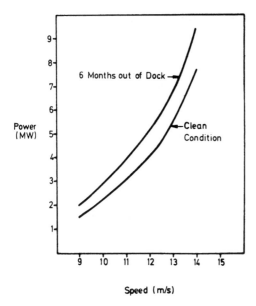

Fig. 11.9 Plot of results

When a ship is moving in shallow water the angle of the line of the maximum height of the diverging waves to the centreline increases, approaching 90 degrees as the ship speed approaches the critical speed. At speeds below the critical, the resistance in shallow water is greater than that in deep water and the resistance increases dramatically as the critical speed is neared. If the ship has enough power to exceed the critical speed then the resistance will fall well below that for deep water at the same speed.

CALCULATION OF WIND RESISTANCE

The fair above-water portion of the main hull experiences less resistance force per unit area than the superstructure. In fact, the resistance per unit of projected area is only about 30 per cent of that of the superstructures. Wind tunnel tests carried out by NPL showed that the resistance offered by the ship could be represented by the equation

$$\text{Resistance} = KBV^2$$

where B = projected area onto a transverse plane of the superstructure plus 30 per cent of the projected area of the above water hull, V = relative wind speed (knots), and K is a coefficient depending on the ship type and the angle of the relative wind to the middle-line of the ship.

It was found that K was fairly constant for angles up to about 15 degrees off the bow and was a maximum for an angle of about 30 degrees off the bow.

If B is in m² and V in knots then, approximately,

$$\text{Resistance} = 48\,KBV^2 \text{ newtons}$$

One useful parameter for comparing results is the *ahead resistance coefficient* (ARC) defined by

$$\text{ARC} = \frac{\text{fore and aft component of wind resistance}}{\frac{1}{2}\rho V_R^2 A_T}$$

In the case of a tanker, the ARC values were reasonably steady for relative winds from ahead to 50 degrees off the bow, the value varying from 0.7 in the light condition to about 0.85 in the loaded condition. Corresponding values for winds up to 40 degrees off the stern were −0.6 and −0.7. Variation with relative wind direction between 50 degrees off the bow and 40 degrees off the stern was approximately linear. Two cargo ships exhibited similar trends but the values of the ARC were about 0.1 lower. The same ARC values apply in the metric system provided consistent units are used.

The results include the effect of the velocity gradient existing in atmospheric winds. (See Chapter 9.) They represent the force experienced by a ship of the size tested (tanker 169 m, cargo ships 149 m and passenger liner 245 m, length overall). Smaller ships will have a greater percentage of their area in the lower regions of the gradient and will suffer proportionately less force. The force can be assumed to vary as the square of the velocity. If the results are to be used to deduce the air resistance experienced by a ship moving ahead with no wind then there is no velocity gradient and the forces deduced must be increased by 25 per cent in the light condition and 40 per cent in the deep load condition for the tanker and cargo ships. The increase for a passenger ship is about 21 per cent.

Wind tunnel tests have been carried out in Japan to help assess the size of mooring rope and the amount of cable veer required by large ships in a strong wind. The problem has since acquired greater urgency due to the increase in ship size and the limitations of available anchorages.

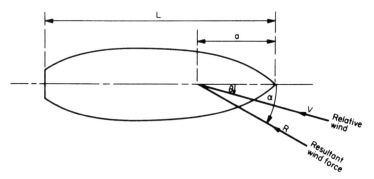

Fig. 11.10 Wind forces on a ship

The Japanese tests showed that the resultant wind force R is given by

$$R = \tfrac{1}{2}\rho C_r V^2 (A \cos^2 \theta + B \sin^2 \theta), \text{newtons}$$

Where $\rho =$ atmospheric density (kgf m^{-4} s^2) $= 0.123$ approx., $A =$ frontal projection area (m^2), $B =$ lateral projection area (m^2), and $V =$ relative wind speed (m/s).

Based on results for a number of ships, it is suggested that C_r values are given by the following relationships:

Cargo ship: $C_r = 1.325 - 0.05 \cos 2\theta - 0.35 \cos 4\theta - 0.175 \cos 6\theta$

Passenger ship: $C_r = 1.142 - 0.142 \cos 2\theta - 0.367 \cos 4\theta - 0.133 \cos 6\theta$

Oil tanker: $C_r = 1.20 - 0.083 \cos 2\theta - 0.25 \cos 4\theta - 0.117 \cos 6\theta$

Approximate values of A and B are given by

$A = B^2[X_A - 0.00475 \, dr]$.   In this case, $B =$ beam

$B = L^2[X_B - 0.0006 \, dr]$

where $dr =$ percentage the actual draught is of the fully-loaded draught and values of X_A and X_B are approximately given in Table 11.4.

The distance aft from the bow at which the resultant wind force acts, expressed as a percentage of ship length, varies uniformly with wind directions between 20° and 160° off the bow and is given by

$$\frac{a}{L} = 0.291 + 0.0023\theta, \quad \theta \text{ in degrees}$$

Table 11.4
Values of X_A and X_B

Ship type	X_A	X_B
3 Island cargo-ship	1.43	0.1195
Cargo-ship with stern machinery	1.225	0.110
Oil tanker	1.095	0.099
Passenger ship	1.455	0.156

Generally, the direction of the resultant wind force, though not the same as the wind direction, changes with it as follows:

$$\alpha = \left\{ 1 - 0.15\left(1 - \frac{\theta}{90}\right) - 0.80\left(1 - \frac{\theta}{90}\right)^3 \right\} \times 90$$

The data above were obtained from measurements on models floating on water. They are therefore subject to a certain unspecified wind gradient. If this is assumed to be comparable to that occurring in nature, then the air resistance force acting on the ship when moving ahead in calm air must be increased as proposed above for the NPL data.

Multiple regression analysis has been used to obtain expressions for the fore and aft wind force component, the lateral wind force component and the

wind-induced yawing moment. Forty-nine sets of experimental data were used
and the equations expressed the force or moment coefficient in the general form

$$a_0 + a_1 \frac{2A_L}{L^2} + a_2 \frac{2A_T}{B^2} + a_3 \frac{L}{B} + a_4 \frac{S}{L} + a_5 \frac{C}{L} + a_6 M$$

where

L and B are overall length and beam

A_L and A_T are the lateral and transverse projected area

S is the length of perimeter of lateral projection of model excluding the
waterline and slender bodies such as masts

C is distance from bow of centroid of lateral projected area

M is number of distinct groups of masts or kingposts seen in lateral projection.

Table 11.5 reproduces the values of a_0 to a_6 for the fore and aft wind force
component together with the corresponding standard error.
Table 11.6 gives typical values of the independent variables to enable approximate values of wind resistance to be calculated in the early design stage.

Table 11.5
Fore and aft component of wind force

$$C_X = a_0 + a_1 \frac{2A_L}{L_{0A}^2} + a_2 \frac{2A_T}{B^2} + a_3 \frac{L_{0A}}{B} + a_4 \frac{S}{L_{0A}} + a_5 \frac{C}{L_{0A}} + a_6 M \pm 1.96 \text{ S.E.}$$

γ_R°	a_0	a_1	a_2	a_3	a_4	a_5	a_6	S.E.
0	2.152	−5.00	0.243	−0.164	—	—	—	0.086
10	1.714	−3.33	0.145	−0.121	—	—	—	0.104
20	1.818	−3.97	0.211	−0.143	—	—	0.033	0.096
30	1.965	−4.81	0.243	−0.154	—	—	0.041	0.117
40	2.333	−5.99	0.247	−0.190	—	—	0.042	0.115
50	1.726	−6.54	0.189	−0.173	0.348	—	0.048	0.109
60	0.913	−4.68	—	−0.104	0.482	—	0.052	0.082
70	0.457	−2.88	—	−0.068	0.346	—	0.043	0.077
80	0.341	−0.91	—	−0.031	—	—	0.032	0.090
90	0.355	—	—	—	−0.247	—	0.018	0.094
100	0.601	—	—	—	−0.372	—	−0.020	0.096
110	0.651	1.29	—	—	−0.582	—	−0.031	0.090
120	0.564	2.54	—	—	−0.748	—	−0.024	0.100
130	−0.142	3.58	—	0.047	−0.700	—	−0.028	0.105
140	−0.677	3.64	—	0.069	−0.529	—	−0.032	0.123
150	−0.723	3.14	—	0.064	−0.475	—	−0.032	0.128
160	−2.148	2.56	—	0.081	—	1.27	−0.027	0.123
170	−2.707	3.97	−0.175	0.126	—	1.81	—	0.115
180	−2.529	3.76	−0.174	0.128	—	1.55	—	0.112

Mean Standard Error 0.103

Table 11.6
Values of independent variables

Variable	$\dfrac{2A_L}{L_{OA}^2}$	$\dfrac{2A_T}{B^2}$	$\dfrac{L_{OA}}{B}$	$\dfrac{S}{L_{OA}}$	$\dfrac{C}{L_{OA}}$	$\dfrac{A_{SS}}{A_L}$	M
Maximum	0.246	2.32	9.75	1.97	0.619	0.595	7
Minimum	0.072	0.88	4.00	1.23	0.401	0.138	1
Mean	0.143	1.78	7.39	1.51	0.506	0.246	4
Ship type							
1	0.192	1.95	7.66	1.44	0.492	0.398	2
2	0.111	1.67	7.80	1.51	0.490	0.258	4
3	0.149	2.04	7.80	1.58	0.489	0.188	4
4	0.122	1.75	7.80	1.51	0.550	0.253	5
5	0.151	2.06	7.80	1.58	0.526	0.175	5
6	0.076	1.03	7.46	1.33	0.547	0.252	3
7	0.117	1.43	7.46	1.40	0.522	0.161	3
8	0.100	1.59	7.46	1.33	0.568	0.211	3
9	0.121	1.68	7.46	1.40	0.537	0.139	3
10	0.166	1.80	6.47	1.45	0.476	0.229	2
11	0.236	1.43	4.05	1.86	0.405	0.396	1

PROPELLER DESIGN

It is possible in this book only to outline the main factors to be considered by the propeller designer. They are:

(a) Shaft revolutions. Apart from the direct influence on propeller efficiency the choice of shaft r.p.m. depends upon the gearing available, critical whirling speeds of shafts and avoidance of the fundamental frequencies of hull vibration;
(b) Number of blades which influences vibration (Fig. 11.11) and cavitation;
(c) Propeller diameter and hence clearance between propeller tips and the hull which has a marked effect on vibration;
(d) Blade area. The greater the blade area for a given thrust the less likely is cavitation;
(e) Boss diameter. Dictated mainly by strength considerations;
(f) Geometry of the blades, e.g. pitch, camber;
(g) The wake in which the propeller is to operate.

The choice of the principal propeller dimensions is an easier problem and is considered below.

CHOICE OF PROPELLER DIMENSIONS

The propeller dimensions are found using methodical series propeller data. This is adequate if the propeller is to be similar in geometry to those forming the methodical series. If it is to differ in some significant way, the dimensions are first approximated to by using the series data and then adjusted as a result of special tests on the propeller so obtained.

It is assumed that the designer has information on P_E, hull efficiency, wake, QPC factor and $(T)_R/(T)_M$ as a result of model experiments and previous ship

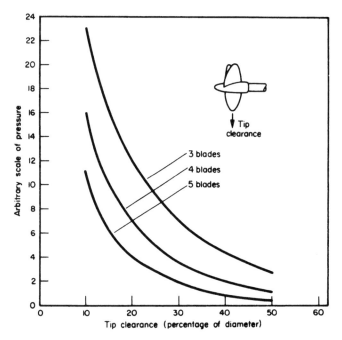

Fig. 11.11 Variation of harmonic pressure with tip clearance

trials. The speed will be stated in the requirements for the ship. The steps in the design process are now:

(*a*) thrust power in open water $= TV_1 = P_T \times \frac{T_k}{T_M}$

from which thrust T can be deduced;
the designer assesses an allowable pressure on the blades if serious cavitation effects are to be avoided. In the absence of other data, it is recommended that a figure of $80\,\text{kN/m}^2$ be used. From this, the blade area to provide the thrust T, and hence **BAR** can be obtained in terms of the diameter of the propeller;

(*c*) for values of **BAR**, for which methodical series data are available, deduce the propeller diameter D. We can now illustrate the method to be used by reference to Fig. 11.5 for a **BAR** of 0.65. The procedure has to be repeated for each **BAR** under consideration;

(*d*) for a series of values of J, calculate the corresponding r.p.m., n;

(*e*) for each n value calculate K_T and by cross plotting find the corresponding P/D ratio and η value;

(*f*) plot n, P and η to a base of J and note the values of n and P corresponding to the maximum η. Note this η value;

(*g*) repeat for a number of **BAR**s and plot n, P and η to base D;

(*h*) read off the values at the optimum value of D. This will either be that for maximum efficiency or, if this diameter is too great, the diameter must be restricted to give a satisfactory tip clearance from the vibration point of view. The propeller diameter may also be limited by consideration of possible damage when the ship is docked or is coming alongside a jetty;

(*i*) the propulsive coefficient can then be deduced using the relationship

$$PC = \left[\frac{\eta_H \times \eta_0 \times \eta_R}{\text{appendage coefficient}}\right] \times QPC \text{ factor, relative to } P_E$$

(*j*) calculate shaft power $= P_E/PC$

EXAMPLE 2. A 174 m long twin-screw ship of 11,684 tonnef similar to that in the example for calculation of P_E had a P_E of 23.49 MW at 28 knots in the deep and dirty condition. Model data for this design suggest the following:

Hull efficiency	$= 0.98$
Relative rotative efficiency	$= 1.00$
Wake	$= 10$ per cent
Appendage coefficient	$= 1.06$

Trial data from a similar ship suggest that with a pressure coefficient of 80 KN/m², the QPC factor is 0.92 and $(T)_R/(T)_M$ is 1.04.

Determine the dimensions and revolutions of the propeller to give maximum efficiency assuming that the maximum diameter from the point of view of docking is 4.27 m. Calculate also the propulsive coefficient at 28 knots and power required at this speed.

Solution: Following the procedure outlined above

$$P_E \text{ per screw} = 11.74 \text{ MW}$$

$$P_T \text{ per screw} = \frac{P_{EA}}{HE} \times \frac{(T)_R}{(T)_M} = \frac{11.74}{0.98} \times 1.04 \times 1.06 = 13.21 \text{ MW}$$

$$V_1 = \frac{28}{1.10} = 25.45 \text{ knots} = 13.1 \text{ m/s}$$

Since $P_T = (\text{thrust}) \times V_1$

Thrust per screw: $\dfrac{13.21}{13.1} = 1.01 \text{ MN}$

Area of blades to restrict pressure loading to 80 kN/m²

$$= \frac{1.01}{80} \times 10^3 = 12.6 \text{ m}^2$$

Hence

$$BAR = \frac{4A_D}{\pi D^2} = \frac{12.6 \times 4}{\pi D^2} = \frac{16.04}{D^2}$$

Gawn (1953) published propeller data for BARs of 0.2, 0.35, 0.5, 0.65, 0.80, 0.95 and 1.1. Corresponding values of D are: 8.94, 6.75, 5.65, 4.96, 4.47, 4.10, 3.81 m.

Because of the limitation imposed on the propeller diameter, we need only consider the last three values of BAR the value 0.80 being used merely to define the trends.

$$J = \frac{V_1}{nD} = \frac{13.1}{nD} = \text{in m, s units}$$

Hence

$$n = \frac{786}{DJ} \text{ r.p.m.}$$

Table 11.7 can now be constructed.

For each value of D and n, K_T can be calculated from

$$K_T = \frac{T}{(\rho/g)n^2 D^4} = \frac{1.01 \times 10^6}{1025 n^2 D^4} = \frac{976}{n^2 D^4}, \ n \text{ in r.p.s.}$$

The calculation can be simplified somewhat by introducing in this expression for K_T the relationship

$$J = \frac{V_1}{nD}$$

$$K_T = \frac{976}{(V_1)^2} \frac{(J)^2 (D)^2}{D^4} = \frac{976 J^2}{V_1^2 D^2}$$

$$= \frac{976}{(13.1)^2} \cdot \frac{J^2}{D^2} = 5.69 \frac{J^2}{D^2}$$

$$= 3.64 \times \frac{1}{D^2} \text{ for } J = 0.8$$

$$= 5.69 \times \frac{1}{D^2} \text{ for } J = 1.0$$

$$= 8.19 \times \frac{1}{D^2} \text{ for } J = 1.2$$

By cross plotting the Gawn data, the values of P/D and η can be obtained for each value of K_T, J and BAR. In order to obtain a reliable value of η, it is recommended that a line be drawn parallel to the base line at the appropriate K_T value, noting the J value appropriate to each P/D and the corresponding η. Plotting η against P/D the η value required can be obtained knowing P/D.

Now by plotting for each BAR the values of n, P/D and η to a base of J the values of P/D (and hence P) and n corresponding to maximum η can be read off. These results together with the η value are now plotted to a base of D.

In this case, the limitation on propeller diameter is the overriding factor, since efficiency is still increasing as the diameter exceeds the 4.27 m value. Corresponding to the 4.27 m diameter

Table 11.7
Propeller calculation

BAR	D	$\dfrac{786}{D}$	J	n	K_T	P/D	η	Values for η_{max}			
								n	P/D	η	P
			0.8	219.9	0.183	1.11	0.658				
0.80	4.47	175.9	1.0	175.9	0.286	1.50	0.665	200	1.24	0.676	18.2
			1.2	146.6	0.410	1.96	0.638				
			0.8	239.5	0.217	1.15	0.648				
0.95	4.10	191.6	1.0	191.6	0.338	1.56	0.640	228	1.24	0.650	16.7
			1.2	159.7	0.487	2.03	0.610				
			0.8	257.8	0.250	1.16	0.615				
1.1	3.81	206.2	1.0	206.2	0.392	1.59	0.610	247	1.24	0.624	15.5
			1.2	171.8	0.563	2.06	0.585				

$$\eta = 0.66$$
$$n = 217 \, \text{r.p.m.}$$
$$P = 5.21 \, \text{m}$$
$$\text{BAR} = \frac{15.9}{D^2} = \frac{15.9}{(4.27)^2} = 0.872$$

Now

$$\text{Propulsive coefficient} = \text{PC} = \left[\frac{\eta_H \times \eta_0 \times \eta_R}{\text{appendage coeff}}\right] \times \text{QPC factor}$$

$$= \left[\frac{0.98 \times 0.66 \times 1.00}{1.06}\right] 0.92 = 0.562$$

$$\text{Shaft power} = \frac{23.49}{\text{PC}} = \frac{23.49}{0.562} = 41.8 \, \text{MW}$$

This is the power which must be developed by the machinery.

PROPELLER DESIGN DIAGRAM

The example above shows that the amount of work involved in designing this one propeller is fairly lengthy, particularly when it is realized that the various cross-plottings necessary in the solution have not been reproduced.

A shorter analysis is possible if use is made of a 3-bladed propeller design diagram based directly on the data published by Gawn. This diagram, developed by the Admiralty Experiment Works is reproduced as Fig. 11.12. It makes use of the parameter K_T/J^2 which depends only on thrust, propeller diameter and speed of advance, since:

$$\frac{K_T}{J^2} = \frac{T}{\rho n^2 D^4} \bigg/ \frac{V_1^2}{n^2 D^2} = \frac{T}{\rho D^2 V_1^2}$$

Thus one unknown, n, has been removed. By calculating the value of K_T/J^2 and drawing a line across the diagram at this level the values of maximum efficiency, J, and P/D can be read off directly at the appropriate BAR value.

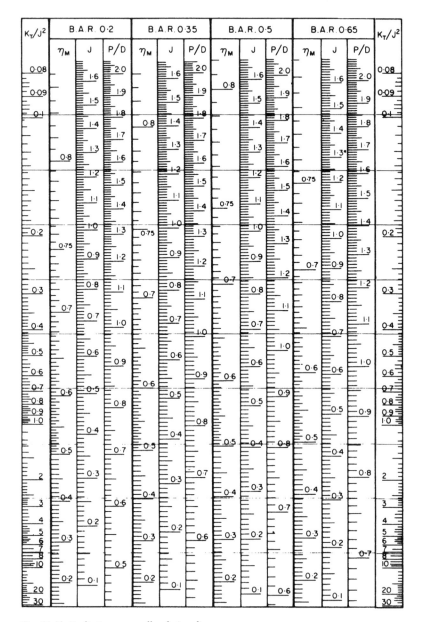

Fig. 11.12 Preliminary propeller design diagram

Fig. 11.12 continued

K_T/J^2	B.A.R. 0·8			B.A.R. 0·95			B.A.R. 1·1			K_T/J^2
	η_M	J	P/D	η_M	J	P/D	η_M	J	P/D	

Diagram gives the maximum peller efficiency, η_M, obtainable any known value of K_T/J^2 corresponding values of advance efficient J and pitch ratio

$$\frac{K_T}{J^2} = \frac{36.13}{D^2 V^2} T$$

$$J = \frac{30.88}{nD}$$

where T = thrust in tonnef
D and P are in metres
n is in r.p.m.
V is in knots

EXAMPLE 3. Use the propeller design diagram to design the propeller in the previous example.

Solution: In the previous example, it was shown that values of D corresponding to BARs of 0.80, 0.95 and 1.1 were 4.47, 4.10 and 3.81 m. Other data calculated in that example were:

$$\text{Thrust per screw} = 1.01 \text{ MN}$$
$$V_1 = 13.1 \text{ m/s}$$

Hence

$$\frac{K_T}{J^2} = \frac{T}{\rho D^2 V_1^2}$$

$$= \frac{1.01}{1025} \times \frac{10^6}{(13.1)^2} \times \frac{1}{D^2} = \frac{5.74}{D^2}$$

A table can now be constructed as in Table 11.8.

Table 11.8
Calculation for worked example using propeller design diagram

BAR	0.80	0.95	1.1
D, m	4.47	4.10	3.81
$\frac{K_T}{J^2}$	0.287	0.341	0.395
η_m	0.676	0.650	0.624
J	0.872	0.842	0.830
P/D	1.24	1.237	1.239
P	5.54	5.07	4.72
DJ	3.90	3.45	3.16
r.p.m.	202	228	249

In Table 11.8, the values of η_m, J and P/D are read directly from the propeller design diagram.

The propeller revolutions, n, can be obtained from

$$n = \frac{V_1 \times 30.88}{DJ} = \frac{786}{DJ}$$

Values of p, n and η_m can be plotted against diameter as before, showing that it is the limit on diameter at 4.27 m which is the governing factor and giving a propeller of the following characteristics

$$D = 4.27 \text{ m} \qquad \text{PC for ship} = 0.565$$
$$P = 5.21 \text{ m} \qquad P_S = 41.8 \text{ MW}$$
$$n = 217 \qquad \text{BAR} = 0.878$$
$$\eta = 0.664$$

Although it is not in general use an interesting form of propeller design chart is presented in Figs 11.13 to 11.15. Figure 11.13 uses vertical and horizontal

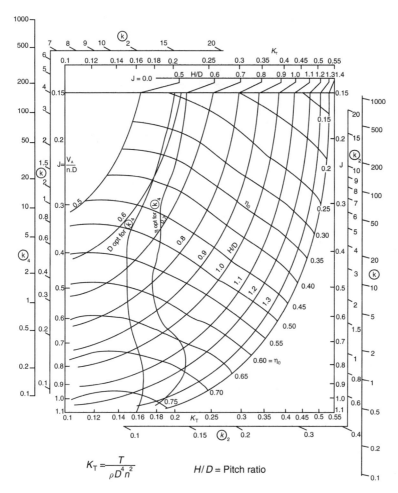

$$K_T = \frac{T}{\rho D^4 n^2}$$ $H/D =$ Pitch ratio

Fig. 11.13 Design diagram for given thrust power

axes of $\log J$ and $\log K_T$. The results of any systematic series can be plotted in the form of curves of constant efficiency and pitch diameter ratio. Those shown are for a Wageningen series 4-bladed screw with a blade area ratio of 0.55. The circular notation used should not be confused with the Froude notation described earlier.

With this type of plot, if

$$\left(K\right)_n = \frac{K_T}{J^n}$$

$$\log \left(K\right)_n = \log K_T - n \log J$$

Thus plots of constant $\left(K\right)_n$ will be straight lines with a slope defined by n. The lines for values of $\left(K\right)_2$ and $\left(K\right)_4$ are represented by short inclined lines around the edges of the diagram.

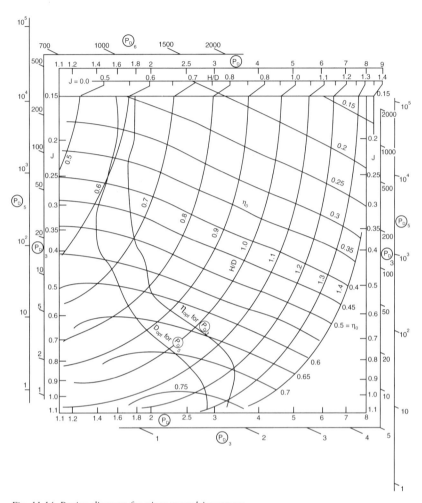

Fig. 11.14 *Design diagram for given propulsion power*

$$\bigcirc\!\!\!\!K_2 = \frac{K_T}{J^2} = T/\rho D^2 V_1^2$$

$$\bigcirc\!\!\!\!K_4 = \frac{K_T}{J^4} = T n^2 / \rho V_1^4$$

Optimum values of n and D for $\bigcirc\!\!\!\!K_2$ and $\bigcirc\!\!\!\!K_4$ respectively are defined by the point at which the propeller efficiency curve is tangential to the $\bigcirc\!\!\!\!K_2$ or $\bigcirc\!\!\!\!K_4$ curve. Thus if T, V_1 and either D or n are known the curves can be used directly to determine the optimum efficiency, pitch ratio and the n or D value.

Figure 11.14 is based on similar arguments and is used where the propulsive power P_D is known. In the diagram

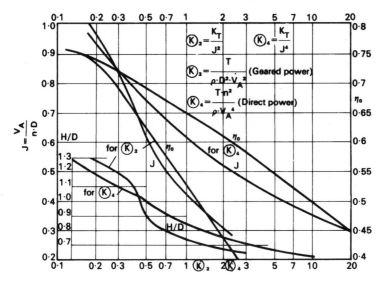

Fig. 11.15 *Optimum curves given T and V_A*

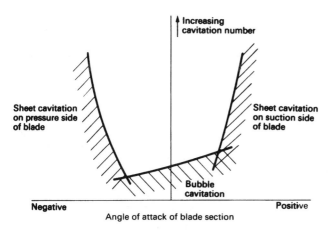

Fig. 11.16

$$\left(P_D\right) = \frac{1000 P_D}{\rho D^5 n^3} = 83{,}776\, K_Q$$

$$\left(P_D\right)_3 = \frac{\left(P_D\right)}{J^3}, \quad \left(P_D\right)_5 = \frac{\left(P_D\right)}{J^5}$$

Figure 11.15 presents optimum curves assuming T and V_1 are known together with propeller revolutions or diameter. Thus either $(K)_2$ or $(K)_4$ is known and the corresponding optimum values of efficiency, advance coefficient and pitch ratio are determined. The value of n or D whichever is the unknown follows from the advance coefficient.

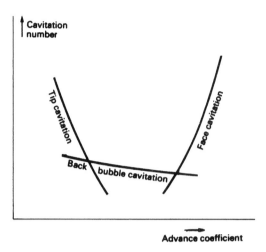

Fig. 11.17

CAVITATION

Figure 11.7 shows the influence of cavitation upon the overall propeller characteristics. In designing a propeller the naval architect needs to know the cavitation characteristics of the chosen blade section. The so called cavitation 'bucket' curve is used as in Fig. 11.16. The shape of the 'bucket' depends upon the blade thickness, camber and angle of attack.

The same characteristic of a 'cavitation bucket' is manifested in the results for the propeller itself as indicated in Fig. 11.17. The wider the bucket the greater the range of J over which the propeller can operate at a given cavitation number without cavitating in one way or another. The curves can be defined as a result of tests carried out in cavitation tunnels although, as has been pointed out in Chapter 10, these can only approximately represent the flow conditions at the propeller in the full-scale ship. Heavy cavitation can reduce the induced velocities ahead of the propeller and the thrust deduction fraction will fall. Cavitation will be more variable and noise and vibratory forces will be greater, if the blades are operating in an unsteady flow compared with the equivalent steady flow. This is in addition to the fall off in propeller efficiency.

INFLUENCE OF FORM ON RESISTANCE

It must be made clear that there is no absolute in terms of an optimum form. The designer has many things to consider besides the powering of the ship, e.g. ability to fit machinery, magazines, etc., seakeeping, manoeuvrability and so on. Even from the point of view of powering, one form may be superior to another at one displacement and over one speed range but inferior at other displacements or speed ranges.

Again, the situation is complicated by the fact that, in general, one parameter cannot be varied without affecting others. For example, to increase length, keeping form coefficients and beam constant, will change displacement and draught. Thus the following comments can be regarded as being valid only in a general qualitative way. Wherever possible, reference should be made to methodical series tests.

Wetted length

Given freedom of choice of length, keeping displacement sensibly constant, a designer will choose a short form for slow speed ships and a long, slender form for high speed ships. This is because an increase in length increases the wetted surface area and hence the skin frictional resistance. At low speed this will more than offset any reduction in wave-making resistance, but for high speeds the possible reduction in wave-making resistance will be all important. Nevertheless, the variation of wave-making resistance with length does not obey a simple law as was explained in Chapter 10. Because of the interference between bow and stern wave systems, there will be optimum bands of length with intermediate lengths being relatively poor.

Prismatic coefficient

This coefficient has little influence on the skin frictional resistance but can have a marked effect on residuary resistance. If possible, reference should be made to methodical series data from models similar to the design under development. Broadly, however, the optimum C_P value increases with increasing Froude number (Fig. 11.18). Since the influence of the prismatic coefficient is mainly related to the residuary resistance, it is not critical for low speed ships. In such ships, the choice of C_P value is much more likely to be governed by the cargo carrying capacity.

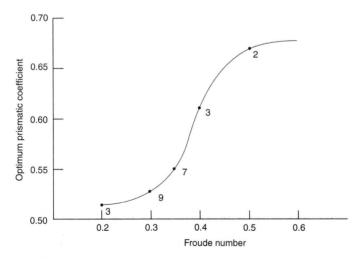

Fig. 11.18 Curve of optimum prismatic coefficient. Figures on the curve are the approximate percentage increases in resistance for ±0.05 change from the optimum C_P

Curve of areas

It is essential that this should be a fair curve with no sudden changes of curvature. Apart from this it is difficult and perhaps dangerous to generalize although it has been shown that very small changes in the curve of areas can produce really large changes in the residuary resistance.

Cross-sectional shapes

Generally, these are not critical but, if other ship requirements permit, U-shaped sections are to be preferred to V-shaped sections forward. This arises principally from the fact that more volume is removed from the vicinity of the waterplane and the wave-making resistance is accordingly reduced. V-shaped sections are used aft for vessels operating at high Froude number.

Centre of buoyancy position

The CB should vary from a few per cent of the length forward of amidships for slow ships to about 10 per cent of the length aft for fast ships.

REDUCING WAVE-MAKING RESISTANCE

Since, physically, wave-making resistance arises from a disturbance of the free surface, it is reasonable to expect that a lower resistance will result from concentrating displacement remote from the waterplane. That is to say, U sections are less resistful than V sections. Other generalizations are dangerous. In one case results for two forms with the same prismatic coefficient and apparently very similar curves of area show that, for one the wave-making resistance is double that of the other for $F_n = 0.23$ with substantial differences for F_n is the range 0.22 to 0.31.

Theory can help to explain why apparently small form changes can lead to large variations in wave-making resistance. It is for this reason that theory can often guide the model experimenter in the search for a better form. It should be emphasized, however, that there is no universal 'optimum' ship form giving minimum resistance at all speeds but rather a best form for a given Froude number.

It has been demonstrated that significant decreases in wave-making resistance occur when the bow and stern wave systems are out of phase. It is therefore reasonable to enquire whether a reduction can be obtained by artificially creating a wave system to interact with the ship system. In fact, this is the principle of the bulbous bow. Depending on its size, a bulb produces a wave system with crests and troughs in positions governed by the fore and aft position of the bulb relative to the bow. Unfortunately, this again can only produce a reduction in resistance over a limited range of F_n and then only at the expense of resistance at other speeds. Where a ship operates for a large percentage of its time at one speed as, for instance, is usually the case for most merchant ships, such a device can be of great benefit and is becoming more extensively used.

Although bulbous bows were introduced to reduce wave-making resistance it is now known that their action is more complex than the above simple explanation. They appear to modify the flow over the ship's hull generally and can lead to reduced resistance in full bodied ships at relatively low speeds. This is probably because they modify the flow over the bilges.

BOUNDARY LAYER CONTROL

Blowing, or more usually suction, offers the possibility of controlling separation of flow and vortex shedding, particularly for full hull forms and on control surfaces. This is similar to the control of flow over aircraft wings.

Another means of reducing drag in streamlined forms is the injection of polymers into the boundary layer. The effect is to reduce turbulence. Royal Navy trials on a surface ship, injecting Polyox at very low concentrations gave reductions in skin friction drag of up to 20 per cent. At 15 knots a concentration of 1.4 ppm gave an overall thrust reduction of 3.5 per cent. Work by the Russian Navy on submarines showed drag reductions of the order of 50 per cent, an increase in maximum speed of 10 per cent and a reduction in hydrodynamic noise level. The addition of the polymer did not affect the thrust deduction factor but the velocity profile in the boundary layer was smoother.

The advantages of such measures must be set against the complications of installing them, including space and weight considerations. The systems also need maintaining and the ejection slots must be kept clear of marine growth. They are most likely to be applied to military vehicles which can benefit from relatively short periods of high speed running, say a total of 100 hours during a mission.

COMPATIBILITY OF MACHINERY AND PROPELLER

Having the geometry of the propeller for the full-power condition, it follows that the thrust and torque variations with shaft revolutions are fully determined. To be satisfactory, the machinery must always be able to develop these torques at the various revolutions, otherwise the machinery will 'lock-up', i.e. as speed is increased the machinery will arrive at a point where its power output is prematurely limited by the torque demanded by the propeller.

If there is no other solution available, it may be possible to solve the problem by fitting a controllable pitch propeller.

STRENGTH OF PROPELLERS

Calculations of propeller strength must take account of the torque and bending moments acting at the blade roots. Stress levels accepted must be such that the propeller will last the life of the ship and must allow for the cyclic variations in loads due to the wake and the increased forces due to ship motions and manoeuvring.

EFFECT OF SPEED ON ENDURANCE

The rate of fuel consumption depends upon:

(*a*) the efficiency of the machinery at various power outputs. This feature is to some extent within the control of the machinery designer, e.g. by designing for optimum efficiency at full power, at cruising power or at some inter- mediate figure in order to balance the two;

(*b*) the power needed to supply the domestic loads of the ship such as lighting, galleys, air-conditioning, etc. This load is often referred to as the 'hotel' load and is independent of the forward speed of the ship.

The hotel load in a modern ship, particularly a passenger ship, can absorb a large proportion of the power generated at low speed. For this reason, the economical speed has tended to increase in recent years. The economic speed is a complicated thing to calculate but factors to be considered are: the fuel bill for covering a given distance at various speeds; the wages bill for the crew; the number of round voyages possible per year; special considerations depending on the payload of the ship, e.g. for a passenger ship a speedier passage may entice passengers away from the airlines. On the other hand, if the journey is between say Southampton and New York there is no attraction in arriving in one of the ports at midnight—if a faster journey is not feasible then the speed might as well be reduced.

In the following example, the influence of speed on the fuel bill is calculated for a typical steam ship of 26.25 MW.

EXAMPLE 4. The shaft power speed curve for a given ship with a total installed power of 26.25 MW, is as shown in Fig. 11.19 and the specific fuel consumption for various percentages of full power are as shown in Fig. 11.20. Calculate the endurance for 1000 tonnef of fuel over a range of speeds and the weight of fuel required for 1000 miles endurance.

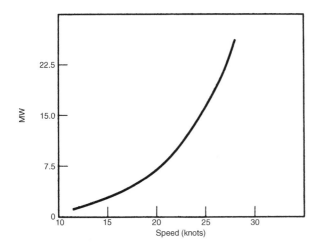

Fig. 11.19 Shaft power/speed curve

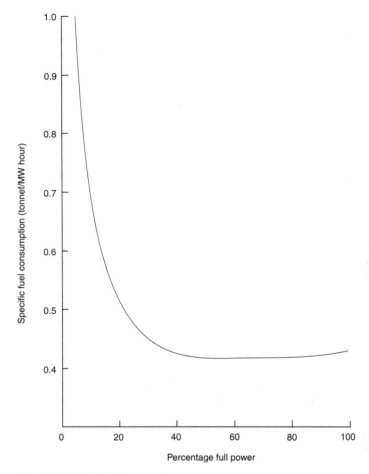

Fig. 11.20 Specific fuel consumption

Solution: At a speed of V knots, the distance travelled in 1 hr is V nautical miles. To travel 1000 miles takes $1000/V$ hr.

If the fuel consumption is S kg per MW hour, the fuel required for 1000 miles at speed V is

$$S \times (\text{power}) \times \frac{1000}{V} \text{ kg}$$

Conversely, the number of hours steaming possible on 1000 tonnef of fuel is given by

$$\text{No. of hours} = \frac{1000 \times 10^3}{S(\text{power})}$$

$$\text{Distance travelled} = \frac{10^6 V}{S(\text{power})} \text{ nautical miles}$$

A table can now be constructed as in Table 11.9.

Table 11.9
Calculation of endurance

V (knots)	Power (MW)	% full power	SFC (tonnef/MW hour)	tonnef of fuel per 1000 miles	Endurance for 1000 tonnef fuel
12	1.43	5.4	0.97	115	8700
14	2.26	8.6	0.74	119	8370
16	3.31	12.6	0.61	126	7920
18	4.80	18.3	0.52	139	7210
20	6.90	26.3	0.46	159	6300
22	9.74	37.1	0.43	190	5250
24	13.65	52.0	0.42	239	4190
26	18.74	71.4	0.42	303	3300
28	26.25	100.0	0.43	403	2480

It will be seen that there is not a lot of penalty in increasing the cruising or endurance speed from 12 to 16 knots. Indeed, economically, the lower salary bill would probably compensate for the increased fuel bill and the ship is a sounder economic proposition because of its increased mileage in a year.

Computational fluid dynamics

With the increasing power of modern computers, computational fluid dynamics (CFD) is becoming a possible tool for the naval architect although it is too complex to cover in this book in other than outline.

For surface ship calculations water can be assumed to be incompressible. This means that the volume of water entering a given elemental space in the vicinity of the hull must be matched by an equal volume of water flowing out. This leads to a continuity equation which the flow must obey. This, taken with the Navier–Stokes equations which define the conservation of momentum of flow, is adequate to characterize the physics for flow around a ship's hull. These equations are not soluble at present in their basic form because of turbulence effects and flow has to be averaged over a period of time. This time period is small compared with overall ship motions but large compared to the turbulence fluctuations. This leads to the Reynolds-averaged Navier–Stokes (RANS) equations

If viscosity, and hence turbulence, is ignored the Euler equations of motion can be used with the continuity equation, although these are not generally recommended for ship applications.

With the additional assumption of irrotational flow the problem reduces to one of potential flow. The three velocity components at any point are now related to the potential, ϕ. The derivative of ϕ in any direction gives the velocity in that direction. The continuity equation then reduces to Laplace's equation for potential flow.

Briefly, then, CFD embraces techniques for solving the equations (RANS, Euler or Laplace) by numerical methods using large numbers of elements around the ship. These techniques include:

- Boundary element methods (BEM), also known as panel methods. These can be used only for tackling potential flow problems. The surface of the hull and the surrounding water surface are divided into discrete elements, or panels. This method is currently the most widely used commercially, with applications such as resistance calculations.
- Finite difference methods (FDM). Unlike BEM, the whole fluid domain is divided into elements. Its name derives from the fact that the derivatives of the fluid equations are approximated by finite differences. The disadvantage of this method is that the mass and momentum of flow may not be fully conserved.
- Finite volume methods (FVM). These also cover the whole fluid domain and use finite differences, but in this case the methods used ensure that mass and momentum are conserved. Most commercial RANSE solvers use FVM. Grid generation is more onerous than for the BEM methods.

Powerful software is available commercially to compute the flow with pressure, velocity components and turbulence at the nodes or centroid of each computational cell. They can also output the forces and moments (pressure and skin friction components) on the hull and appendages. Software flow visualization tools enable visual checks to be made at each stage of calculation.

Although CFD can be used for seakeeping and manoeuvring assessments, strip methods which are now well established tend to be used in preference.

The capability of CFD is gradually improving. The techniques are not easy to apply and to date most applications have been by researchers. Early work was related to deeply submerged bodies with calculations covering the hull, control surfaces and propulsors. Deeply submerged bodies, of course, avoid the complication presented by the sea/air interface on which vessels other than deeply submerged submarines must operate. However, methods can solve the free surface interface although not yet for all hull forms.

CFD techniques can be used in the design process looking at a range of design options in conjunction with semi-empirical methods and model tests. In many cases the theoretical results need to be adjusted to fit observed trial data. Typical applications have been to simulate resistance and, increasingly, propulsion. Its use in wake prediction and propeller design is established. For propellers it can predict the thrust in fully developed cavitating and non-cavitating conditions in steady and unsteady flow. It is not so good for predicting torque, the onset of cavitation or the hull/propeller interaction. More work is needed to take account of the effects of air and impurities entrained in the water.

As an example of what can be achieved, in one application MIT compared propellers behind a tapered and full stern axisymmetric submarine hull. A vortex lifting-line method was used to give a preliminary propeller design, providing three-dimensional blade geometries. Then a 3-D vortex-lattice lifting-surface propeller blade program, coupled to a RANSE solver, was used to find an optimized propeller geometry and the resulting forces for the propeller/hull combination. The combination of methods took account of the propeller/hull interactions giving thrust deduction and the effective inflow to the propeller.

Although not conclusive this study suggested that a full stern hull may have advantages.

Grid generation is a key element in CFD. To generate a good grid for dealing efficiently with any given problem is not easy at present. It demands considerable experience of the methods. Also, grid generation accounts for a major part of the time needed for a CFD investigation, perhaps 80 per cent. In future it is to be expected that more user friendly methods will become available, making use of expert systems computer techniques to guide the less experienced practitioner.

As CFD methods become easier to use, and with more experience of comparing theoretical results with model and full-scale data, it is to be expected that they will become more common as part of the design process.

Summary

The methods of presentation and calculation presented enable the general principles established in Chapter 10 to be applied to the calculation of the shaft power required to propel a new design at the required speed. Allowance can be made for air resistance and for hull fouling as a result of marine growth which increases with time out of dock. As part of the process, a suitable design of screw propeller is obtained from methodical series data.

This is, of course, only a beginning. While it enables a suitable design of hull and propeller to be effected, it does not describe many of the refinements to the process. These the student must pursue through the transactions of the learned societies.

Problems

1. A ship of 50 MN displacement is driven at a speed of 12 knots. A ship of 65 MN of similar form is being designed. At what speed of the larger ship should its performance be compared with the 50 MN ship?
2. A ship of length 64.6 m, 6.02 m beam, 1.98 m draught, wetted surface 369 m² and displacement 4.26 MN has a resistance of 35 kN at 15.8 knots.

    Deduce the dimensions and effective power of a ship of similar form 233 m long at the corresponding speed.
3. With the definitions of C and C_T given in the text show that

$$C_T = \frac{8\pi}{1000} \frac{C}{S}$$

4. A 5 m model of a 180 m long ship is towed in a ship tank at a speed of 1.2 m/sec. The towing pull is 11.77 newtons. Assuming that 60 per cent of the resistance force is due to skin friction, calculate the corresponding speed for the ship in knots and the P_E at this speed assuming a wetted surface area of 3600 m² in the ship.

5. A destroyer 97.5 m × 1473 tonnef has a full speed of 35 knots with clean bottom. The $\textcircled{C}$ value of a 4.88 m model of the same form is 1.75 at the corresponding speed. Estimate the proportion of the ship resistance at full speed attributable to skin friction using
   (1) Froude analysis
   (2) ITTC ship-model correlation line.
   NOTE: Use Question 3 to change $\textcircled{C}$ to C_T.
6. A model of a vessel, 122 × 20 × 7.3 m draught, of 8697 tonnef displacement, is run, and the curve of P_E on a base of speed of ship is 2.42, 3.01, 3.74, 4.62 and 5.71 MW for 16, 17, 18, 19 and 20 knots respectively. Make an estimate of the P_E of a ship of 16,250 tonnef, of similar form, for speeds of 20 and 21 knots and give the dimensions of the new ship.
7. The data below relates to the 4.88 m model of a ship 120 m long, 4064 tonnef displacement. Assuming $\textcircled{S} = 7.15$ and a propulsive coefficient of 0.57 plot the P_S-speed curve for the ship when 6 months out of dock. Assume that the skin frictional resistance increases by $\frac{1}{4}$ of a per cent per day out of dock.
   What speed is likely in this condition with 30 MW?

$\textcircled{K}$	1.6	2.0	2.5	3.0	3.5	4.0	4.5
$\textcircled{C}$	1.150	1.176	1.245	1.418	1.500	1.770	2.085

8. The following data relates to a 4.88 m model of a ship 195 m long and 14,453 tonnef displacement.

$\textcircled{K}$	1.2	1.5	2.0	2.3	2.5	3.0	3.1	3.2	3.3	3.4
$\textcircled{C}$	1.205	1.170	1.145	1.145	1.155	1.160	1.170	1.181	1.210	1.253

Compute and plot the P_S-speed curve for the ship 6 months out of dock assuming $\textcircled{S} = 7.40$, and a propulsive coefficient of 0.55.
   What power is required for 20 and 28 knots?
9. A ship of 6096 tonnef displacement is required to have a maximum speed of 25 knots. The total effective power including appendages is 11.19 MW. Other data are:

Hull efficiency	= 1.0
Mean wake	= 1.4 per cent
Relative rotative efficiency	= 0.98
Quasi-propulsive coefficient factor	= 0.88
Pressure coefficient	= 7.65 tonnef/m^2

Use the propeller design diagram to find the propeller diameter and shaft power required for shaft revolutions of 200 r.p.m. Calculate the efficiency, pitch and r.p.m. of a 3.96 m diameter propeller.

10. A propeller is found by calculation to have a K_T/J^2 value of 0.2225 when developing a P_T of 3.73 MW at a speed of advance through the water of 25 knots. What is the diameter of the propeller?

   Use the propeller design diagram to find the maximum efficiency possible and the corresponding pitch.

11. What is *cavitation* and what is its effect on torque, thrust and efficiency of a propeller? Explain why experiments on models to determine the effects of cavitation cannot be carried out in an open ship tank.

   A 4 m propeller has been designed for a destroyer to give a top speed of 30 knots at 250 r.p.m. It is desired to run a 50 cm model of the propeller in a cavitation tunnel at a water speed of 6 m/s. At what water pressure and r.p.m. must the model propeller be run, to simulate ship conditions? C.L. of propeller below surface = 4 m. Atmospheric pressure = 10^5 N/m^2, water vapour pressure = 1700 N/m^2. Froude wake factor = −0.01.

12. Estimate the r.p.m. and expected thrust and torque of a model propeller, 229 mm in diameter, fitted behind a 5.49 m model of a ship 137 m long. The model is to be run in fresh water. The following data are available for the ship (in salt water, 0.975 m^3/tonnef):

shaft r.p.m.	= 110
ship speed	= 8.03 m/s
Froude wake factor	= 0.47
P_D	= 4.77 MW
P_T	= 2.97 MW

13. What is meant by the terms quasi-propulsive coefficient and quasi-propulsive coefficient factor? Explain how these quantities are obtained from model experiments and full-scale trials.

   The following results were obtained from experiments on a 5 m model of a new design of frigate: Length 130 m, displacement 3600 tonnef,

Ship Ⓒ at 27 knots	= 1.59
Hull efficiency	= 0.95
Relative rotative efficiency	= 0.98
Openwater screw efficiency	= 0.69
Appendage coefficient	= 1.07

   On trials, the frigate achieved a speed of 27 knots at a measured shaft power of 22 MW. Calculate the quasi-propulsive coefficient and the quasi-propulsive coefficient factor.

14. Assuming that the curve of K_T against J is a straight line over the working range, show that, for a given speed of advance, the thrust (T) developed by a propeller varies with its r.p.m. (N) in the following manner:

$$T = AN^2 - BN$$

where A and B are constants for a given propeller and speed of advance.

For a particular 50.8 cm model propeller, running at 168 m/min, the constants have the following values: $A = 75.6$; $B = 289$, when N is measured in hundreds of r.p.m., and T is measured in newtons.

Calculate the thrust of a geometrically similar propeller of 3.05 m diameter at 200 r.p.m. and a speed of advance of 12 knots in sea water.

15. Resistance experiments are to be run on a 6.1 m model of a new warship design. Estimate the model resistance in newtons at a speed corresponding to the ship's full speed, given the following design information:

Length, 177 m; displacement, 11,379 tonnef; estimated P_S at full speed of 30 knots, 55.18 MW; estimated propulsive coefficient, 0.54; wetted surface, 3530 m².

Use the ITTC line to calculate skin frictional resistance for model and ship; applying a roughness correction of 0.0004 for the ship only.

The appropriate kinematic viscosities are:

fresh water, 1.229×10^{-6} m²/s
salt water, 1.113×10^{-6} m²/s

16. Describe, briefly, the three main causes of 'wake' when considering a ship moving through the water.

How is the hull efficiency related to wake and augment?

The thrust–r.p.m. curve for a model propeller, run in open water at an advance speed of 2.5 m/s, is given by the equation:

$$T = 53\, N^2 - 225\, N.$$

where T is in newtons, N is in hundreds of r.p.m.

This curve coincides with the 'behind thrust' curve for the propeller-ship model combination when run at a speed of advance of 2.8 m/s.

The augmented resistance can be approximated to the straight line:

$$T = 45\, N - 145$$

and the model hull resistance, when towed without the propellers and at a speed of 2.8 m/s is 51 newtons.

Determine the propeller revolutions for model self propulsion and hence find the wake, augment and hull efficiency for the propeller-ship combination.

17. Describe the methods by which the hull efficiency elements may be deduced from model experiments.

Why can good comparisons be made between various sizes of propellers by only maintaining the advance coefficient constant, when running deeply submerged in open water?

The results of tests on a 30 cm diameter propeller run at 500 r.p.m. are shown below.

What will be the maximum efficiency obtainable for this propeller, and the appropriate speed of advance?

J	0	0.2	0.4	0.6	0.8	1.0	1.2
K_T	0.715	0.62	0.50	0.37	0.25	0.14	0.03
K_Q	0.13	0.114	0.094	0.072	0.05	0.032	0.014

18. Very briefly, describe methods by which hull-efficiency elements may be determined from model experiments.

After towing a model, it is deduced that the effective power required to move the ship's hull and appendages at a speed of 28 knots would be 17.9 MW of which 39 per cent would be due to skin friction. Skin friction for the model would account for 43.5 per cent of the total resistance.

Experiments were then conducted with the model and propellers combined at a model speed of 2.44 m/s which corresponds to the required ship's speed. At self-propulsion the shaft speed was 14.0 rev/s and thrust and torque provided were 32.7 and 4.3 newtons respectively. Open water experiments with the same propellers gave the same thrust and revs when advancing at 2.33 m/s. Determine the augment and wake fractions and the propeller efficiency behind the ship.

Hence, assuming RRE and appendage coefficient to be both 1.0, deduce the P_S required for 28 knots and a QPC factor of 0.94.

19. A new design has a displacement of 115 MN and length 174 m. Corresponding values of ⓒ and ⓚ for a 4.88 m model are:

ⓚ	2.2	2.3	2.4	2.5	2.6	2.7	2.8
ⓒ	1.130	1.130	1.132	1.138	1.140	1.141	1.150
ⓚ	2.9	3.0	3.1	3.2	3.3	3.4	
ⓒ	1.153	1.156	1.172	1.193	1.235	1.283	

Assuming ⓢ $= 7.205$ and a propulsive coefficient of 0.55, plot the curve of power (MW) against speed for the clean ship. Assuming that the skin frictional resistance increases by $\frac{1}{4}$ per cent per day out of dock, plot the corresponding curve for the ship 6 months out of dock.

Assuming 33.56 MW installed power, calculate the maximum speeds (*a*) in the clean condition and (*b*) 6 months out of dock.

12 Seakeeping

Seakeeping qualities

The general term sea worthiness must embrace all those aspects of a ship design which affect its ability to remain at sea in all conditions and to carry out its specified duty. It should, therefore, include consideration of strength, stability and endurance, besides those factors more directly influenced by waves. In this chapter, the term seakeeping is used to cover these more limited features, i.e. motions, speed and power in waves, wetness and slamming.

The relative importance of these various aspects of performance in waves varies from design to design depending upon what the operators require of the ship, but the following general comments are applicable to most ships.

Motions

Excessive amplitudes of motion are undesirable. They can make shipboard tasks hazardous or even impossible, and reduce crew efficiency and passenger comfort. In warships, most weapon systems require their line of sight to remain fixed in space and to this end each system is provided with its own stabilizing system. Large motion amplitudes increase the power demands of such systems and may restrict the safe arcs of fire.

The phase relationships between various motions are also important. Generally, the phasing between motions is such as to lead to a point of minimum vertical movement about two-thirds of the length of the ship from the bow. In a passenger liner, this area would be used for the more important accommodation spaces. If it is desirable to reduce the vertical movement at a given point, then this can be achieved if the phasing can be changed, e.g. in a frigate motion at the flight deck can be the limiting factor in helicopter operations. Such actions must inevitably lead to increased movement at some other point. In the frigate, increased movement of the bow would result and wetness or slamming might then limit operations.

Speed and power in waves

When moving through waves the resistance experienced by a ship is increased and, in general, high winds mean increased air resistance. These factors cause the ship speed to be reduced for a given power output, the reduction being aggravated by the less favorable conditions in which the propeller is working. Other unpleasant features of operating in waves such as motions, slamming and wetness are generally eased by a reduction in speed so that an additional speed reduction may be made voluntarily.

Wetness

When the relative movement of the bow and local wave surface becomes too great, water is shipped over the forecastle. At an earlier stage, spray is driven over the forward portion of the ship by the wind. Both conditions are undesirable and can be lessened by increasing freeboard. The importance of this will depend upon the positioning of upper deck equipment and its sensitivity to salt spray. Spray rails, flare angles and knuckles may all influence the troublesome nature of spray which, in cold climates, causes ice accretion.

Slamming

Under some conditions, the pressures exerted by the water on a ship's hull become very large and slamming occurs. Slamming is characterized by a sudden change in the vertical acceleration of the ship followed by a vibration of the ship girder in its natural frequencies. The conditions leading to slamming are high relative velocity between ship and water, shallow draught and small rise of floor. The area between 10 and 25 per cent of the length from the bow is the area most likely to suffer high pressures and to sustain damage.

Ship routing

Since the ship behaviour depends upon the wave conditions it meets, it is reasonable to question whether overall performance can be improved by avoiding the more severe waves. This possibility has been successfully pursued by some authorities. Data from weather ships are used to predict the speed loss in various ocean areas and to compute the optimum route. In this way, significant savings have been made in voyage times, e.g. of the order of 10–15 hours for the Atlantic crossing.

Importance of good seakeeping

No single parameter can be used to define the seakeeping performance of a design. In a competitive world, a comfortable ship will attract more passengers than a ship with a bad reputation. A ship with less power augment in waves will be able to maintain tighter schedules or will have a lower fuel bill. In extreme cases, the seakeeping qualities of a ship may determine its ability to make a given voyage at all.

Good seakeeping is clearly desirable, but the difficulty lies in determining how far other design features must, or should, be compromised to improve seakeeping. This will depend upon each particular design, but it is essential that the designer has some means of judging the expected performance and the effect on the ship's overall effectiveness. Theory, model experiment and ship trial all have a part to play. Because of the random nature of the sea surface in which the ship operates, considerable use is made of the principles of statistical analysis.

Having improved the physical response characteristics of a ship in waves the overall effectiveness of a design may be further enhanced by judicious siting of critical activities and by fitting control devices such as anti-roll stabilizers.

As with so many other aspects of ship design a rigorous treatment of sea-keeping is very complex and a number of simplifying assumptions are usually made. For instance, the ship is usually regarded as responding to the waves as a rigid body when assessing motions and wetness although its true nature as an elastic body must be taken into account in a study of structure. In the same way it is instructive, although not correct, to study initially the response of a ship to regular long-crested waves ignoring the interactions between motions, e.g. when the ship is heaving the disturbing forces will generate a pitching motion. This very simple approach is now dealt with before considering coupled motions.

Ship motions

It was seen in Chapter 4, that a floating body has six degrees of freedom. To completely define the ship motion it is necessary to consider movements in all these modes as illustrated in Fig. 12.1. The motions are defined as movements of the centre of gravity of the ship and rotations about a set of orthogonal axes through the c.g. These are space axes moving with the mean forward speed of the ship but otherwise fixed in space.

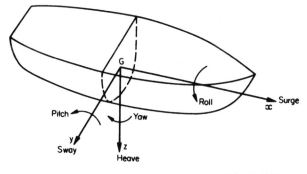

Translation or rotation	Axis	Description	Positive sense
Translation	Along x	Surge	Forwards
	Along y	Sway	To starboard
	Along z	Heave	Downwards
Rotation	About x	Roll	Starboard side down
	About y	Pitch	Bow up
	About z	Yaw	Bow to starboard

Fig. 12.1 Ship motions

It will be noted that roll and pitch are the dynamic equivalents of heel and trim. Translations along the x- and y-axis and rotation about the z-axis lead to no residual force or moment, provided displacement remains constant, as the ship is in neutral equilibrium. For the other translation and rotations, movement is opposed by a force or moment provided the ship is stable in that mode. The magnitude of the opposition increases with increasing displacement from the equilibrium position, the variation being linear for small disturbances.

This is the characteristic of a simple spring system. Thus, it is to be expected that the equation governing the motion of a ship in still water, which is subject to a disturbance in the roll, pitch or heave modes, will be similar to that governing the motion of a mass on a spring. This is indeed the case, and for the undamped case the ship is said to move with simple harmonic motion.

Disturbances in the yaw, surge and sway modes will not lead to such an oscillatory motion and these motions, when the ship is in a seaway, exhibit a different character to roll, pitch and heave. These are considered separately and it is the oscillatory motions which are dealt with in the next few sections. It is convenient to consider the motion which would follow a disturbance in still water, both without and with damping, before proceeding to the more realistic case of motions in waves.

UNDAMPED MOTION IN STILL WATER

It is assumed that the ship is floating freely in still water when it is suddenly disturbed. The motion following the removal of the disturbing force or moment is now studied for the three oscillatory motions.

Rolling

Let ϕ be the inclination of the ship to the vertical at any instant. The moment, acting on a stable ship, will be in a sense such as to decrease ϕ. For small values of ϕ,

$$\text{moment} = -\Delta \overline{GM}_T \phi$$

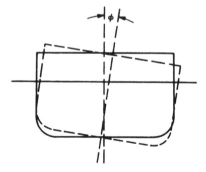

Fig. 12.2 Rolling

Applying Newton's laws of motion

$$\text{moment} = (\text{moment of inertia about } 0x)(\text{angular acceleration})$$

i.e.

$$-\Delta \overline{GM}_T \phi = +\frac{\Delta}{g} k_{xx}^2 \frac{\mathrm{d}^2 \phi}{\mathrm{d}t^2}$$

i.e.

$$\frac{d^2\phi}{dt^2} + \left(g\frac{\overline{GM_T}}{k_{xx}^2}\right)\phi = 0$$

This is the differential equation denoting simple harmonic motion with period T_ϕ where

$$T_\phi = 2\pi\left(\frac{k_{xx}^2}{g\overline{GM_T}}\right)^{\frac{1}{2}} = \frac{2\pi k_{xx}}{(g\overline{GM_T})^{\frac{1}{2}}}$$

It will be noted that the period of roll is independent of ϕ and that this will hold as long as the approximation $\overline{GZ} = \overline{GM_T}\phi$ applies, i.e. typically up to ±10 degrees. Such rolling is termed *isochronous*.

In practice k_{xx} must be increased to allow for what are usually termed 'added mass' effects due to motion induced in the water although this does not mean that a specific body of water actually moves with the ship. Added mass values vary with frequency but this variation can often be ignored to a first order. Typically the effect increases k_{xx} by about 5 per cent.

Hence

$$T_\phi \propto \frac{1}{(\overline{GM_T})^{\frac{1}{2}}}$$

Thus the greater is $\overline{GM_T}$, i.e. the more stable the ship, the shorter the period and the more rapid the motion. A ship with a short period is said to be 'stiff'—compare the stiff spring—and one with a long period is said to be 'tender'. Most people find a long period roll less unpleasant than a short period roll.

Pitching

This is analogous to roll and the motion is governed by the equation

$$\frac{d^2\theta}{dt^2} + \left(\frac{g\overline{GM_L}}{k_{yy}^2}\right)\theta = 0$$

and the period of the motion is

$$T_\theta = \frac{2\pi k_{yy}}{(g\overline{GM_L})^{\frac{1}{2}}} \text{ for very small angles of pitch.}$$

Heaving

Let z be the downward displacement of the ship at any instant. The force acting on the ship tends to reduce z and has a magnitude F_z given by

$$F_z = -\frac{A_W z}{u}$$

where u is the reciprocal weight density of the water.

Hence, the heaving motion is governed by the equation

$$\frac{\Delta}{g}\frac{d^2 z}{dt^2} = -\frac{A_W z}{u}$$

or

$$\frac{d^2 z}{dt^2} + \frac{g A_W}{u\Delta} z = 0$$

from which

$$\text{period} = 2\pi \left(\frac{u\Delta}{g A_W}\right)^{\frac{1}{2}}$$

Δ may be effectively increased by a significant amount (perhaps doubled) by the 'added mass' effect.

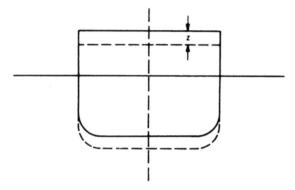

Fig. 12.3 Heaving

DAMPED MOTION IN STILL WATER

Now consider what happens when the motion is damped. It is adequate to illustrate the effect of damping on the rolling motion.

Only the simplest case of damping is considered here namely, that in which the damping moment varies linearly with the angular velocity. It opposes the motion since energy is always absorbed.

Allowing for the entrained water the equation for rolling in still water becomes

$$\frac{\Delta}{g} k_{xx}^2 (1 + \sigma_{xx})\ddot{\phi} + B\dot{\phi} + \Delta \overline{GM}_T \phi = 0$$

where

$$\frac{\Delta k_{xx}^2}{g}\sigma_{xx} = \text{augment of rolling inertia of ship due to entrained water}$$

$$B = \text{damping constant.}$$

This can be likened to the standard differential equation

$$\ddot{\phi} + 2k\omega_0\dot{\phi} + \omega_0^2\phi = 0$$

where

$$\omega_0^2 = \frac{g\overline{GM}_T}{k_{xx}^2(1 + \sigma_{xx})} \quad \text{and} \quad k = \frac{Bg}{2\omega_0\Delta k_{xx}^2(1 + \sigma_{xx})}$$

which in turn defines the effective period T_ϕ of the motion as

$$T_\phi = \frac{2\pi}{\omega_0}(1 - k^2)^{-\frac{1}{2}} = 2\pi k_{xx}\left(\frac{1 + \sigma_{xx}}{g\overline{GM}_T}\right)^{\frac{1}{2}}(1 - k^2)^{-\frac{1}{2}}$$

When the damping is not proportional to the angular velocity the differential equation is no longer capable of ready solution.

APPROXIMATE PERIOD OF ROLL

Of the various ship motions the roll period is likely to vary most from design to design and, because of the much greater amplitudes possible, it is often the most significant. Various approximate formulae have been suggested for calculating the period of roll including:

$$T_\phi = 2\pi\frac{K}{(g\overline{GM}_T)^{\frac{1}{2}}}$$

Suggested values of K for merchant ships and warships are given by the respective expressions:

Merchant ships

$$\left(\frac{K}{B}\right)^2 = F\left[C_B C_u + 1.10 C_u(1 - C_B)\left(\frac{H}{T} - 2.20\right) + \frac{H^2}{B^2}\right]$$

where

$$C_u = \text{upper deck area coeff.} = \frac{1}{LB}\,(\text{deck area})$$
$$H = \text{effective depth of ship} = D + A/L_{pp}$$
$$A = \text{projected lateral area of erections and deck}$$
$$L_{pp} = \text{L.B.P.}$$
$$T = \text{mean moulded draught}$$
$$F = \text{constant} = 0.125 \text{ for passenger and cargo ship,}$$
$$= 0.133 \text{ for oil tankers,}$$
$$= 0.177 \text{ for whalers.}$$

Warships

$$\left(\frac{K}{B}\right)^2 = F\left[C_B C_u + 1.10 C_e (1 - C_B)\left(\frac{H_n}{T} - 2.20\right) + \frac{H_n^2}{B_u^2}\right]$$

where

B_u = max. breadth under water;

C_e = exposed deck area coeff.;

$H_n = D + A_n/L_{pp}$;

D = depth from top of keel to upper deck;

A_n = sum of the projected lateral areas of forecastle, under bridge
and gun;

F = constant ranging from 0.172 for small warships to 0.177 for large
warships.

MOTION IN REGULAR WAVES

In Chapter 9, it is explained that the irregular wave systems met at sea can be regarded as made up of a large number of regular components. A ship's motion record will exhibit a similar irregularity and it can be regarded as the summation of the ship responses to all the individual wave components. Theoretically, this super-position procedure is valid only for those sea states for which the linear theory of motions is applicable, i.e. for moderate sea states. It has been demonstrated by several authorities, however, that provided the basic data is derived from relatively mild regular components, the technique can be applied, with sufficient accuracy for most engineering purposes, to more extreme conditions. Thus, the basic element in ship motions is the response of the ship to a regular train of waves. For mathematical convenience, the wave is assumed to have a sinusoidal profile. The characteristics of such a system were dealt with in Chapter 9.

In the simple approach, it is necessary to assume that the pressure distribution within the wave system is unaffected by the presence of the ship. This is one of the assumptions made by William Froude in his study of ship rolling and is commonly known as 'Froude's Hypothesis'.

Rolling in a beam sea

The equation for rolling in still water is modified by introducing a forcing function on the right-hand side of the equation. This could be obtained by calculating the hydrodynamic pressure acting on each element of the hull and integrating over the complete wetted surface.

The resultant force acting on a particle in the surface of a wave must be normal to the wave surface. Provided the wave-length is long compared with the beam of the ship, it is reasonable to assume that the ship is acted on by a resultant force normal to an 'effective wave surface' which takes into account all the sub-surfaces interacting with the ship. Froude used this idea and further assumed that the 'effective wave slope' was that of the sub-surface passing through the centre of buoyancy of the ship.

With this assumption it can be shown that, approximately, the equation of motion for undamped rolling motion in beam seas becomes

$$\frac{\Delta}{g} k_{xx}^2 (1 + \sigma_{xx})\ddot{\phi} + \Delta \overline{GM}_T (\phi - \phi') = 0$$

where $\phi' = \alpha \sin \omega t$; α = maximum slope of the surface wave; ω = frequency of the surface wave.

If ϕ_0 and ω_0 are the amplitude and frequency of unresisted rolling in still water, the solution to this equation takes the form

$$\phi = \phi_0 \sin(\omega_0 t + \beta) + \frac{\omega_0^2 \alpha}{\omega_0^2 - \omega^2} \sin \omega t$$

The first term is the free oscillation in still water and the second is a forced oscillation in the period of the wave train.

The amplitude of the forced oscillation is

$$\frac{\omega_0^2 \alpha}{\omega_0^2 - \omega^2}$$

When the period of the wave system is less than the natural period of the ship ($\omega > \omega_0$), the amplitude is negative which means that the ship rolls into the wave (Fig. 12.4(a)). When the period of the wave is greater than the natural period of the ship, the amplitude is positive and the ship rolls with the wave (Fig. 12.4(b)). For very long waves, i.e. ω very small, the amplitude tends to α and the ship remains approximately normal to the wave surface. When the frequencies of the wave and ship are close the amplitude of the forced oscillation becomes very large.

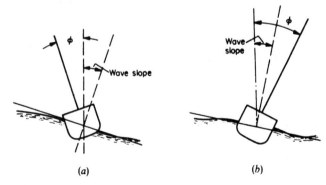

(a)   (b)

Fig. 12.4

The general equation for rolling in waves can be written as:

$$\ddot{\phi} + 2k\omega_0 \dot{\phi} + \omega_0^2 \phi = \omega_0^2 \alpha \cos \omega t$$

The solution to this differential equation is

$$\phi = \mu a \cos(\omega t - \varepsilon)$$

where

$$\tan \varepsilon = \frac{2k\Lambda}{1 - \Lambda^2}$$

$\Lambda = $ *tuning factor* $= \omega/\omega_0$; $\mu = $ *magnification factor* $= 1/\{(1 - \Lambda^2)^2 + 4k^2\Lambda^2\}^{\frac{1}{2}}$.

Plots of the phase angle ε and magnification factor are presented in Fig. 12.5. It will be appreciated that these expressions are similar to those met with in the study of vibrations.

The effect of damping is to cause the free oscillation to die out in time and to modify the amplitude of the forced oscillation. In an ideal regular sea, the ship would oscillate after a while only in the period of the waves. In practice, the maximum forced roll amplitudes occur close to the natural frequency of the ship, leading to a ship at sea rolling predominantly at frequencies close to its natural frequency.

Pitching and heaving in waves

In this case, attention is focused on head seas. In view of the relative lengths of ship and wave, it is not reasonable to assume, as was done in rolling, that the wave surface can be represented by a straight line. The principle, however, remains unchanged in that there is a forcing function on the right-hand side of the equation and the motions theoretically exhibit a natural and forced oscillation. Because the response curve is less peaked than that for roll the pitch and heave motions are mainly in the frequency of encounter, i.e. the frequency with which the ship meets successive wave crests.

Another way of viewing the pitching and heaving motion is to regard the ship/sea system as a mass/spring system. Consider pitching. If the ship moved extremely slowly relative to the wave surface it would, at each point, take up an equilibrium position on the wave. This may be regarded as the static response of the ship to the wave and it will exhibit a maximum angle of trim which will approach the maximum wave slope as the length of the wave becomes very large relative to the ship length. In practice, the ship hasn't time to respond in this way, and the resultant pitch amplitude will be the 'static' angle multiplied by a magnification factor depending upon the ratio of the frequencies of the wave and the ship and the amount of damping present. This is the standard magnification curve used in the study of vibrations. Provided the damping and natural ship period are known, the pitching amplitude can be obtained from a drawing board study in which the ship is balanced at various points along the wave profile.

Having discussed the basic theory of ship motions, it is necessary to consider in what form the information is presented to the naval architect before proceeding to discuss motions in an irregular wave system.

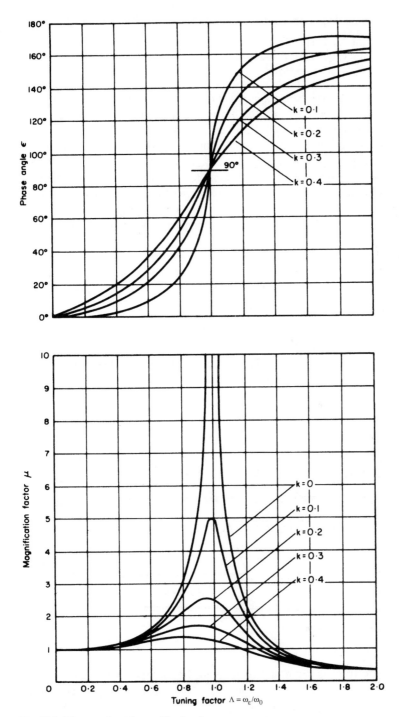

Fig. 12.5 *Phase angle and magnification factor*

PRESENTATION OF MOTION DATA

It is desirable that the form of presentation should permit ready application to ships of differing sizes and to waves of varying magnitude. The following assumptions are made:

(a) Linear motion amplitudes experienced by geometrically similar ships are proportional to the ratio of the linear dimensions in waves which are geometrically similar and in the same linear ratio. That is, the heave amplitude of a 200 m ship in waves 150 m long and 6 m high will be double that of a 100 m ship in waves 75 m × 3 m; $V/\sqrt{L}$ constant;
(b) Angular motion amplitudes are the same for geometrically similar ship and wave combinations, i.e. if the pitch amplitude of the 200 m ship is 2 degrees, then the pitch amplitude for the 100 m ship is also 2 degrees;
(c) For a ship in a given wave system all motion amplitudes vary linearly with wave height;
(d) Natural periods of motions for geometrically similar ships vary with the square root of the linear dimension, i.e. the rolling period of a ship will be three times that of a one-ninth scale model.

These assumptions follow from the mathematical analysis already outlined.
A quite common plot for motions in regular waves, is the amplitude, expressed non-dimensionally, to a base of wave-length to ship length ratio for a series of $V/\sqrt{L}$ values. The ordinates of the curve are referred to as *response amplitude operators* (See Fig. 12.6).

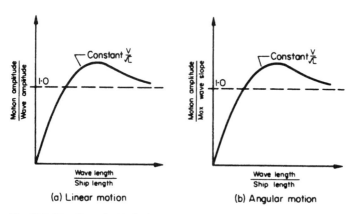

Fig. 12.6 Non-dimensional plotting

This system of plotting is non-dimensional, but a slight complication arises with angular motions when using wave spectra which are in terms of wave height. Since wave height is proportional to wave slope, the data can be presented as in Fig. 12.7 with no need to differentiate between linear and angular motions, although the curves are no longer non-dimensional.

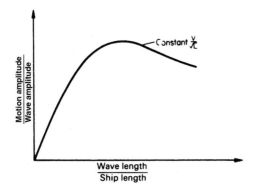

Fig. 12.7 Presentation of data for spectral analysis

Some typical response curves are reproduced in Fig. 12.8.

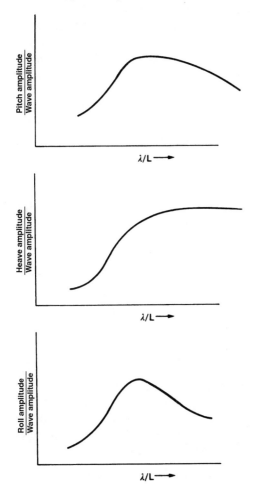

Fig. 12.8 Typical response curves

MOTION IN IRREGULAR SEAS

The foundations for the study of ship motions in irregular seas were laid in 1905 by R. E. Froude when he wrote in the context of regular wave experiments:

> 'Irregular waves such as those commonly met with at sea—are only a compound of a number of regular systems (individually of a comparatively small magnitude) of various periods, ranging through the whole gamut (so to speak) represented by our diagrams, and more. And the effect of such a compound wave series on the models would be more or less a compound of the effects proper to the individual units composing it.'

It has been seen that for regular waves the motion data can be presented in the form of response amplitude operators (RAO), for various ship speeds in waves of varying dimension relative to the ship length. Generally, a designer is concerned with a comparison of two or more designs so that, if one design showed consistently lower RAOs in all waves and at all speeds, the conclusion to be reached would be clear cut. This is not usually the case, and one design will be superior to the other in some conditions and inferior in other conditions. If it is known, using data such as that presented in Chapter 9, that on the intended route, certain waves are most likely to be met then the design which behaves better in these particular waves would be chosen.

Of more general application is the use of the concept of wave spectra. It was shown in Chapter 9 that, provided phase relationships are not critical, the apparently irregular sea surface can be represented mathematically by a spectrum of the type

$$S(\omega) = \frac{A}{\omega^5} \exp\left(-\frac{B}{\omega^4}\right)$$

where ω = circular frequency in radians per second.

A and B are constants which can be expressed in terms of the characteristic wave period and/or the significant wave height.

Since

$$\lambda = \frac{2\pi g}{\omega^2} \quad \frac{\mathrm{d}\lambda}{\mathrm{d}\omega} = -\frac{4\pi g}{\omega^3}$$

If λ is to be used as the base for the spectrum instead of ω then the requirement that the total spectral energy is constant leads to

$$S(\lambda) = S(\omega)\frac{\mathrm{d}\omega}{\mathrm{d}\lambda} = -\frac{A}{4\pi g\omega^2}\exp\left(-\frac{B}{\omega^4}\right)$$

For a known ship length the wave spectrum can be replotted to a base of λ/L to correspond to the base used above for the motion response amplitude operators. Then the wave spectrum and motion data such as that presented in Fig. 12.9 for heave can be combined to provide the energy spectrum of the motion.

Various motion parameters can then be derived from the spectral characteristics as for the waves themselves. See Chapter 9.

For example average heave amplitude $= 1.25\sqrt{m_0}$

where m_0 is the area under the heave spectrum.

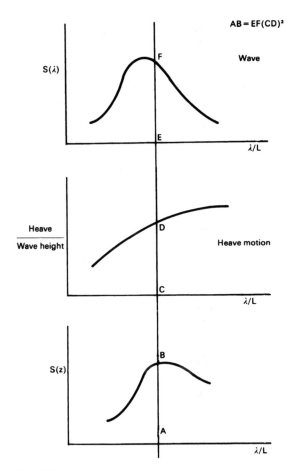

AB = EF(CD)²

Wave

$S(\lambda)$

λ/L

$\dfrac{\text{Heave}}{\text{Wave height}}$

Heave motion

λ/L

$S(z)$

λ/L

Fig. 12.9

In some cases motion data is presented to a base of frequency of encounter of the ship with the wave. The same process can be followed to arrive at the motion spectrum but noting that the wave spectrum is derived from an analysis of the variation of the surface elevation at a fixed point. In this case then it must be modified to allow for the effective or encounter spectrum as experienced by the ship.

If the ship is moving at velocity V at an angle ψ to the direction of advance of the wave system the wave spectrum as experienced by the ship is obtained by multiplying

(a) abscissae by $\left(1 - \dfrac{\omega V}{g}\cos\psi\right)$

(b) ordinates by $\left(1 - \dfrac{2\omega V}{g}\cos\psi\right)^{-1}$

When the ship is moving directly into the wave system $\cos\psi = -1$.

The effect of ship speed on the shape of the wave spectrum is illustrated in Fig. 12.10 which shows a spectrum appropriate to a wind speed of 30 knots and ship speeds of 0, 10, 20 and 30 knots.

To illustrate the procedure for obtaining the motion spectra, consider one speed for the ship and assume that the encounter spectrum for that speed is as shown in Fig. 12.11(*a*). Also, assume that the amplitude response operators for heave of the ship, at that same speed, are as shown in Fig. 12.11(*b*). The ordinate of the wave energy spectrum is proportional to the square of the amplitude of the component waves. Hence, to derive the energy spectrum for the heave motion as shown in Fig. 12.11(*c*), the following relationship is used

$$S_z(\omega_E) = [Y_{z\zeta}(\omega_E)]^2 S_\zeta(\omega_E)$$

i.e.

$$RC = (RB)^2(RA)$$

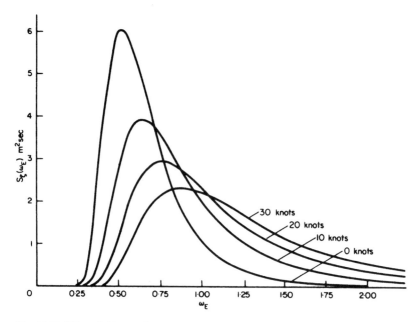

Fig. 12.10 *Effect of ship speed on encounter spectrum*

If the area under the motion energy spectrum is obtained by integration, the significant heave amplitude, etc., can be deduced by using the same multiplying factors as those given in Chapter 9 for waves.

For example, if m_0 is the area under the roll spectrum

average roll amplitude $= 1.25\sqrt{m_0}$

significant roll amplitude $= 2\sqrt{m_0}$

average amplitude of $\dfrac{1}{10}$ highest rolls $= 2.55\sqrt{m_0}$.

Any of these quantities, or the area under the spectrum, can be used to compare designs at the chosen speed. The lower the figure the better the design and the single numeral represents the overall response of the ship at that speed in that wave system. The process can be repeated for other speeds and other spectra. The actual wave spectrum chosen is not critical provided the comparison is made at constant significant wave height and not constant wind speed.

EXAMPLE 1. A sea spectrum for the North Atlantic is defined by the following table, $S_\zeta(\omega)$ being in m^2 s.

ω	0.3	0.4	0.5	0.6	0.7	0.8	0.9	1.0	1.1	1.2
$S_\zeta(\omega)$	0.20	2.00	4.05	4.30	3.40	2.30	1.50	1.00	0.70	0.50

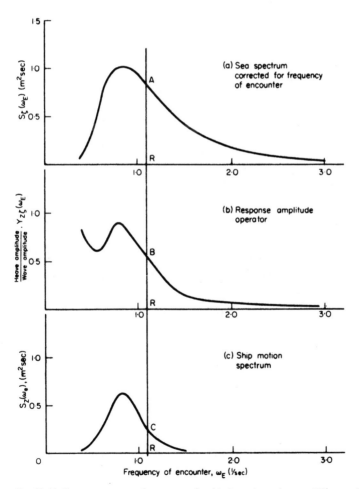

Fig. 12.11 Energy spectra and response of a ship in an irregular sea (illustrated for heave)

Calculate the encounter spectra for a ship heading directly into the wave system at speeds of 10,20 and 28 knots.

Assuming that the heave response of a ship, 175 m in length, is defined by Fig. 12.12 deduce the heave spectra for the three speeds and hence the probability curves for the motion.

Solution: It has been shown for the wave spectra, that

$$\omega_E = \omega \left(1 + \frac{\omega V}{g} \right)$$

For 10 knots;

$$V = 10 \times \frac{1852}{3600} = 5.14\,\mathrm{m/s} \quad g = 9.807\,\mathrm{m/s^2} \quad \therefore \ \omega_E = \omega(1 + 0.525\omega)$$

similarly for 20 and 28 knots ω_E is equal to $\omega(1 + 1.05\omega)$ and $\omega(1 + 1.47\omega)$ respectively.

Figure 12.12 is used by calculating the wave-length appropriate to each ω value. λ and λ/L are tabulated below with the response amplitude operators from Fig. 12.12. Since curves show response at each speed the RAOs apply to the appropriate ω_E.

It has also been shown that ordinates of the spectrum must be multiplied by

$$\left(1 + \frac{2\omega V}{g} \right)^{-1} = (1 + 1.05\omega)^{-1} \text{ for } 10\,\mathrm{knots}$$

$$(1 + 2.10\omega)^{-1} \text{ for } 20\,\mathrm{knots}$$

$$(1 + 2.94\omega)^{-1} \text{ for } 28\,\mathrm{knots}$$

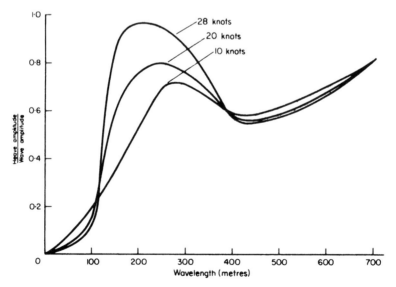

Fig. 12.12

The calculations can be carried out in tabular fashion as below for 10 knots and repeated for 20 knots and 28 knots.

ω	$1 + 0.525\omega$	ω_E	$S_\zeta(\omega)$	$1 + 1.05\omega$	$S_\zeta(\omega_E)$
0.3	1.158	0.347	0.20	1.315	0.15
0.4	1.210	0.484	2.00	1.420	1.41
0.5	1.263	0.632	4.05	1.525	2.65
0.6	1.315	0.789	4.30	1.630	2.64
0.7	1.368	0.958	3.40	1.735	1.96
0.8	1.420	1.136	2.30	1.840	1.25
0.9	1.473	1.326	1.50	1.945	0.77
1.0	1.525	1.525	1.00	2.050	0.49
1.1	1.578	1.736	0.70	2.155	0.32
1.2	1.630	1.956	0.50	2.260	0.22

			RAO		
ω	λ(m)	λ/L	10 knots	20 knots	28 knots
0.3	689	3.97	0.80	0.80	0.80
0.4	387	2.23	0.60	0.60	0.60
0.5	247	1.42	0.69	0.80	0.95
0.6	171	0.985	0.44	0.69	0.93
0.7	126	0.730	0.28	0.40	0.29
0.8	96.6	0.556	0.18	0.15	0.10
0.9	76.5	0.440	0.12	0.08	0.05
1.0	61.6	0.354	0.10	0.06	0.04
1.1	51.2	0.295	0.08	0.05	0.03
1.2	43.0	0.250	0.07	0.04	0.03

The ordinates of the heave motion spectrum at each speed are obtained by multiplying the wave spectrum ordinate by the square of the RAO as in the table below:

ω	10 knots			20 knots			28 knots		
	$S_\zeta(\omega_E)$	*RAO*	$S_Z(\omega_E)$	$S_\zeta(\omega_E)$	*RAO*	$S_Z(\omega_E)$	$S_\zeta(\omega_E)$	*RAO*	$S_Z(\omega_E)$
0.3	0.15	0.80	0.098	0.12	0.80	0.079	0.11	0.80	0.068
0.4	1.41	0.60	0.508	1.09	0.60	0.392	0.92	0.60	0.330
0.5	2.65	0.69	1.262	1.98	0.80	1.270	1.64	0.95	1.480
0.6	2.64	0.44	0.511	1.91	0.69	0.910	1.56	0.93	1.350
0.7	1.96	0.28	0.153	1.38	0.40	0.207	1.11	0.29	0.093
0.8	1.25	0.18	0.041	0.86	0.15	0.019	0.69	0.10	0.007
0.9	0.77	0.12	0.011	0.52	0.08	0.003	0.41	0.05	0.001
1.0	0.49	0.10	0.005	0.32	0.06	0.001	0.25	0.04	—
1.1	0.32	0.08	—	0.21	0.05	—	0.17	0.03	—
1.2	0.22	0.07	—	0.14	0.04	—	0.11	0.03	—

The heave spectra can now be plotted and the areas under each obtained to give m_0. Values of m_0 so deduced are

10 knots: $m_0 = 0.37$ and $\sqrt{(2m_0)} = 0.86$

20 knots: $m_0 = 0.62$ and $\sqrt{(2m_0)} = 1.11$

28 knots: $m_0 = 0.72$ and $\sqrt{(2m_0)} = 1.20$

The probability that at a random instant of time the heave exceeds some value z is given by $P(z) = 1 - \text{erf}(z/\sqrt{(2m_0)})$. The error function, erf, is obtained from standard mathematical tables.

MOTION IN OBLIQUE SEAS

The procedure outlined above for finding the motion spectra can be applied for the ship at any heading provided the appropriate encounter spectrum is used and the response amplitude operators are available for that heading.

In a regular wave system, as the ship's course is changed from directly into the waves, two effects are introduced, viz.:

(a) the effective length of the wave is increased and the effective steepness is decreased;
(b) the frequency of encounter with the waves is decreased as already illustrated.

An approximation to motions in an oblique wave system can be obtained by testing in head seas with the height kept constant but length increased to $\lambda/\cos\psi$ and with the model speed adjusted to give the correct frequency of encounter. This is a reasonable procedure for vertical motions but it is only an approximation.

SURGE, SWAY AND YAW

As already explained, these motions exhibit a different character from that of roll, pitch and heave. They are not subject to the same theoretical treatment as these oscillatory motions but a few general comments are appropriate.

Surge

At constant power in still water a ship will move at constant speed. When it meets waves there will be a mean reduction in speed due to the added resistance and changed operating conditions for the propeller. The speed is no longer constant and the term surge or surge velocity is used to define the variation in speed about the new mean value. Several effects are present. There is the orbital motion of the wave particles which tends to increase the speed of the ship in the direction of the waves at a crest and decrease it in a trough. In a regular wave system, this speed variation would be cyclic in the period of encounter with the waves. In an irregular sea, the height and hence the resistance of successive waves varies giving rise to a more irregular speed variation. This is superimposed upon the orbital effect which is itself irregular in this case. The propellers will also experience changing inflow conditions due to the waves and the ship's responses. The thrust will vary, partly depending upon the dynamic characteristics of the propulsion machinery and transmission system. The resulting surge is likely to be highly non-linear.

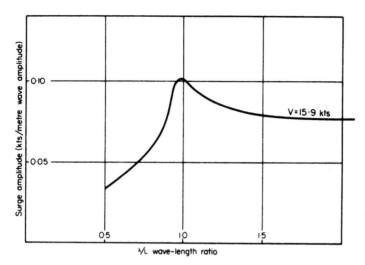

Fig. 12.13 Surging in head sea

The surge experienced by a vessel of length 146.15 m is shown in Fig. 12.13.

The maximum response occurs in waves approximately equal in length to the ship. In waves of this length and 5 m high, the speed oscillation is about ±0.25 knots. The effect varies approximately linearly with speed.

Sway

When the wave system is other than immediately ahead or astern of the ship, there will be transverse forces arising from similar sources to those causing the surging motion. In a regular sea, these would lead to a regular motion in the period of encounter with the waves but, in general, they lead to an irregular athwartships motion about a mean sideways drift. This variation about the mean is termed sway. It is also influenced by the transverse forces acting on the rudder and hull due to actions to counteract yaw which is next considered.

Yaw

When the wave system is at an angle to the line of advance of the ship the transverse forces acting will introduce moments tending to yaw the ship. Corrective action by the rudder introduces additional moments and the resultant moments cause an irregular variation in ship's heading about its mean heading. This variation is termed yawing. In a regular sea with an automatic rudder control system, the motion would exhibit a regular period depending on the period of encounter and the characteristics of the control equation (see Chapter 13). In general, however, the motion is quite irregular.

Some of the difficulty of maintaining course in rough weather is indicated in Fig. 12.14 which is for a ship of 146 m.

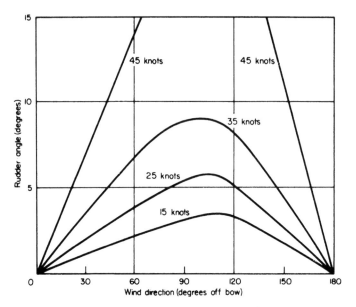

Fig. 12.14 Rudder angles for different wind speeds and directions

Large amplitude rolling

Linear theory shows that large angles of roll can occur when the wave encounter frequency of a beam sea is close to the ship's natural frequency of roll. The amplitude reached will depend upon the degree of damping and whether any stabilizing devices, such as active fins, are employed.

Linear theory asumes a steady metacentric height but when a ship is moving through waves this height is a dynamic quantity not a static one. As the wave surface moves along the length of the ship the shape of the underwater form changes, particularly at the bow and stern, an effect accentuated by heave and pitching motions. These changes lead to variations in the effective metacentric height. When a ship is in a following sea metacentric height variations are long period. Particularly in ships with flat transom sterns there may be a loss of stability and the resulting roll amplitudes can be very large.

Another non-linear effect which causes rolling occurs when the dominant encounter period approximates half the natural period of roll in head or following seas. If associated with fairly large stability variations, large roll angles can result. This phenomenon is often called *half cycle* or *parametric* rolling. It starts quite unexpectedly and quickly reaches very large amplitudes. Model tests, conducted at MARIN, on a 240 m cruise ship suffered roll amplitudes of 40 degrees. It was found that below a certain wave height threshold the rolling was negligible, above the threshold a fairly regular roll motion builds up. The threshold wave height depends upon the ship's heading, the peak period and ship's speed. Zero speed proved the most severe test condition. In following seas a significant wave height as low as 2 m was sufficient to trigger the rolling. In head seas the threshold was 2.75 m. Above the threshold the effect of increasing wave height was dramatic. The threshold wave height increased with increasing ship speed.

Limiting seakeeping criteria

The ability of a ship to carry out its intended mission efficiently may be curtailed by a number of factors. There is a correspondingly wide range of limiting seakeeping criteria. The limit may be set by the ability of the ship itself, or its systems, to operate effectively and safely, or by the comfort or proficiency of passengers or crew. In so far as equipment or personnel performance is degraded when motions (e.g. vertical acceleration) exceed a certain level, careful siting of the related activity within the ship in an area of lesser motions may extend the range of sea conditions in which operation is acceptable. Other features such as slamming or propeller emergence are dependent on overall ship geometry and loading although here again the design of the ship (e.g. its inherent strength in the case of slamming) can determine the acceptable level before damage occurs or conditions become unsafe.

There is a potential danger in applying 'standard' acceptance levels of any criterion to a new design. There must be a judicious choice, both of criteria and acceptance levels, to reflect the particular design, its function and its similarity to previous designs for which operating experience is available. Thus a new design may have been specifically strengthened forward to enable it safely to withstand high slamming loads. Nevertheless guideline figures applicable to general ship types are useful in preliminary design development. Some performance parameters can be assessed in different ways. This may lead to different absolute values of criteria. Hence in using criteria values it is important they be computed for a new design using the same method as that adopted in establishing the acceptable levels.

It is now appropriate to review briefly the seakeeping parameters most frequently used as potential limiting criteria. They are speed and power in waves, slamming, wetness, propeller emergence and impairment of human performance.

SPEED AND POWER IN WAVES

As a wave system becomes more severe, the power needed to drive the ship through it at a given speed increases. The difference arises mainly from the increased resistance experienced by the hull and appendages, but the overall propulsive efficiency also changes due to the changed conditions in which the propeller operates. If the propulsion machinery is already producing full power, it follows that there must be an enforced reduction in speed. Past a certain severity of waves, the motions of the ship or slamming may become so violent that the captain may decide to reduce speed below that possible with the power available. This is a voluntary speed reduction and might be expected to be made in merchant ships of fairly full form at Beaufort numbers of 6 or more. The speed reduction lessens as the predominant wave direction changes from directly ahead to the beam (Fig. 12.15).

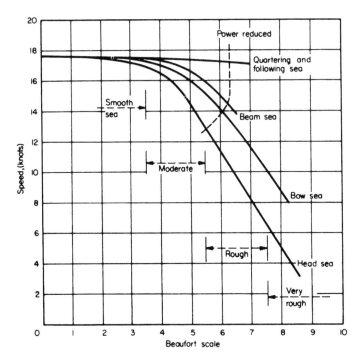

Fig. 12.15

Figure 12.16 shows how the power required for various speeds increases with increasing sea state as represented by the Beaufort number. The figure applies to a wave system 10 degrees off the bow and to a ship 150 m long with a longitudinal radius of gyration equal to 22 per cent of the length. Decreasing the longitudinal moment of inertia decreases the additional power required and also results in drier decks forward.

Figure 12.17 shows the reduction in speed which occurs at constant power (5.83 MW) for the ship in the same conditions and shows the significance of varying the longitudinal radius of gyration. The effect of the variation is less significant in large ships than in small. It is associated with a reduction in natural pitching period.

Other ship design features conducive to maintaining higher speed in rough weather are a low displacement–length ratio, i.e. $\Delta/(L/100)^3$, and fine form forward. Increased damping by form changes or the deliberate introduction of a large bulbous bow can also help. When it is realized that the passage times of ships in rough weather may be nearly doubled, it is clearly of considerable importance to design the ship, both above and below water, so that it can maintain as high a speed as possible. Wetness is a significant factor influencing the need to reduce speed, and this is dealt with later.

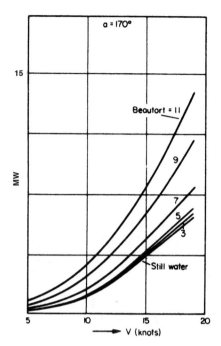

Fig. 12.16 Power in waves for a 150 m long ship

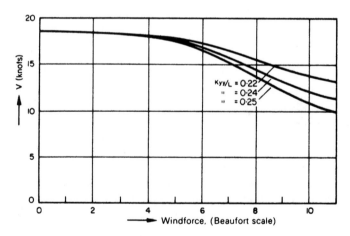

Fig. 12.17 Variation in speed at constant power

SLAMMING

Slamming is a high frequency transient vibratory response of a ship's hull to wave impact. It occurs at irregular intervals. Each blow causes the ship to shudder and is followed by a vibration of the ship's structure. The impact may be large enough to cause physical damage to the ship, the most vulnerable area being between 10 and 25 per cent of the ship's length from the bow. It is the possibility of this damage occurring that causes an experienced captain to reduce speed when his ship begins to slam badly. This always leads to reduced severity of slamming. Lightly loaded cargo ships are particularly liable to slam and the enforced speed reduction may be as much as 40 per cent. Slamming is likely to occur when the relative velocity between the ship's bottom and the water surface is large (usually when the ship is nearly level and the bow has greatest downward velocity); the bow has emerged from and is re-entering the water with a significant length of the bottom roughly parallel to the local water surface; there is a low rise of floor forward increasing the extent of the ship's bottom parallel with the local sea surface. A high relative velocity between the waves and a heavily flared bow can cause a similar, but generally less severe, effect. The heavy flare may throw the water clear of the bow reducing green seas but possibly increasing spray.

The slamming impact lasts for about $1/30$ s and does not perceptibly modify the downward movement of the bow. It can be detected as a disturbance in the acceleration record and by the ensuing vibration which can last for about 30 s. Slamming pressures as high as $0.7 \, \text{N/mm}^2$ have been recorded in low speed ships. Pressures in high speed ships are generally less because of their finer form forward. The pressure can be shown to be proportional to the square of the relative velocity of impact and inversely proportional to the square of the tangent of the angle of deadrise (Fig. 12.18). This is based on analyses of the similar problem of a seaplane landing.

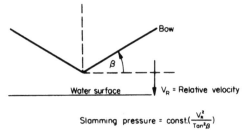

$$\text{Slamming pressure} = \text{const.}\left(\frac{V_R^2}{\text{Tan}^2\beta}\right)$$

Fig. 12.18 Slamming pressure

When β is very small, the pressures obtained would tend to infinity on the above simple analysis. In practice, pressures are limited by the elastic response of the ship structure. The peak pressure moves higher in the ship section as the bow immersion increases.

If, in a given case, the conditions for high pressure apply over a limited area, the blow is local and may result only in local plate deformation. If the pressure acts over a larger area, the overall force acting on the ship is able to excite vibrations of the main ship girder. Since the duration of this vibration, typically of the order of 30 s, is long compared with the stress cycle induced by the main wave, the stress record will be as in Fig. 12.19 assuming the slam occurs at time T_0. The primary ship girder stresses may be increased by 30 per cent or more.

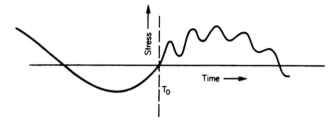

Fig. 12.19 *Slamming vibration superimposed on regular stress cycle*

In considering the limiting seakeeping criteria associated with slamming, it is assumed that the relative motion between the ship and wave surface is sinusoidal.

$$r = r_0 \sin(\omega t - \delta), r_0 = \text{peak relative motion}$$

differentiating

$$\dot{r} = \omega r_0 \cos(\omega t - \delta)$$

If a slam occurs at time $t = 0$

$$\delta = \sin^{-1}\left[-\frac{T}{r_0}\right], T = \text{local draught}$$

Assuming the probability of exceeding a peak relative motion of r_0 follows a Rayleigh distribution

$$P_r(r \geq r_0) = \exp\left(-\frac{r_0^2}{2m_0}\right), m_0 = \text{variance of relative motion}$$

Assuming that a peak relative motion close to r_0 occurs once in N oscillations

$$P_r = \frac{1}{N} = \frac{2\pi}{T_s \omega}, T_s = \text{arbitrary sample time}$$

$$\therefore r_0 = \left\{-2m_0 \log_e\left[\frac{2\pi}{T_s \omega}\right]\right\}^{\frac{1}{2}}$$

The slamming pressure is given by

$$p = \frac{1}{2}\rho \dot{r}^2 F(\beta), \beta = \text{local deadrise angle}$$

An average plot for $F(\beta)$ against β is given in Fig. 12.20. The pressure is limited to $\rho C_w \dot{r}$ where C_w is the velocity of sound in water.

The total force acting on unit length of the ship at a given point is the sum of the forces on the flat of keel and that on the two sides of the hull with deadrise. In each case the force arises from a pressure which is assumed to reach its peak

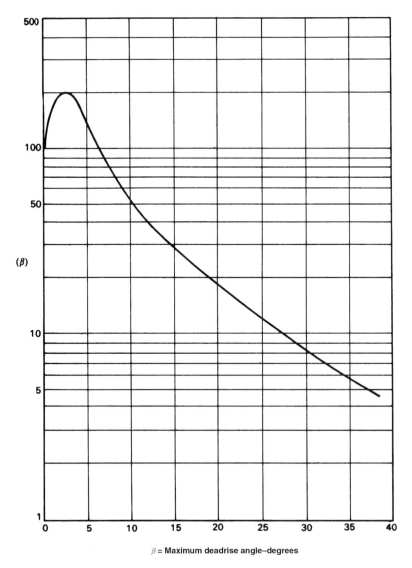

β = **Maximum deadrise angle–degrees**

Fig. 12.20 Slamming pressure coefficient

instantaneously and then decay exponentially. Little is known about the longitudinal distribution of impact loads but a conservative approach is to assume simultaneous impact at hull sections for which keel emergence is predicted in an arbitrary sample period.

WETNESS

By wetness is meant the shipping of spray or green seas over the ship and, unless otherwise qualified, refers to wetness at the bow.

It is not possible at present to calculate wetness accurately but it may be assessed by:

(a) calculating the relative vertical movement of the bow and water surface and assuming that the probability of deck wetness is the same as that of the relative motion exceeding the freeboard at the stem head;

(b) running a model in waves and noting for each of a range of sea conditions the speed at which the model is wet and assuming that the ship will behave in a similar way. Model tests at the right F_n can represent green seas but not spray effects.

Methods based on (a) are usually adopted. If it is valid to assume that the relative motion between ship and waves is sinusoidal and that the probability of deck immersion follows a Rayleigh distribution:

$$P_r = \exp\left[-\frac{F^2}{2m_0}\right], F = \text{freeboard}$$

The average time interval between the deck being wet at a given station is

$$t_w = \frac{2\pi}{P_r \omega} = \frac{2\pi}{\omega} \exp\left[\frac{F^2}{2m_0}\right]$$

where $\omega =$ average frequency of the relative motion

$$= \sqrt{\frac{m_2}{m_0}}$$

$m_2 =$ variance of the relative velocity

Besides trying to reduce the incidence of wetness the naval architect should:

(a) design decks forward so that water clears quickly;

(b) avoid siting forward any equipments which may be damaged by green seas or which are adversely affected by salt water spray.

A bulwark can be fitted to increase freeboard provided it does not trap water. The sizes of freeing ports required in bulwarks are laid down in various international regulations.

PROPELLER EMERGENCE

Using an arbitrary assumption that the propeller should be regarded as having emerged when a quarter of its diameter, D, is above water, criteria corresponding to those used in wetness follow, viz:

$$P_r = \exp\left[-\left(T_p - \frac{D}{4}\right)^2 \middle/ 2m_0\right]$$

$T_p =$ depth of propeller boss below the still waterline

Average time interval between emergencies

$$t_p = \frac{2\pi}{P_r \omega}, \quad \omega = \sqrt{\frac{m_2}{m_0}}$$

DEGRADATION OF HUMAN PERFORMANCE

Besides reducing comfort, motions can reduce the ability and willingness of humans to work and make certain tasks more difficult. Thus in controlling machinery, say, motions may degrade the operator's ability to decide what he should do and, having decided, make the execution of his decision more difficult. There is inadequate knowledge of the effects of motion on human behaviour but in broad terms it depends upon the acceleration experienced and its period. These can be combined in a concept of subjective motion. In this, the combinations of acceleration and frequency are determined at which the subjects feel the motion to have the same intensity. Denoting this level as subjective magnitude *(SM)* with a value of 10, other combinations of acceleration and frequency are assessed as of $SM = 10n$ when they were judged to be n times as intense as the original base *SM*. It is found that

$$SM = A(f)a^{1.43}, a = \text{acceleration amplitude in 'g'}$$

With the frequency, f, in Hz, $A(f) = 30 + 13.53(\log_e f)^2$
Assuming the sinusoidal results can be applied to random motions

$$a = \frac{2}{g}\sqrt{m_{4a}}, \quad f = \frac{1}{2\pi}\sqrt{\frac{m_{2a}}{m_{0a}}}$$

where m_{0a}, m_{2a} and m_{4a} are the variances of the absolute motion, velocity and acceleration respectively in SI units.
    Then

$$SM = \left[3.087 + 1.392\left\{\log_e \frac{1}{2\pi}\sqrt{\frac{m_{2a}}{m_{0a}}}\right\}^2\right]m_{4a}^{0.715}$$

The motions experienced by any individual will depend upon their position in the ship. An overall figure for a ship can be obtained by applying a weighting curve representing the distribution of personnel in the ship. Alternatively for a localized activity (e.g. on the bridge) the *SM* for that one location can be obtained. Unfortunately no clear-cut limiting *SM* could be proposed although a figure of 15 has been suggested as an absolute maximum. Five actions a designer can take to prevent or mitigate the adverse affects of ship motion, especially sea sickness, are

(a) Locate critical activities near the effective centre of rotation.
(b) Minimize head movements.
(c) Align operator position with the ship's principal axes.
(d) Avoid combining provocative sources.
(e) Provide an external visual frame of reference.

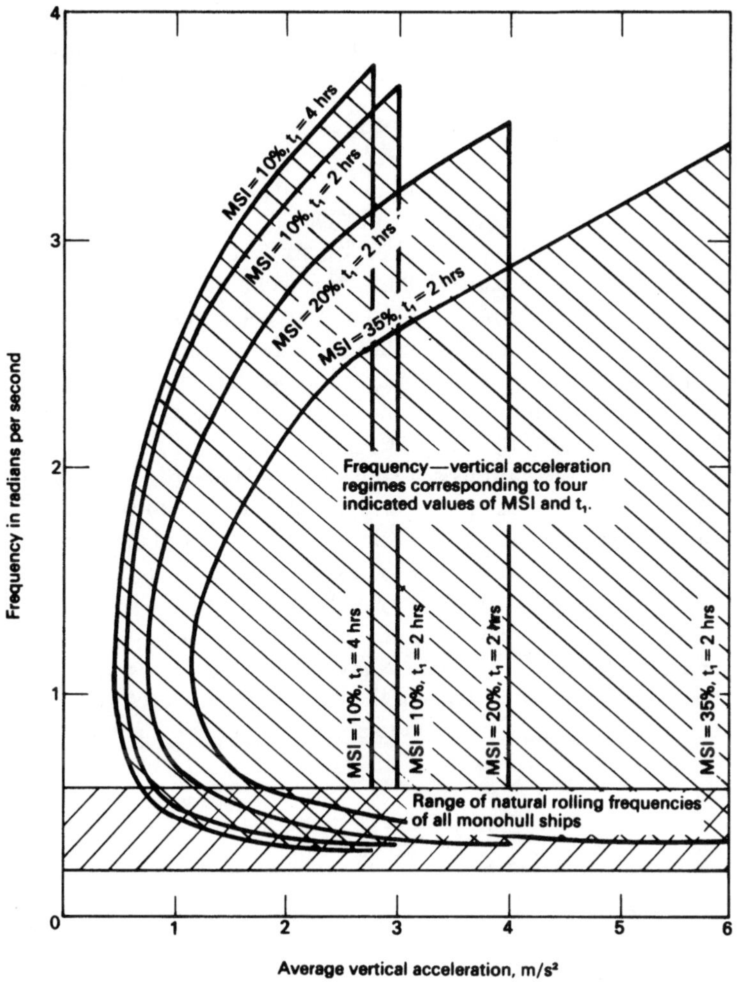

Frequency in radians per second
Average vertical acceleration, m/s²
0
1
2
3
4
5
6
MSI=10%, t₁=4 hrs
MSI=10%, t₁=2 hrs
MSI=20%, t₁=2 hrs
MSI=35%, t₁=2 hrs
Frequency—vertical acceleration regimes corresponding to four indicated values of MSI and t₁.
Range of natural rolling frequencies of all monohull ships

Fig. 12.21 Motion sickness index

Another approach to the effect of motions on personnel is the concept of *motion sickness incidence* (MSI). The MSI is the percentage of individuals likely to vomit when subject to the given motion for a given time. Plots can be made as in Fig. 12.21. The limitations of this approach are that the data relate to unacclimatized subjects, sinusoidal motions and the fact that human performance may be degraded long before vomiting occurs.

Overall seakeeping performance

A number of possible limiting seakeeping criteria have just been discussed. Their variety and the range of sea conditions expected in service mean that no single performance parameter is likely to be adequate in defining a design's

overall seakeeping performance. This applies even within the restricted definition of seakeeping adopted in this chapter. However a methodology is developing which permits a rational approach, the steps of which are now outlined.

(a) The *sea states* in which the ship is to operate are established. The need may
be specific in the sense that the ship will operate on a particular route at
certain seasons of the year, or it may be as general as world-wide operations
all the year round. Ocean wave statistics can be used to determine the
ranges of wave height, period and direction likely to be met for various
percentages of time. As described later, the technique of wave climate
synthesis can be used to improve the reliability of predictions based on
observed wind and wave data. This establishes the number of days a year
the ship can be expected to experience various wave conditions and these
can be represented by appropriate wave spectra, e.g. by adopting the
formulation recommended by the ITTC.

(b) The *ship responses* in the various sea states can be assessed from a knowledge of its responses in regular waves. Even in long-crested seas the ship
response depends upon the severity of the sea, the ship speed and the ship's
heading relative to the wave crest line. Thus motions can be represented by
a polar diagram, such as Fig. 12.22, in which contours are drawn for given
values of response for each of a range of significant wave heights. Assuming
a linear dependency the contours can be expressed as response operators.
    If it is desired to compare designs on the basis of their relative motions at
various speeds (or Froude numbers) the areas within the polar plot can be

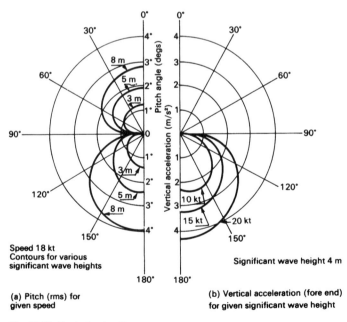

Fig. 12.22 *Typical polar diagrams*

used. This will average out variations with heading. If it is known that the ship will transit on some headings more frequently than others the polar plots can be adjusted by means of suitable weighting factors. For vessels which are symmetrical about their centre-line plane in geometry and loading the polar plots will be symmetrical about the vertical axis.

(c) *Limiting conditions.* It is not usually the motion amplitudes *per se* which limit the ability of a ship to carry out its intended mission. More often it is a combination of motions and design features leading to an undesirable situation which can only be alleviated by reduction in speed or a change of heading. That is to say the ship's freedom of action is restricted.

The usual action is to reduce speed as this has the effect of avoiding synchronism with wave components other than short waves which have less effect on motions anyway. A change of course is only effective when there is a predominant wave direction and often can only be adopted for relatively short periods of time.

Various limiting seakeeping criteria have been discussed above. For any chosen criterion, the speed above which the agreed acceptable limit of the criterion is expected to be exceeded can be plotted on a polar diagram. As with motions, the area within the plot, adjusted if necessary by weighting

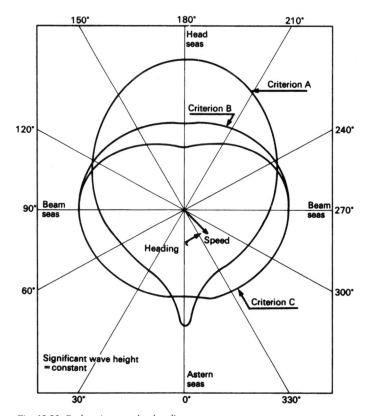

Fig. 12.23 Seakeeping speed polar diagram

factors for different headings, can be used as an overall measure of the design's performance in terms of the selected criterion in the given sea state. The greater the area the greater the range of speeds and headings over which the vessel can operate.

Plots, and hence areas, will be required for each wave spectrum of interest, each of which can be characterized by its significant wave height.

Another concept is that of a *stratified measure of merit*. The time the ship is expected to encounter various wave conditions has been established from ocean wave statistics. The product of this time and the area under the polar plot for the chosen seakeeping criterion is plotted against the significant wave height appropriate to the wave spectrum. The area under the resulting curve represents the overall performance of the design with respect to the selected criterion, over the period of time covered by the wave data used. This could be a single voyage, a year or the whole life of the ship.

(*d*) The *operational* ability of a design will not always be limited by the same criterion. Thus the design's overall potential must be assessed in relation to all the possible limiting criteria. If all criteria of interest are plotted on a common polar plot the area within the inner curve at each heading represents the overall limiting performance of that design in the selected sea state. The measure of merit concept can be used as for a single criterion.

This operational ability assumes a common mission throughout the life of the ship. If it is known in advance that the ship will have different missions at different times the method will need to be modified to reflect the different influence of the various criteria on the missions concerned.

In practice a captain must judge the operational importance of maintaining speed against the risk to the ship. The captain of a warship is more likely to reduce speed on a peacetime transit than in a wartime operation.

This general approach is one method of assessing the relative seagoing performance of competitive vehicle types. Other 'scoring' methods suggested are:

(*a*) The percentage time a given vehicle in a given condition of loading can perform its function in a specified area, in a given season at a specified speed without any of a range of chosen seakeeping criteria exceeding agreed values.

(*b*) The time a vehicle needs to transit between two specified locations in calm water divided by the time the vehicle would require to travel between the same locations in rough weather without any of the selected criteria value being exceeded.

Acquiring data for seakeeping assessments

It will be appreciated from the foregoing that two things are necessary to enable an assessment to be made of seakeeping performance, viz. a knowledge of:

(*a*) wave conditions for the area to which the assessment is related and specifically how the total energy of the wave system is distributed with respect to frequency;

(*b*) the responses of the ship in regular sinusoidal waves covering the necessary frequency band. These responses are normally defined by the appropriate response amplitude operators in the form of response per unit wave height.

SELECTION OF WAVE DATA

Chapter 9 gave information on the type of data available on sea conditions likely to be met in various parts of the world. Much of this is based on visual observations, both of waves and winds. As such they involve an element of subjective judgment and hence uncertainty. In particular visual observations of wave periods are likely to be unreliable. Care is necessary therefore in the interpretation and analysis of wave data if sound design decisions are to be derived from them.

The National Maritime Institute, (now BMT Ltd), has developed a method, known as wave climate synthesis, for obtaining reliable long-term wave data from indirect or inadequate source information. This approach can be used when instrumented wave measurements are not available. Essentially relationships derived from corresponding sets of instrumented and observed data are used to improve the interpretation of observed data. Various sources of data and of methods of analysing it are available such as the Marine Information and Advisory Service of the Institute of Oceanographic Sciences and agencies of the World Meteorological Organisation. Much of the data is stored on magnetic tape.

The NMI analysis uses probabilistic methods based on parametric modelling of the joint probability of wave height and wind speed. Important outputs are:

(a) Wave height

When a large sample is available raw visual data provide reasonable probability distributions of wave height. However, comparisons of instrumented and visual data show that better distributions can be derived using best fit functional modelling to smooth the joint probability distributions of wave height and wind speed. This is illustrated in Fig. 12.24 for OWS *India* in which the 'NMIMET Visual' curve has been so treated.

Analysis of joint probabilities for wave height and wind speed from measured data leads to the relationship

$$\text{Mean wave height} = H_{\text{r}} = \left[(aW_{\text{r}}^n)^2 + H_2^2\right]^{\frac{1}{2}}, \ W_{\text{r}} = \text{wind speed.}$$

Standard deviation of the scatter about the mean is

$$\sigma_{\text{r}} = H_2(b + cW_{\text{r}})$$

The joint probability distribution is given by a gamma distribution

$$P(H_{\text{s}}/H_{\text{r}}, \sigma_{\text{r}}) = \frac{q^{p+1}}{\Gamma(p+1)} H_{\text{s}}^p \exp(-qH_{\text{s}})$$

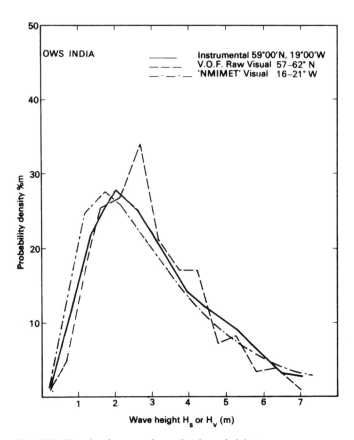

Fig. 12.24 *Visual and measured wave height probabilities*

where $p = \dfrac{H_r^2}{\sigma_r^2} - 1$

$q = H_r/\sigma_r^2$

H_s = significant wave height

H_2, a, b, c and n are the model parameters for which, in the absence of more specific data, suitable standard values may be used. The following values have been recommended on the basis of early work using instrumental data from a selection of six stations. In quoting them it should be noted that they are subject to review in the light of more recent work and meanwhile should be regarded as only valid for use with measured wind speeds up to a limit of about 50 knots. It should also be noted that the numerical values cited are to be used in association with units of metres for wave height and knots for wind speed.

	H_2 (metres)	a	b	c	n
Open ocean	2.0	0.033	0.5	0.0125	1.46
Limited fetch	0.5	0.023	0.75	0.0188	1.38

The wave height probabilities follow from the wind speed probabilities using:

$$P(H_s) = \sum_r (H_s/H_r, \sigma_r) \times P(W_r)$$

Wave directionality data can be obtained if the joint probability distributions of wave height and wind speed are augmented by corresponding joint probabilities of wave height and period and wind speed and direction.

(b) Wave periods

Reliability of visual observations of wave period is poor. NMI adopted a similar approach to that used for wave height but using a different functional representation. Based on analysis of instrumented wave height/period data, wave height and period statistics can be synthesized when reliable wave height data are available using:

$$P(T) = \sum_r P(T/H_r) \times P(H_r)$$

where

$$P(T/H_r) = F_1(\mu_h, \sigma_h, \mu_t, \sigma_t, \rho)$$

$$= [2\pi(1-\rho^2)\sigma_t^2]^{-\frac{1}{2}} \exp\left\{ \frac{-1}{2(\sqrt{(1-\rho^2)}\sigma_t)^2} \left[(t - \mu_t) - \rho\frac{\sigma_t}{\sigma_h}(h - \mu_h) \right]^2 \right\}$$

$$P(H_r) = F_2(\mu_h, \sigma_h)$$

$$= \frac{1}{\sqrt{2\pi}\sigma_h} \exp\left\{ -\frac{(h - \mu_h)^2}{2\sigma_h^2} \right\} \left\{ 1 - \frac{C_s}{6} \left[3\left(\frac{h - \mu_h}{\sigma_h}\right) - \left(\frac{h - \mu_h}{\sigma_h}\right)^3 \right] \right\}$$

where μ_h = mean value of h

$\quad\quad \sigma_h$ = standard deviation of h

$\quad\quad \mu_t$ = mean value of $t = \ln \mu_T - \sigma_t^{3/2}$

$\quad\quad \sigma_t$ = standard deviation of $t = 0.244 - 0.0225\mu_H$

$\quad\quad \rho$ = correlation coefficient = $0.415 + 0.049\mu_H$

$\quad\quad C_s$ = skewness parameter = $E\left(\frac{[h - \mu_h]^3}{\sigma_h^3}\right)$

In these expressions h and t are the logarithmic values of H and T respectively. μ_h, σ_h and C_s follow from the given probability distribution of H as does μ_H the mean wave height.

$$\mu_T = 3.925 + 1.439\mu_H$$

The numerical values of the coefficients in the formulae for σ_t, ρ and μ_t were derived by regression analysis of over 20 sets of instrumental data.

(c) Extreme wave height

Sometimes the designer needs to estimate the most probable value of the maximum individual wave height in a given return period. After the probabilities of H_s are obtained the corresponding cumulative probabilities are computed and plotted on probability paper.

The methods used by NMI Ltd for analysing these cumulative probabilities for H_s differed from those described in Chapter 9 and are suitable for use when, as in the case of visual data, wave records are not available.

The data define exceedance probabilities for H_s up to a limiting level $1/m$ where m is the number of H_s values (or visual estimates of height) available. It is commonly required to extrapolate these to a level $1/M$ corresponding to an extreme storm of specified return period, R years, and duration, D hours, and in this case $M = 365 \times 24 \times R/D$.

In the NMI method this extrapolation is achieved by use of a 3-parameter Weibull distribution, the formula for the cumulative probability being:

$$P(x > H_s) = \exp - \left[\frac{(H_s - H_0)^n}{b} \right]$$

with values of the parameters n, b and H_0 determined numerically by least square fitting of the available data. The most probable maximum individual wave height H_{max} corresponding to the significant height H_{sM} for the extreme storm having exceedance probability $1/M$ is then estimated by assuming a Rayleigh distribution of heights in the storm, so that $H_{max} \doteq (\frac{1}{2} \ln N)^{\frac{1}{2}} H_{sM}$, where N is an estimate of the number of waves in the storm given by $N = 3600D/T$, where T is an estimated mean wave period.

OBTAINING RESPONSE AMPLITUDE OPERATORS

It has been shown that response amplitude operators are convenient both in presenting the results of regular motions in non-dimensional form and as a means of deducing overall motion characteristics in irregular seas. How are these RAOs to be obtained for a given ship? If it is a new design then calculation or model experiments must be used. If the ship exists then ship trials are a possibility. As with other aspects of ship performance the naval architect makes use of all three approaches. Theory helps in setting up realistic model tests which in turn help to develop the theory indicating where simplifying assumptions, e.g. that of linearity of response, are acceptable. Full-scale trials provide evidence of correlation between ship and model or ship and theory.

As knowledge has built up confidence in theory, and as more powerful computers have facilitated more rigorous but lengthy calculations, theory has become the favoured approach to assessing ship motions at least in the early design stages and for conventional forms. Models can be used for the final form to look at deck wetness and rolling in quartering seas for which the theory is less reliable, or to confirm data for unusual hull configurations.

Theory

To outline how theoretical predictions are made, consider the simple case of a ship heading directly into a regular series of long-crested waves at constant speed, i.e. the ship's heading is normal to the line of wave crests. If the ship is symmetrical about its longitudinal centreplane its longitudinal motions, i.e. pitch and heave, will be uncoupled from its lateral motions, sway, yaw and roll (see Fig. 12.1). It is further assumed that pitch and heave are unaffected by any surge.

The fundamental equations follow from Newton's second law of motion

$$m\ddot{z} = F$$

$$J\ddot{\theta} = M$$

In studying ship motion it is usual to consider only the changes in force and moment between the ship moving in calm water and in waves. Thus F and M are the summations of the fluid forces and moments acting on the ship due to the relative motion of ship and wave. Coupling will exist between pitch and heave and the general equations of motion will be of the form

$$(m + a)\ddot{z} + b\dot{z} + cz + d\ddot{\theta} + e\dot{\theta} + g\theta = F_0 \cos(\omega_e t + \alpha)$$

$$(I_{yy} + A)\ddot{\theta} + B\dot{\theta} + C\theta + D\ddot{z} + E\dot{z} + Gz = M_0 \cos(\omega_e t + \beta)$$

where ω_e is the frequency of encounter with the waves.

It is necessary now to obtain expressions for the various coefficients a, b, etc., and A, B, etc., in these equations. Most modern approaches are based on what is known as the 'strip theory' or 'slender body theory'.

The methods differ in detail but all make the following simplifying assumptions:

- Ship responses are small, varying linearly with wave height.
- The ship is slender, its length being much larger than its beam or draught.
- The hull is rigid.
- Ship sections are wall-sided.
- Speed is moderate so that no significant planing lift is produced.
- The water depth may be regarded as deep.
- The presence of the hull has no effect on waves. This is often referred to as the *Froude–Kriloff hypothesis*.

The vessel is considered as composed of a number of thin transverse slices or strips. The flow about each is assumed to be two-dimensional and the same as would exist if the section were part of an infinitely long oscillating cylinder of equal cross-section. This is clearly a simplification of the real state of affairs in that it ignores any interaction between the flow around adjacent sections and the three-dimensional flow that must exist particularly at the ends of the ship. In the case of a vessel pitching and heaving but with no forward velocity the assumption seems not unreasonable, with the motion of each strip considered as a combination of the two motions.

When the ship has forward motion the assumption of two-dimensional flow seems more debatable. The water mass remains approximately steady in space. Although there are orbital motions of water particles in waves the water so disturbed does not move forward with the ship. Cross-sections of the ship are taken relative to a set of axes fixed in space and account is taken of the changing form at any transverse plane due to the forward velocity of the ship. Some authorities regard the theory as unsatisfying but it has been found to provide good comparisons between theory and experiment for ships down to 40 metres in length and for that reason is widely used. Clearly accuracy depends upon, *inter alia*, the values of 'added mass' and damping assumed and much of the development has centred on such factors.

The analysis methods used begin by considering an infinitely long semi-circular cylinder heaving, swaying and rolling in a water free surface. Flow is assumed to be inviscid and incompressible so that potential theory can be used. Conformal transformation techniques are then used to extend the semi-circular results into those for realistic ship shapes. This technique was first used by Lewis (1929) for vibration studies. He found that a truncated transformation yielded acceptable results and he defined his sections in terms of beam/draught ratio and a cross-section area coefficient. The method has been expanded upon by later workers in the field.

In still water the vessel is in equilibrium with buoyancy equal to its weight since it is assumed that the hull does not develop any planing lift. Only changes relative to this still water condition are considered.

Concentrating for the moment on the vertical movement of the strip, at any instant the relative vertical displacement of a point on the hull relative to the water surface will be composed of terms due to heave, pitch and wave surface elevation, i.e.

$$z_r = z - x\theta - \zeta$$

Axes are taken with origin in the still water plane vertically above the centre of gravity and the transverse plane under consideration is a distance x from that

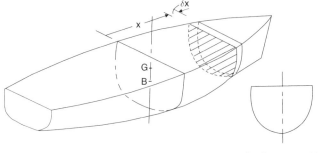

Section at x, δx thick

Fig. 12.25 Strip theory

origin. As described above the transverse planes are taken at fixed spatial positions. Thus they move relative to the ship based axes such that:

$\dot{x} = -u$ where $u =$ ahead velocity of ship.

Differentiating z_r with respect to time

$$\dot{z}_r = w_r = \dot{z} - (x\dot{\theta} + \dot{x}\theta) - \dot{\zeta}$$
$$= \dot{z} - x\dot{\theta} + u\theta - \dot{\zeta}$$

Differentiating again

$$\ddot{z}_r = \dot{w}_r = \ddot{z} - x\ddot{\theta} - \dot{x}\dot{\theta} + \dot{u}\theta + u\dot{\theta} - \ddot{\zeta}$$
$$= \ddot{z} - x\ddot{\theta} + 2u\dot{\theta} - \ddot{\zeta} \text{ since } \dot{u} = 0$$

The forces acting on a 'strip' with these motions would normally be:

(*a*) an inertial force opposing the acceleration of that section of the ship;
(*b*) hydrodynamic forces due to the relative acceleration and velocity of hull and water;
(*c*) a hydrostatic force due to the increased immersion, z_r.

The so-called 'added mass' term varies with time so that the total effect is one of momentum change

$$\frac{d(a_n w_r)}{dt} = w_r \frac{da_n}{dt} + a_n \dot{w}_r$$

The subscript n is used to denote the particular strip of the ship. If the mass of the ship itself over this strip is m_n then the vertical force δF on a strip of length δx will be of the form

$$\frac{\delta F_n}{\delta x} = -m_n(\ddot{z} - x\ddot{\theta}) - a_n \ddot{z}_r - \left(b_n + \frac{da_n}{dt}\right)\dot{z}_r - c_n z_r$$

a_n, b_n and c_n will at least depend upon hull form and the first two may also vary with motion frequency.

Additional moments, and forces, will arise from yawing and rolling.

The added mass or *acceleration* term, a_n, is obtained by reference to an oscillating circular cylinder. Such a cylinder deeply immersed in a perfect fluid and oscillating normal to its axis has an added mass equal to the displaced water mass. Hence in the more general case of a non-circular cylinder it is reasonable to use an expression of the form:

added mass coefficient $= \rho(\text{constant})(\text{cross-sectional area})$
$$= \rho k_2 A$$

When the cylinder oscillates in a free surface a standing wave system is formed which modifies the added mass term by a second factor, k_4, which depends upon the frequency of oscillation. In this case:

added mass coefficient $= a_n = \rho k_2 k_4 A$

Suitable values for k_2 and k_4 for typical ship forms may be found in published papers for a range of beam/draught ratios and for a number of section shapes at each ratio value. Values are available for rectangular and triangular shapes. Coefficient values for a ship are obtained by comparing each section with the sections presented in the reference. Hence water inertia mass per unit length is calculated for each section, the water inertia curve plotted and added to the normal weight distribution curve. The water inertia is effectively independent of forward speed.

Lewis found that due to the flat bottom amidships the greater part of the added mass is amidships. Also he found that the water inertia is nearly independent of draught so that natural frequency will vary only slowly with the displacement.

The *relative velocity* coefficient is the sum of two components, viz:

(*a*) a dissipative damping component, b_n, due to radiated waves

$$b_n = \rho g^2 \bar{A}^2 / \omega_e^3$$

where ω_e = frequency of radiated wave
$= $ frequency of wave encounter.

$\bar{A}$ = ratio of amplitudes of radiated waves and the relative vertical motion.
$\bar{A}$ values can be found from the literature;

(*b*) a dynamic damping component arising from the rate of change of a_n.

Thus this term involves the way in which the added mass coefficient varies along the length of the ship.

Havelock (1956) compared the damping coefficients obtained by strip theory and three-dimensional calculation. A completely immersed body was assumed and Havelock argued the relative magnitudes would be a guide to the relative values for a freely floating body. Vossers in discussion supported this view based on calculations for a Michell ship. Figure 12.26 gives figures for a spheroid of L/B ratio of 8. For free oscillations of typical ship forms the values calculated by the two methods are approximately equal but for forced oscillations at lower frequencies the values can differ significantly, particularly for pitch.

The *vertical displacement* term, c_n, is the added buoyancy due to increased immersion. This depends upon the waterplane area in the ship

$$c_n = \rho g \delta A_W$$
$$= \rho g B_n \text{ per unit length, } B_n = \text{local beam}$$

The value of δF acting on each element can be obtained from the above and its moment about the vessel's centre of gravity. Integrating along the length of the ship gives the overall vertical force and moment.

Strip theory is used widely for prediction of rigid body motions and gives reasonable results for most purposes except for rolling. Unfortunately the damping coefficients for rolling are non-linear and depend upon the motion

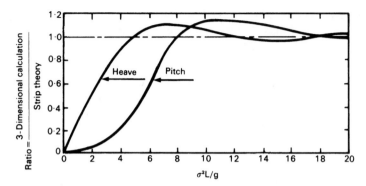

Fig. 12.26 Ratios of damping coefficients for heaving and for pitching

amplitude. This arises from the importance of eddy shedding at the bilges as the ship rolls and skin friction, which are due to viscosity, ignored in strip theory. To circumvent this problem a device can be adopted of an equivalent linear damping coefficient. The predominant rolling motions occur at the natural roll frequency. The equivalent linear roll damping coefficient is based on that frequency and the energy dissipated during the roll. It will have components due to eddy making, skin friction and drag forces on appendages.

Strip theory can be used to calculate ship motions in regular waves. The results can then be applied to motions in irregular seas by summation of the responses to a large number of regular components by which the irregular sea can be represented.

The same basic approach can be used where the ship is moving at an angle to the line of wave crests and to the other degrees of freedom. However, although early applications of strip theory to pitching and heaving were successful, predictions of sway, roll and yaw were less good. Correlation of theoretical results with model or full-scale data showed significant errors in roll, particularly for higher-speed ships. The problem arises principally from the difficulty of assessing roll damping. This has been attributed to inadequate treatment of hull appendages. Better prediction of roll damping and response can be obtained using strip theory to compute added mass, wave making damping and exciting forces including:

(a) lifting surface contributions to damping and exciting forces. The various hull appendages act as lifting surfaces when the ship is underway, their contribution varying with speed. The forces depend upon the geometry of the appendages and the resulting moments upon their position relative to the centre of gravity of the ship;

(b) viscous roll damping, which has contributions from the bilge keels and other appendages, the eddy-making resistance of the hull and hull friction;

(c) hull circulatory effects.

These methods permit fair prediction of lateral motions. Studies show that there is negligible coupling of roll and yaw into sway, the sway being characterized by an exponential decay with a 90 degrees phase lead, and that yaw amplitudes are small. The theory correctly predicts the substantial reduction in roll response with increasing ship speed and the variation in roll response with ship's heading relative to the waves.

Model experiment

Several methods of model testing are in essence available, viz:

(*a*) measuring the response of a model in regular waves and deducing a set of response amplitude operators for use in the superpositioning theory to predict performance in an irregular sea. For reliable results such tests must be conducted in relatively moderate waves, typically with wave length to height ratio of 40:1. The test facilities can be relatively simple but a large number of runs will be required to cover adequately the range of speed and wavelength involved;

(*b*) running the model in a standard irregular pattern and analysing the data to provide the response amplitude operators. Several test runs will generally be necessary for each speed to obtain sufficient motion cycles to provide adequate confidence levels in the subsequent statistical analysis. The wavemakers must be capable of creating an irregular wave pattern with the desired spectral characteristics;

(*c*) *transient wave* or *impulse* wave testing. This method can be regarded as a special case of method (*b*). The wavemaker starts at high frequency, slows down and then stops. Thus it initially produces short waves which are gradually overtaken by the later, longer waves. The model starts in calm water, passes through a short wave sequence and finishes its run in calm water. By analysing the complete wave and motion records a full picture is obtained of the model responses with a considerable reduction in testing time. The range of frequencies present in the wave sequence must cover the range for which response operators are required, the waves must not become so steep that they break and the records must not be affected by reflections from the end of the tank;

(*d*) creating a representation of an actual (recorded on ship trial, say,) irregular wave pattern and running the model in it to record motion, wetness, power, etc. In practice it would be very difficult to ensure the model experienced the same wave pattern as the ship which would be the only way to compare directly the pattern of wetness and slamming. However, if the spectral form is correct the model and ship performance can be compared on a statistical basis and thus provide some check on the adequacy of the linear superpositioning assumption.

As with wave records discussed in Chapter 9, it must be remembered that the results of a particular series of experiments can only be regarded as a sample of all possible experimental outcomes. To illustrate this, consider a single run in an irregular wave system. The actual surface shape repeats itself only after a very long period of time, if at all. The actual model data obtained from a particular run depends on when, during that period, the experiment is run. Tests for statistical significance can be applied.

Ship trials

In principle it is the ship trial which should provide the final check on the adequacy of theoretical and experimental predictions of ship behaviour. Unfortunately the actual ocean never exhibits a standard long-crested sinusoidal wave pattern. Thus no direct measurement of individual response operators is possible for comparison with predictions. Even truly long-crested irregular seas are never met. On occasion the sea surface may approximate to the long crested form but waiting for good conditions can be expensive of time and money. If they are met the sea and motions can be measured and response operators deduced. Comparison of ship with prediction can be made on the basis of these response operators or on the motion parameters predicted for the sea spectrum as measured.

Unfortunately, it is often the more extreme conditions that are of most concern and it is for these that the usual assumptions of linear superposition are likely to yield the greatest inaccuracies. These would be additional to those arising due to inaccuracies in recording waves and motions, sea variations over a recording period and differences in wind and tide. Thus a comparison of two ships as a result of trials conducted at different times and in different sea conditions has a number of limitations, particularly under limiting conditions for operations. Some of these are avoided, or reduced, if the two ships can go on trials in company. The sea condition will then be the same for both ships although it must be remembered that the sea condition may favour one design, particularly if the ships are of different length. It will be appreciated, therefore, that all trials data must be used intelligently with due allowance for the above factors.

The conduct of ship trials is discussed later.

Non-linear effects

Most of the remarks in this chapter, and, indeed, in the book, relate to simple linear equations of motion. This is because it is the simple approach, and

- it is a desirable step, to give a basic understanding, before going on to consider non-linear equations;
- it is an assumption that has served the naval architect with believable accuracy over many years when the tools were not available for more precise studies.

To some extent the success of the linear approach in the past has been due to the use of conservative design methods, or factors of safety that may have been unreasonably high. The key word here is 'may' because, in truth, the naval architect did not know. Designs to these methods had performed well in service but had they been over designed? Some would argue that because ships are still lost, they were inadequately designed, but perhaps there were other reasons for the losses. Or, the successful ships may have been lucky in that they did not meet the more extreme conditions for which they were designed; chance governs much of our lives.

Unfortunately, it is for the more extreme sea conditions and ship responses that non-linearities are most important. It is these conditions that are of greatest concern in safety. Extreme loading in a seaway and broaching-to are two examples of situations when it is desirable, indeed necessary, to take account of non-linear effects. How then do these effects arise?

- Some are due to the variation in the underwater form as the ship moves through, and responds to, the waves. This will arise even for moderate angles due to flare, but become severe when the deck edge becomes immersed or the bow leaves the water completely.
- Cross-coupling of motions will occur. Heave and roll will induce pitching, for example because the ship's shape and therefore forces upon it differ forward and aft.
- Some arise from the physics of the situation. Taking roll damping, the forces on appendages at high speed are proportional to roll velocity but viscous damping forces vary as the square of the roll velocity.

Software is available nowadays to enable the designer to solve non-linear, coupled, six degrees of freedom motion calculations. The exciting forces and moments arise from hydrostatics, from movement through the water (resistance and propulsion), from control surface movements (rudders and stabilizers), wind and wave forces. The computations are more time consuming and costly and should only be used where their greater accuracy is justified. The more realistic the equations the more closely a design's real-life behaviour will be predicted. Any remaining factors of safety can be more specific. For instance, factors will still be needed in structural design to allow for factors such as plate rolling tolerances, built-in distortions and defects. It is not possible in this book to cover non-linear analyses but the student should be aware of the limitations of the linear approach.

Frequency domain and time domain simulations

It has been shown how a ship's response amplitude operators can be used, in conjunction with a wave spectrum for a given sea condition, to produce a corresponding motion spectrum for each motion. This can be done for a range of ship speeds, headings and sea states. From the motion spectrum one can deduce statistical information about the motion by assuming the spectrum to be of a certain form; say, a Gaussian or Rayleigh distribution. Thus the significant pitch angle, the probability of exceeding a given roll angle, and so on, can be calculated. This is usually quite adequate for assessing the general seakeeping performance of a design. Extreme motion probabilities are difficult to assess because of doubts as to the actual form of the ends of the spectrum.

It is also possible to generate from the spectrum a time sequence for any motion. By dividing the spectrum into a large number of small frequency intervals the motion amplitude at a large number of frequencies are obtained. These individual motions can be assumed to be sinusoidal and can be combined in random phase angle to produce the variation of the motion with time. Velocities and accelerations are obtained by differentiating with respect to time. This simulation of the motion, because it derives from the spectral frequency, is known as a frequency domain simulation. While relatively easy to generate, such simulations suffer from a number of weaknesses:

- The simulation must be continued for a long time in order to ensure that a good representation of motions in any sea state is obtained. If too short the designer may find that the simulation is of a period of relative quiet in the motion, or one of relative violence. This latter may, of course, be what the designer is seeking if it is extreme motions which are being studied.
- It is a linear representation. All motions will be proportional to wave height. This may not be too critical if the spectra used were based on motion levels typical of those being studied. It is not acceptable for many investigations into extreme motions and their effects. See the remarks on non-linear effects, above.
- Cross-coupling of motions is ignored except insofar as it is implicit in the spectra used.

The frequency domain simulation can produce reasonable representations of motions for many purposes but is inadequate for extreme conditions. One limitation is that all motions responses are assumed proportional to wave height. Another form of simulation reproduces how a ship behaves in a specific train of waves. The wave form and the ship form are defined in the *time domain*. By calculating the forces and moments on the ship as it moves through the waves the true combinations of motion can be obtained. The constantly changing shape of the immersed hull, and the forces acting on it, follow from the wave form and the motions of the ship. Cross-coupling effects will be built in. A more accurate picture of factors such as change of freeboard will result and be allowed for. Again the analysis must be continued for long enough if a full representation of the general motion is required.

The accuracy of the motions obtained will depend upon the methods used to calculate the forces involved. *Non-linear effects* due to the changing form of the immersed volume will be implicit in the method. Other non-linearities such as the variation of viscous damping with amplitude of motion (viscous roll damping is proportional to the square of the roll velocity) can be introduced into the equations used. Responses are no longer linearly related to wave height. Programs have been developed for this type of simulation. While in principle the more advanced methods of *computational fluid dynamics* could be used, in practice it has been found adequate to use potential flow methods. They do, as one would expect, require considerable computing power which is compounded by the length of simulation needed. Such computations are becoming more common as the power of computers increases and costs reduce.

Another advantage of a time domain simulation is that so-called 'memory' effects are included. These arise from the fact that the forces at any instant depend, to a degree, upon the preceding motions. Memory effects are important to some design calculations. For instance, in the design of securing systems it is necessary to know the time history of the loads as well as the maximum loads experienced. To illustrate how memory effects may be significant, consider a ship responding to rudder movements. When the rudder is put over the ship begins to turn and the flow around the hull and rudder change. The immediate effects of further rudder movement will depend upon the water flow which will be in transition until the ship takes up a steady rate of turn. In a seaway the flow around the hull is constantly changing in a pattern which does not repeat itself.

Improving seakeeping performance

It has been seen that overall seakeeping performance is limited not so much by motions *per se* as by by the interplay between motions and other design features. Thus overall performance can be improved by such actions as:

(*a*) siting critical activities in less-affected areas of the ship. Examples are the siting of passenger accommodation towards the position of minimum vertical motion; placing helicopter operations aft in frigates and placing only very rugged equipment forward on the forecastle;
(*b*) rerouting of ships to avoid the worst sea conditions;
(*c*) providing local stabilization for certain equipments such as radars.

To an extent these can only be regarded as palliatives and it is necessary to consider how the motions themselves can be reduced. Care is needed to ensure the reduction is 'useful'. For example, if human performance is a limiting factor significant reductions in vertical acceleration over a wide range of higher frequencies is counter-productive if it is won at the expense of even a small increase in acceleration in the frequency band critical to humans. Bearing this reservation in mind there are a number of ways open to the naval architect:

Use can be made of a radically different hull form. Examples, some of which will be discussed in a later chapter, are:

(*a*) the Small Waterplane Area Twin Hull (SWATH) ship. Essentially the use of a small waterplane area reduces the exciting forces and moments, the twin hull restoring the desired static stability qualities and weather deck area.
(*b*) the semi-submersible. The concept is similar to the SWATH. A major part of the vessel is well below the still waterplane so that the waves exert little force on it. This configuration is used for oil drilling rigs where a stable platform is essential and it must be held accurately in position over the seabed.
(*c*) hydrofoil craft. With suitable height sensors and foil incidence control systems a hydrofoil can provide a high speed, steady platform in sea states

up to those in which the waves impact the hull. This depends upon foil separation from the hull.

Special ship types can be very effective in specific applications but usually there are penalties which means that the vast majority of ships are still based on a conventional monohull. In this case the designer can either improve performance by detail form changes or stabilize the whole ship. These are now considered.

INFLUENCE OF FORM ON SEAKEEPING

It can be dangerous to generalize on the effect of varying form parameters on the seakeeping characteristics of a design. A change in one parameter often leads to a change in other parameters and a change may reduce motions but increase wetness. Again, the trend arising from a given variation in a full ship may not be the same as the trend from the same variation in a fine ship. This accounts for some apparently conflicting conclusions from different series of experiments. It is essential to consult data from previous similar ships and particularly any methodical series data that is available covering the range of principal form parameters applicable to the new design.

With the above cautionary remarks in mind the following are some general trends based on the results of methodical series data:

Length is an important parameter in its own right. This can be appreciated by considering the response of a ship to a given wave system. If the ship is long compared with the component waves present, it will pitch and heave to a small extent only, e.g. a large passenger liner is hardly affected by waves which cause a 3000 tonnef frigate to pitch violently. With most ships there does come a time when they meet a wave system which causes resonance but the longer the ship the less likely this is.

Forward waterplane area coefficient. An increase reduces the relative motion at the bow but can lead to increased vertical wave bending moment.

Length to beam ratio has little influence on motions although lower L/B ratios are preferable.

Length to draught ratio. High values lead to resonance with shorter waves and this effect can be quite marked. Because of this, high L/T ratios lead to lower amplitudes of pitch and heave in long waves and greater amplitudes in short waves. A high L/T ratio is more conducive to slamming.

Block coefficient. Generally the higher the block coefficient the less the motions and the greater the increase in resistance, but the influence is small in both cases.

Prismatic coefficient. The higher the C_P value the less the motion amplitudes but the wetter the ship. High C_P leads to less speed loss at high speed and greater speed loss at low speed.

Beam to draught ratio. Higher values reduce vertical acceleration but may lead to greater slamming.

Longitudinal radius of gyration. In waves longer than the ship, a small radius of gyration is beneficial in reducing motions.

A bulbous bow generally reduces motions in short waves but can lead to increased motions in very long waves.

Forward sections. U-shaped sections usually give less resistance in waves and a larger longitudinal inertia. V-shaped sections usually produce lower amplitudes of heave and pitch and less vertical bow movement. Above-water flare has little effect on motion amplitudes but can reduce wetness at the expense of increased resistance and possible slamming effects.

Freeboard. The greater the freeboard the drier the ship.

It will be noted that a given change in form often has one effect in short waves and the opposite effect in long waves. In actual ocean conditions, waves of all lengths are present and it would not be surprising, therefore, if the motions, etc., in an irregular wave system showed less variation with form changes. Research has shown that, for conventional forms, the overall performance of a ship in waves is not materially influenced by variations in the main hull parameters. A large ship will be better than a small one.

Local form changes can also assist in reducing the adverse consequences of motion, e.g. providing finer forms forward with large deadrise angle can reduce slamming forces.

SHIP STABILIZATION

There is a limit to the extent to which amplitudes of motion can be reduced in conventional ship forms by changes in the basic hull shape. Fortunately, considerable reductions in roll amplitudes are possible by other means, roll being usually the most objectionable of the motions as regards comfort. In principle, the methods used to stabilize against roll can be used to stabilize against pitch but, in general, the forces or powers involved are too great to justify their use.

Stabilization systems

These fall naturally under two main headings:

(*a*) Passive systems in which no separate source of power is required and no special control system. Such systems use the motion itself to create moments opposing or damping the motion. Some, such as the common bilge keel, are external to the main hull and with such systems there is an added resistance to ahead motion which has to be overcome by the main engines. The added resistance is offset, partially at least, by a reduction in resistance of the main hull due to the reduced roll amplitude.

Other passive systems, such as the passive anti-roll tanks, are fitted internally. In such cases, there is no augment of resistance arising from the system itself.

The principal passive systems (discussed presently) fitted are:

Bilge keels (and docking keels if fitted)
Fixed fins

Passive tank system
Passive moving weight system.

(*b*) Active systems in which the moment opposing roll is produced by moving masses or control surfaces by means of power. They also employ a control system which senses the rolling motion and so decides the magnitude of the correcting moment required. As with the passive systems, the active systems may be internal or external to the main hull.

The principal active systems fitted are:

Active fins
Active tank system
Active moving weight
Gyroscope.

Brief descriptions of systems

The essential requirement of any system is that the system should always generate a moment opposing the rolling moment.

(*a*) With active fins a sensitive gyro system senses the rolling motion of the ship and sends signals to the actuating system which, in turn, causes the fins to move in a direction such as to cause forces opposing the roll. The actuating gear is usually electrohydraulic. The fins which may be capable of retraction into the hull, or may always protrude from it, are placed about the turn of bilge in order to secure maximum leverage for the forces acting upon them. The fins are usually of the balanced spade type, but may incorporate a flap on the trailing edge to increase the lift force generated.

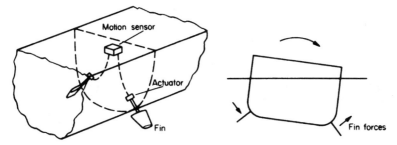

Fig. 12.27 *Active fin system*

The capacity of a fin system is usually expressed in terms of the steady angle of heel it can cause with the ship moving ahead in still water at a given speed. Since the force on a fin varies in proportion to the square of the ship speed, whereas the $\overline{GZ}$ curve for the ship is, to a first order, independent of speed, it follows that a fin system will be more effective the higher the speed. Broadly speaking, a fin system is not likely to be very effective at speeds below about 10 knots.

(*b*) Active weights systems take a number of forms, but the principle is illustrated by the scheme shown in Fig. 12.28. If the weight W is attached to a

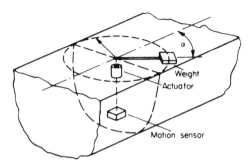

Fig. 12.28 Active weight system

rotating arm of radius R then, when the arm is at an angle α to the centre line of the ship and on the higher side,

Righting moment $= WR\sin\alpha$

Such a system has the advantage, over the fin system, that its effectiveness is independent of speed. It involves greater weight and power, however, and for these reasons is not often fitted.

(*c*) Active tank systems are also available in a variety of forms as illustrated in Fig. 12.29. The essential, common, features are two tanks, one on each side of the ship, in which the level of water can be controlled in accord with the dictates of the sensing system. In scheme (*a*), water is pumped from one tank to the other so as to keep the greater quantity in the higher tank. In scheme (*b*), the water level is controlled indirectly by means of air pressure above the water in each tank, the tanks being open to the sea at the bottom. Scheme (*b*) has the advantage of requiring less power than scheme (*a*). In scheme (*c*), each tank has its own pump but otherwise is similar to scheme (*a*).

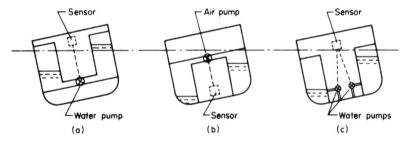

Fig. 12.29 Active tank systems

(*d*) All active stabilizing systems depend upon gyroscopes as part of their control system. If the gyroscope is massive enough, use can be made of the torque it generates when precessed to stabilize the ship. Such systems are not commonly fitted because of their large space and weight demands.

(*e*) Bilge keels are so simple and easy to fit that very few ships are not so fitted. They typically extend over the middle half to two-thirds of the ship's length at

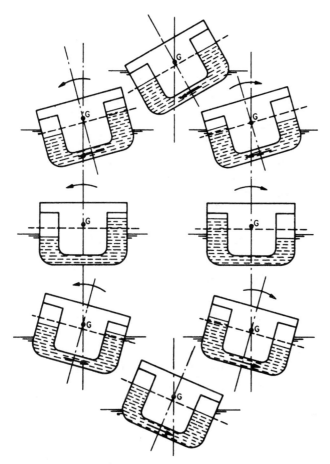

Fig. 12.30 Passive tank system

the turn of bilge. Compared with a ship not fitted, bilge keels can produce a reduction of roll amplitude of 35 per cent or more. They are usually carefully aligned with the flow around the hull in calm water so as to reduce their resistance to ahead motion. Unfortunately, when the ship rolls the bilge keels are no longer in line with the flow of water and can lead to significant increases in resistance. For this reason, some large ships may be fitted with a tank stabilizing system and dispense with bilge keels.

(*f*) Fixed fins are similar in action to bilge keels except that they are shorter and extend further from the ship's side. An advantage claimed for them is that, by careful shaping of their cross-section, the lift generated at a given ahead speed can be increased compared with the drag they suffer. A disadvantage is that, projecting further from the hull, they are more susceptible to damage. They are generally less effective at low speed.

(*g*) Passive tank systems use the roll of the ship itself to cause water in the tanks to move in such a way as to oppose the motion. Starting from rest with water level in the two tanks, if the ship rolls to starboard water flows from port to starboard until the maximum angle of roll is reached. As the ship now tries to recover, the water will try to return but will nevertheless lag and the moment due to the water will oppose the roll velocity. Also, if the resistance of the duct is high the water will not be able to return before the ship is rolling to port, i.e. the level of water in the tanks can be made to lag the roll motion. By carefully adjusting the resistance of the duct the system can be 'tuned' to give maximum stabilizing effect. This will be when the phase lag is 90 degrees.

One limitation of such a scheme is that the system can only be 'tuned' to one frequency. This is chosen as the natural period of roll because it is at this period that the really large angles of roll can be built up. At other frequencies the passive tank system may actually lead to an increase in roll angle above the 'unstabilized' value, but this is not usually serious because the roll angles are small anyway. A more sophisticated system is one in which the resistance in the duct can be varied to suit the frequency of the exciting waves. In this way roll damping is achieved in all wave lengths.

(*h*) Passive moving weight systems are similar in principle to the passive tank systems but are generally less effective for a given weight of system.

Comparison of principal systems

Table 12.1 compares the principal ship stabilizing systems. The most commonly fitted, apart from bilge keels, are the active fin and passive tank systems.

Performance of stabilizing systems

The methods of predicting the performance of a given stabilizer system in reducing motion amplitudes in irregular seas are beyond the scope of this book. A common method of specifying a system's performance is the roll amplitude it can induce in calm water, and this is more readily calculated and can be checked on trials.

When the ship rolls freely in still water, the amplitude of each successive swing decreases by an amount depending on the energy absorbed in each roll. At the end of each roll the ship is momentarily still and all its energy is stored as potential energy. If ϕ_1 is the roll angle, the potential energy is $\frac{1}{2}\Delta\overline{GM}\phi_1^2$. If, on the next roll, the amplitude is ϕ_2 then the energy lost is

$$\frac{1}{2}\Delta\overline{GM}(\phi_1^2 - \phi_2^2) = \Delta\overline{GM}\left(\frac{\phi_1 + \phi_2}{2}\right)(\phi_1 - \phi_2) = \Delta\overline{GM}\phi\delta\phi$$

where ϕ = mean amplitude of roll.

The reduction in amplitude, $\delta\phi$, is called the *decrement* and in the limit is equal to the slope of the curve of amplitude against number of swings at the mean amplitude concerned. That is,

$$\delta\phi = \left(-\frac{\mathrm{d}\phi}{\mathrm{d}n}\right)_\phi$$

Table 12.1
Comparison of stabilizer systems. (Figures are for normal installations)

Type	Activated fin	Passive tank	Active tank	Massive gyro (active)	Moving weight (active)	Moving weight (passive)	Bilge keel	Fixed fin
Percentage roll reduction	90%	60–70%	No data	45%	No data	No data	35%	No data
Whether effective at very low speeds	No	Yes	Yes	Yes	Yes	Yes	Yes	No
Reduction in deadweight	1% of displacement	1–4% of displacement	Comparable with passive tank	2% of displacement	Comparable with passive tank		Negligible	
Any reduction in statical stability	No	Yes	Yes*	No	Yes*	Yes	No	No
Any increase in ship's resistance	When in operation	No	No	No	No	No	Slight	Slight
Auxiliary power requirement	Small	Nil	Large	Large	Large	Nil	Nil	Nil
Space occupied in hull	Moderate generally less than tanks	Moderate	Moderate	Large	Moderate	Less than tanks	Nil	Nil
Continuous athwartships space	No	Generally	Yes	No	Yes	Yes	No	No
Whether vulnerable to damage	Not when retracted	No	No	No	No	No	Yes	Very
First cost	High	Moderate	Probably high†	Very high	Probably high†	Probably high†	Low	Moderate†
Maintenance	Normal mechanical	Low	Normal mechanical	Probably high	Normal mechanical	Normal mechanical	Often high	Probably high

* There is an effective reduction in statical stability, since allowance must be made for the possibility of the system stalling with the weight all on one side.
† These systems have not been developed beyond the experimental stage and the cost comparison is based on general consideration.

This means that when stabilizers are rolling a ship to a steady amplitude ϕ, the energy lost to damping per swing is

$$\Delta\overline{\mathrm{GM}}\phi\left(\frac{\mathrm{d}\phi}{\mathrm{d}n}\right)_\phi$$

and this is the energy that must be provided by the stabilizers.

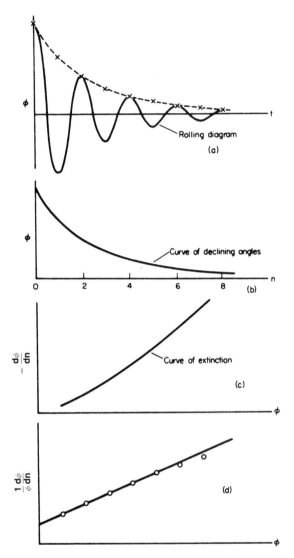

Fig. 12.31 Decrement curve

The value of $d\phi/dn$ can be derived from model or full-scale experiments by noting successive amplitudes of roll as roll is allowed to die out naturally in otherwise still water. These amplitudes are plotted to base n (i.e. the number of swings) and the slope measured at various points to give values of $-d\phi/dn$ at various values of ϕ. (See Fig. 12.31 (*a*), (*b*) and (*c*).)

In most cases, it is adequate to assume that $d\phi/dn$ is defined by a second order equation. That is

$$-\frac{d\phi}{dn} = a\phi + b\phi^2$$

or

$$-\frac{1}{\phi}\frac{d\phi}{dn} = a + b\phi$$

By plotting $(1/\phi)\, d\phi/dn$ against ϕ as in Fig. 12.31(d), a straight line can be drawn through the experimental results to give values of a and b.

Considering forcing a roll by moving weight, the maximum amplitude of roll would be built up if the weight could be transferred instantaneously from the depressed to the elevated side at the end of each swing as shown in Fig. 12.32.

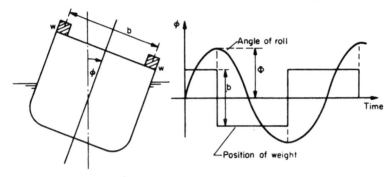

Fig. 12.32 *Instantaneous weight transfer*

The change in potential energy of the weight at each transfer is $wb\sin\phi$. Hence, approximately

$$wb\phi = \Delta\overline{GM}\phi\left(-\frac{d\phi}{dn}\right)$$

or

$$-\frac{d\phi}{dn} = \frac{w}{\Delta}\frac{b}{\overline{GM}}$$

It follows that the moving weight can increase the roll amplitude up to the value appropriate to this value of $d\phi/dn$.

EXAMPLE 2. Assuming that for a rolling ship, the slope of the curve of declining angles is $-d\phi/dn = a\phi + b\phi^2$, find the values of a and b given the following corresponding values of $-d\phi/dn$ and ϕ in degree units:

ϕ	5	10	15	20
$-\dfrac{d\phi}{dn}$	0.75	2.00	3.75	6.00

The above figures apply to a 3000 tonne frigate with a metacentric height of 1 m. A mass of 15 tonnes is made to move across the deck in simple harmonic motion with an amplitude of 6 m. Find the steady rolling angle which can be set up by the weight.

Solution: To find the values of a and b the values of $-(1/\phi)\,d\phi/dn$ must be plotted against ϕ

ϕ	5	10	15	20
$-\dfrac{1}{\phi}\dfrac{d\phi}{dn}$	0.15	0.20	0.25	0.30

This is a straight line and gives $a = 0.10$, $b = 0.01$, i.e.

$$-\frac{d\phi}{dn} = 0.10\phi + 0.01\phi^2$$

In this case, the weight is moving with simple harmonic motion. (This would be the effect if the weight were moving in a circle on the deck at constant speed.) If the angle of roll at any instant is given by

$$\phi = \Phi \sin \frac{2\pi t}{T_0}$$

then the distance of the weight from the centre line is

$$\frac{b}{2} \sin\left(\frac{2\pi t}{T_0} + \frac{\pi}{2}\right)$$

where b is the double amplitude of motion.

The movement of ship and weight must be 90 degrees out of phase for an efficient system.

The work done by the weight in moving out to out (i.e. per swing) is

$$\int_{-b/2}^{+b/2} w\phi\, d\left\{\frac{b}{2} \sin\left(\frac{2\pi t}{T_0} + \frac{\pi}{2}\right)\right\}$$

$$= w\Phi \frac{b}{2}\frac{2\pi}{T_0} \int_{-T_0/2}^{T_0/2} \sin\frac{2\pi t}{T_0} \cos\left(\frac{2\pi t}{T_0} + \frac{\pi}{2}\right) dt$$

$$= \frac{\pi}{T_0} wb\Phi \int_0^{T_0/2} \sin^2\frac{2\pi t}{T_0}\, dt$$

$$= \frac{\pi}{T_0} wb\Phi \frac{1}{2}\left[t - \frac{T_0}{4\pi}\sin\frac{4\pi t}{T_0}\right]_0^{T_0/2}$$

$$= \frac{\pi}{4} wb\Phi$$

i.e.

$$\frac{\pi}{4} wb\Phi = \Delta\overline{GM}\Phi\left(-\frac{d\phi}{dn}\right)$$

From which

$$-\frac{d\phi}{dn} = \frac{\pi}{4}\frac{wb}{\Delta\overline{GM}} = \frac{\pi}{4}\frac{mb}{\Sigma\,\overline{GM}}$$

Hence, if ϕ is the steady rolling angle produced

$$0.1\phi + 0.01\phi^2 = \frac{\pi}{4} \times \frac{15}{3000} \times \frac{2(6)}{1} \times \frac{180}{\pi}$$

whence,

$$\phi = 12.2 \text{ degrees}$$

Experiments and trials

TEST FACILITIES

Seakeeping experiments can be conducted in the conventional long, narrow ship tanks usually used for resistance tests provided they are fitted with a wavemaker at one end and a beach at the other. Unfortunately in such tanks it is only possible to measure the response of the model when heading directly into or away from the waves.

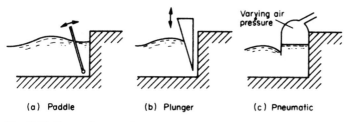

(a) Paddle   (b) Plunger   (c) Pneumatic

Fig. 12.33 Types of wave-maker

Various forms of wavemaker have been employed (Fig. 12.33). Beaches also take a number of forms but are essentially devices for absorbing the energy of the incident waves. They reduce the amplitude of the reflected waves which would otherwise modify the waves experienced by the model. There are a number of basins specially designed for seakeeping experiments. They permit the model to be run at any heading relative to, or to manoeuvre in, the waves. Short crested wave systems can be generated. The basin at Haslar is depicted in Fig. 12.34. It is 122 m long by 61 m wide and uses a completely free remotely-controlled model. In some facilities, e.g. that at the David Taylor Model Basin, Carderoc, the basin is spanned by a bridge so that models can be run under a carriage either free or constrained. This is also a feature of the 170 m by 40 m Seakeeping and Manoeuvring Facility at MARIN which opened in 1999, in which realistic short crested wave conditions can be created.

Whilst a free model has no guides that can interfere with its motion, the technique presents many difficulties. The model must contain its own propulsion system, power supplies, radio-control devices as well as being able to record much of the experimental data. It must be ballasted so as to possess the correct stability characteristics and scaled inertias in order that its response as a dynamic system will accurately simulate that of the ship. All this must be achieved in a relatively small model as too large a model restricts the effective

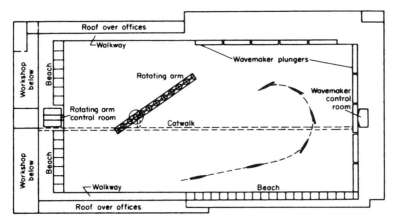

Fig. 12.34 Seakeeping basin at Haslar

length of each run. This is very important for experiments where data have to be presented statistically.

CONDUCT OF SHIP TRIALS

Ship trials are carried out for a variety of reasons, including:

(*a*) to confirm that the ship meets her design intention as regards performance;
(*b*) to predict performance during service;
(*c*) to prove that equipment can function properly in the shipboard environment;
(*d*) to provide data on which future ship designs can be based;
(*e*) to determine effect on human performance.

   Whilst resistance and propulsion trials are usually carried out in calm water, those concerned with motions must by definition be carried out in rough water. Stabilizer performance is a special case and is discussed later. Two types of trial are possible:

(*a*) short duration trials in which the ship responses to a measured sea system
are recorded;
(*b*) prolonged period trials in which statistical data is built up of ship response
in a wide range of sea conditions.

   The first type of trial is essential if it is wished to compare ship with model or calculated response operators over a range of ship speeds and headings. Then the likely long-term behaviour of the ship on voyage can be deduced as described earlier.
   The second type of trial provides a comparison of actual and assessed behaviour during a voyage or over a period of time. Differences may be due either to the ship not responding to the wave systems as predicted or to the wave systems experienced not being those anticipated so the data is of limited

value in assessing prediction methods. The longer the time period the better the measure of a ship's 'average' performance.

Short duration trials are expensive and the opportunity is usually taken to record hull strains, motions, shaft torque and shaft thrust at the same time. Increasing attention is being paid to the performance of the people on board. The sea state itself must be recorded and this is usually by means of a wave recording buoy. For the second type of trial a simpler statistical motion recorder is used, often restricted to measurement of vertical acceleration. No wave measurements are made but sea states are observed. Statistical strain gauges may also be fitted. Satellites can be used to measure the wave system in which the trial ship is operating and can help record the ship's path.

Although various methods have been proposed for measuring a multi-directional wave system it is a very difficult task. Good correlation has been achieved between calculated and measured sea loadings in some trials by applying the cosine squared spreading function proposed by the ITTC (see Chapter 9) to a spectrum based on buoy measurements.

In the earliest trials the waves were recorded by a shipborne wave recorder but nowadays a freely floating buoy is used. Signals are transmitted to the trials ship over a radio link or recorded in the buoy for recovering at the end of the trial. Vertical motions of the buoy are recorded by an accelerometer and movement of the wave surface relative to the buoy by resistive probes. Roll, pitch and azimuth sensors monitor the attitude of the buoy. For studying complex wave systems in detail several buoys may be used.

A typical sequence for a ship motion trial is to:

(*a*) carry out measured mile runs at the start of the voyage to establish the ship's smooth water performance and to calibrate the log;
(*b*) carry out service trials during passage to record sample ship motions and propulsive data under normal service conditions;
(*c*) launch the recording buoy, record conditions and recover buoy, when conditions are considered suitable, i.e. waves appear to be sufficiently long-crested;
(*d*) carry out a manoeuvre of the type shown in Fig. 12.35 recording motions and waves for each leg.

Figures denote time in minutes spent on each leg. The accuracy of the analysis depends upon the number of oscillations recorded. For this reason, the legs running with the seas are longer than those with ship running head into the waves. The overall time on the manoeuvre has to be balanced against the possibility of the sea state changing during the trial. The two sets of buoy records and a comparison of the results from the initial and final legs provides a guide to the stability of the trial conditions. The remaining steps of the sequence are:

(*e*) launch buoy for second recording of waves;
(*f*) repeat (*c*), (*d*) and (*e*) as conditions permit;
(*g*) carry out service trials on way back to port;
(*h*) carry out measured mile runs on return.

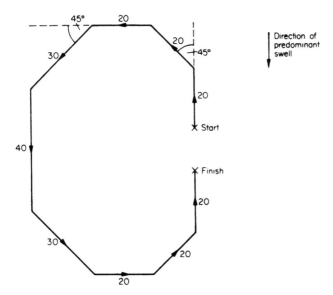

Fig. 12.35 *Typical seakeeping manoeuvre*

On return to harbour, a very lengthy analysis is required. Indeed, this type of trial would not be feasible without computer aid.

STABILIZER TRIALS

Stabilizers, fitted to reduce rolling in a seaway, can be specified directly in terms of roll under stated conditions but such performance can never be precisely proven on trial. As an alternative a designer may relate performance to the steady angle of heel that can be generated by holding the fins over in calm water at a given speed. Heel can be measured directly to establish whether or not such contractual requirements have been met. Forced rolling trials in calm water can be used to study the performance of shipborne equipment under controlled conditions. It is often difficult to distinguish the effects of the stabilizers from the cross-coupling effect of the rudders; indeed, it is possible to build up a considerable angle of roll in calm water by the judicious use of rudder alone.

Problems

1. The mast of a small floating raft, due to the passage of a train of deep water waves, is observed to oscillate with a period of 7 s and amplitude from the vertical of ±8 degrees. Find the height, length and velocity of the waves in metric units.
2. A box-shaped vessel has a length of 50 m, beam 12 m, draught 3 m, free-board 6 m. The height of the c.g. above the bottom is 4.5 m. Assuming that the weight is uniformly distributed throughout the length and section of the vessel, and neglecting the effects of associated water, calculate the free periods of roll, pitch and heave in salt water.

3. A vessel, length 200 m, whose periods of free roll and pitch are $12\frac{1}{2}$ and 10 s respectively, is steaming at 20 knots in a sea of wave-length 250 m. Calculate the headings on which the greatest rolling and pitching are likely.

4. A ship is rolling with a constant amplitude, the rolling being maintained by moving a weight across the deck, i.e. the energy put into the ship by the moving weight just balances the energy dissipated by damping.

   Show, by general arguments, that the ideal case is one in which the weight is transferred instantly from the depressed side to the elevated side at the end of each swing.

   Compare the relative weights required to maintain a given small rolling amplitude assuming

   (a) instantaneous transfer
   (b) weight moving with constant velocity
   (c) weight moving with s.h.m.

   The amplitude of the movement is the same in each case.
   [*Note*: Assume rolling motion is given by $\phi = \Phi \sin(2\pi t/T)$.]

5. Show that, neglecting damping forces, the rolling of a ship to small angles in still water, without ahead motion, is simple harmonic. Hence, derive an expression for the natural period of oscillation of the ship in terms of the radius of gyration (k) and the $\overline{\text{GM}}$ of the ship. What is the effect of entrained water? How do rolling considerations affect the choice of $\overline{\text{GM}}$ for a passenger ship?

   During rolling trials on an aircraft carrier, a natural period of roll of 14 s was recorded. The displacement was 50,000 tonnef and the $\overline{\text{GM}}$ was 2.5 m. The inertia coefficient, allowing for the effect of entrained water is 20 per cent. Calculate the radius of gyration of the aircraft carrier.

6. A ship, 4000 tonnef displacement, 150 m length, 15 m beam and 1 m metacentric height has a rolling period of 10.5 s and a decrement equation

$$-\frac{\mathrm{d}\phi}{\mathrm{d}n} = 0.20\phi + 0.15\phi^2 \ (\phi \text{ in radians})$$

   If the ship is to be rolled to an amplitude of 10° estimate the weight required to be moved instantaneously across the deck assuming that it can be moved through 12 m.

7. The differential equation for the rolling motion of a ship in regular waves can be expressed in the form:

$$\ddot{\phi} + 2k\omega_0\dot{\phi} + \omega_0^2\phi = \omega_0^2 F_0 \sin \omega_E t$$

   Explain the significance of the terms in this equation.

   The equation of the rolling motion of a particular ship in regular waves can be expressed in the form:

$$\ddot{\phi} + 0.24\dot{\phi} + 0.16\phi = 0.48 \sin \omega_E t$$

where ϕ is the roll angle in degrees.

Calculate the amplitudes of roll when ω_E is equal to 0.2, 0.4 and 0.8, commenting upon their relative magnitudes. What would be the period of damped rolling motion in calm water?

8. A ship motion trial is carried out in a long-crested irregular wave system. The spectrum of the wave system as measured at a stationary point is defined by the following table:

$S_\zeta(\omega)$, (wave height)$^2/\delta\omega$ (m^{2}s)	1.2	7.6	12.9	11.4	8.4	5.6
ω, frequency (1/s)	0.3	0.4	0.5	0.6	0.7	0.8

The heave energy spectrum obtained from accelerometers in the ship, when moving at 12 knots on a course of 150 degrees relative to the waves, is defined as follows:

$S_Z(\omega_E)$, (heave)$^2/\delta\omega_E$ (m^{2}s)	0.576	1.624	1.663	0.756	0.149	0.032
ω_E, frequency of encounter (1/s)	0.4	0.5	0.6	0.7	0.8	0.9

Derive the response curve, in the form of heave/wave height, for the ship at this speed and heading, over the range of frequencies of encounter from 0.4 to 0.9.

9. The successive maximum angles in degrees recorded in a model rolling experiment are:

Port   15 (start)   10.4   7.7   5.9
Starboard   12.3   8.9   6.7

What are the '*a*' and '*b*' coefficients? What maximum angle would you expect to be attained at the end of the tenth swing?

10. A vessel, unstable in the upright position, lolls to an angle α. Prove that, in the absence of resistance, she will roll between $\pm\phi$ or between ϕ and $\sqrt{(2\alpha^2 - \phi^2)}$ according as ϕ is greater or less than $\alpha\sqrt{2}$. All angles are measured from the vertical.

Explain how the angular velocity varies during the roll in each case.

11. A rolling experiment is to be conducted on a ship which is expected to have '*a*' and '*b*' extinction coefficients of 0.08 and 0.012 (degree units).

The experiment is to be conducted with the displacement at 2134 tonnef and a metacentric height of 0.84 m. The period of roll is expected to be about 9 s.

A mechanism capable of moving a weight of 6.1 tonnef in simple harmonic motion 9.14 m horizontally across the vessel is available.

Estimate:

(a) the maximum angle of roll likely to be produced,
(b) the electrical power of the motor with which the rolling mechanism should be fitted (assume an efficiency of 80 per cent).

12. A vessel which may be regarded as a rectangular pontoon 100 m long and 25 m wide is moving at 10 knots into regular sinusoidal waves 200 m long

and 10 m high. The direction of motion of the vessel is normal to the line of crests and its natural (undamped) period of heave is 8 s.

If it is assumed that waves of this length and height could be slowed down relative to the ship, so that the ship had the opportunity of balancing itself statically to the wave at every instant of its passage, the ship would heave in the effective period of the wave and a 'static' amplitude would result. With the wave at its correct velocity of advance relative to the ship a 'dynamic' amplitude will result which may be regarded as the product of the 'static' amplitude and the so-called 'magnification factor'.

Calculate the amplitude of heave of the ship under the conditions described in the first paragraph, making the assumption of the second paragraph and neglecting any Smith correction.

The linear damping coefficient k, is 0.3.

13. A ship, 4000 tonnef displacement, 140 m long and 15 m beam has a transverse metacentric height of 1.5 m. Its rolling period is 10.0 s and during a rolling trial successive (unfaired) roll amplitudes, as the motion was allowed to die down, were:

    11.3, 8.6, 6.8, 5.6, 4.5, 3.7 and 3.1 degrees

    Deduce the 'a' and 'b' coefficients, assuming a decrement equation of the form

    $$-\frac{d\phi}{dn} = a\phi + b\phi^2$$

14. The spectrum of an irregular long-crested wave-system, as measured at a fixed point, is given by:

$S\zeta(\omega)$, (wave amplitude)$^2/\delta\omega$ (m^2 s)	0.3	1.9	4.3	3.8
ω, (frequency) (1/s)	0.3	0.4	0.5	0.6

    A ship heads into this wave system at 30 knots and in a direction such that the velocity vectors for ship and waves are inclined at 120 degrees. Calculate the wave spectrum as it would be measured by a probe moving forward with the speed of the ship.

    Discuss how you would proceed to calculate the corresponding heave spectrum. Illustrate your answer by calculating the ordinate of the heave spectrum at a frequency of encounter of 0.7 s.

    The relationship between amplitude of heave and wave amplitude at this frequency of encounter for various speeds into regular head seas of appropriate length should be taken as follows:

heavy amplitude / wave amplitude	0.71	0.86	0.92	0.95	0.96
speed (knots)	20	40	60	80	100

    Assume that, to a first approximation, the heave amplitude of a ship moving at speed V obliquely into long-crested waves is the same as the heave amplitude in regular head seas of the same height and of the same 'effective length' (i.e. the length in the direction of motion) provided the speed V_1 is adjusted to give the same frequency of encounter.

15. Assuming that a ship heaves in a wave as though the relative velocity of wave and ship is very low, show that the 'static' heave is given by

$$\frac{\text{heave amplitude}}{\text{wave amplitude}} = -\frac{\sin n\pi}{\pi n} \sin \frac{2\pi t}{T_E}$$

for a rectangular waterplane, where

$$n = \frac{\text{length of ship}}{\text{length of wave}} = \frac{L}{\lambda}$$

and the wave is defined by

$$\zeta = \frac{H}{2}\sin\left(\frac{2\pi n x}{L} - \frac{2\pi t}{T_E}\right)$$

Show that there is zero 'static' response at $n = 1.0$.

13 Manoeuvrability

General concepts

All ships require to be controllable in direction in the horizontal plane so that they can proceed on a straight path, turn or take other avoiding action as may be dictated by the operational situation. They must further be capable of doing this consistently and reliably not only in calm water but also in waves or in conditions of strong wind. In addition, submarines require to be controllable in the vertical plane, to enable them to maintain or change depth as required whilst retaining control of fore and aft pitch angle.

Considering control in the horizontal plane, a study of a ship's manoeuvrability must embrace the following:

(a) the ease with which it can be maintained on a given course. The term steering is commonly applied to this action and the prime factor affecting the ship's performance is her directional or dynamic stability. This should not be confused with the ship's stability as discussed in Chapter 4;
(b) the response of the ship to movements of her control surfaces, the rudders, either in initiating or terminating a rate of change of heading;
(c) the response to other control devices such as bow thrusters;
(d) the ability to turn completely round within a specified space.

With knowledge on these factors the designer can ensure that the ship will be controllable; calculate the size and power of control surfaces and/or thrusters to achieve the desired standards of manoeuvrability; design a suitable control system—autopilot or dynamic positioning system; provide the necessary control equations for the setting up of training simulators.

For control in the vertical plane, it is necessary to study:

(a) ability to maintain constant depth, including periscope depth under waves;
(b) ability to change depth at a controlled pitch angle.

Submarine stability and control is dealt with in more detail in a later section.

It is clear from the above, that all ships must possess some means of directional control. In the great majority of cases, this control is exercised through surfaces called rudders fitted at the after end of the ship. In some cases, the rudders are augmented by other lateral force devices at the bow and, in a few special applications, they are replaced by other steering devices such as the vertical axis propeller. This chapter is devoted mainly to the conventional, rudder steered, ship but the later sections provide a brief introduction to some of the special devices used.

It is important to appreciate that it is not the rudder forces directly in themselves that cause the ship to turn. Rather, the rudder acts as a servo-system which causes the hull to take up an attitude in which the required forces and moments are generated hydrodynamically on the hull. Rudders are fitted aft in a ship because, in this position, they are most effective in causing the hull to take up the required attitude and because they benefit from the increased water velocity induced by the propellers. At low speed, when the rudder forces due to the speed of the ship alone are very small, a burst of high shaft revolutions produces a useful side force if the propellers and rudders are in line.

In the early days of man's movements on water, directional control was by paddle as in a canoe today. That is to say, the heading was controlled by applying a force either on the port or starboard side of the craft. As vessels grew in size, the course was changed by means of an oar over the after end which was used to produce a lateral force. Later again, this was replaced by a large bladed oar on each quarter of the ship and, in turn, this gave way to a single plate or rudder fitted to the transom. The form of this plate has gradually evolved into the modern rudder. This is streamlined in form to produce a large lift force with minimum drag and with leading edge sections designed to reduce the loss of lift force at higher angles of attack. In some cases the single rudder has given way to twin or multiple rudders.

DIRECTIONAL STABILITY OR DYNAMIC STABILITY OF COURSE

It was seen in Chapter 4, that when disturbed in yaw there are no hydrostatic forces tending either to increase or decrease the deviation in ship's head. In this mode, the ship is said to be in a state of neutral equilibrium. When under way, hydrodynamic forces act on the hull which can have either a stabilizing or destabilizing effect. However, in the absence of any external corrective forces being applied, the ship will not return to its initial line of advance when subject to a disturbance. Hence, *directional stability* cannot be defined in terms precisely similar to those used for transverse stability. A ship is said to be directionally stable when, having suffered a disturbance from an initial straight path, it tends to take up a new straight line path.

In the study of ships, directional stability usually refers to the situation without any control forces being applied. That is to say, what may be termed the inherent course stability of the ship is considered. If the degree of instability were small the vessel could probably be steered by a helmsman (or autopilot) but at the expense of continuous control surface movements which would involve greater resistance and more wear and tear on the machinery. A high degree of instability would mean the ship was uncontrollable. At the other end of the scale too high a level of course stability would make the ship hard to control because of lack of response to rudder movements.

Figure 13.1 shows an arrow with a large tail area well aft of its centre of gravity. Consider a small disturbance which deflects the arrow through a small angle ψ relative to its initial trajectory. The velocity of the arrow is still substantially along the direction of the initial path and the tail surfaces, being now at an angle of attack ψ, develop a lift force F which is in the direction

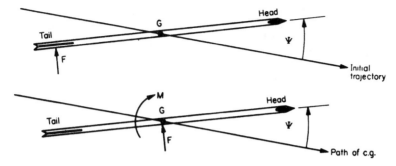

Fig. 13.1 Arrow suffering a disturbance

shown. This force is clearly acting in such a manner as to reduce ψ since it is equivalent to a side force F acting at, and a moment M acting about, the c.g. Provided the tail surfaces are large enough, the forces acting on the rest of the arrow will be negligible compared with F. Other forces will be acting on the head and shank of the arrow so that F should, more precisely, be regarded as the resultant force acting on the arrow.

The sideways force acting at G will have the effect of changing the direction of movement of the c.g. but, as M causes ψ to decrease, M and F will decrease becoming zero and then negative as the axis of the arrow passes through the path of the c.g. Thus, the arrow will oscillate a little and then settle down on a new straight path, ignoring gravitational pull. The deviation from the original path will depend upon the damping effect of the air and the relative magnitudes of M and F. It follows that to maintain a near constant path, the arrow should have large tail surfaces as far aft of G as possible. Directional stability of this very pronounced type is often referred to as 'weather-cock' stability by an obvious analogy.

Application to a ship

It is not possible to tell merely by looking at the lines of a ship whether it is directionally stable or not. Applying the principle enunciated above, it can be argued that for directional stability the moment acting on the hull and its appendages must be such that it tends to oppose any yaw caused by a disturbance, i.e. the resultant force must act aft of G. The point at which it acts is commonly called the *centre of lateral resistance*. As a general guide, therefore, it is to be expected that ships with large skegs aft and with well rounded forefoot will tend to be more directionally stable than a ship without these features but otherwise similar. Also, as a general guide, long slender ships are likely to be more directionally stable than short, tubby forms.

The sign convention used in this chapter is illustrated in Fig. 13.2.

It is necessary to differentiate between 'inherent' and 'piloted' controllability. The former represents a vessel's open loop characteristics and uses the definition that when, in a given environment, a ship can attain a specified manoeuvre with some steering function, that ship is said to be manoeuvrable. This ability

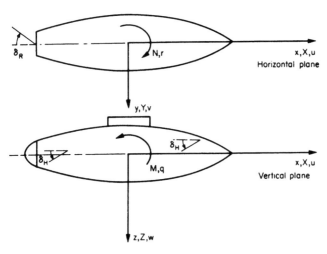

Fig. 13.2 Sign convention

depends upon the environment so that some situations could arise in which the ship becomes unmanoeuvrable. Piloted manoeuvrability reflects the ability of a ship, when controlled by a human operator or an autopilot, to perform a manoeuvre such that deviations from a preset mission remain within acceptable limits. In deciding whether a ship is manoeuvrable in this sense the mission must be specified and the limits within which it is to be achieved.

STABILITY AND CONTROL OF SURFACE SHIPS

For a surface ship we need only consider linear motions along the x and y axes and angular motion about the z axis, the axes used being body axes. If the ship is disturbed from its straight line course in such a way that it has a small sideways velocity v it will experience a sideways force and a yawing moment which can be denoted by Y_v and N_v respectively. If this was the only distur-bance the ship would exhibit directional stability if the moment acted so as to reduce the angle of yaw and hence v. In the more general case the disturbed ship will have an angular velocity, angular and linear accelerations and will be subject to rudder actions. All these will introduce forces and moments. Con-sidering only small deviations from a straight path so that second order terms can be neglected, the linear equations governing the motion become

$$(m - Y_{\dot{v}})\dot{v} = Y_v v + (Y_r - m)r + Y_{\delta R}\delta_R$$
$$(I - N_{\dot{r}})\dot{r} = N_v v + N_r r + N_{\delta R}\delta_R$$

where subscripts v, r and δ_R denote differentiation with respect to the lateral component of velocity (radial), rate of change of heading and rudder angle respectively, i.e. $Y_v = \partial Y/\partial v$, etc. Y denotes component of force on ship in y direction and N the moment of forces on ship about z-axis. m is the mass of the ship.

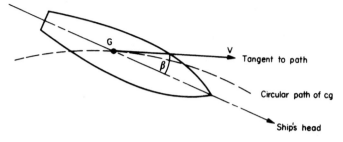

Fig. 13.3

Equations are needed only for motion along the transverse axis and about the vertical axis as it is assumed that the ship has a steady forward speed. Put into words the equations are saying no more than that the rate of change of momentum in the y axis direction is equal to the force in that direction and that that force is the sum of all such terms as (rate of change of Y with lateral velocity) × (lateral velocity).

The equations can be expressed non-dimensionally by

$$(m' - Y'_{\dot{v}})\dot{v}' = Y'_v v' + (Y'_r - m')r' + Y'_{\delta_R}\delta'_R$$
$$(I' - N'_{\dot{r}})\dot{r}' = N'_v v' + N'_r r' + N'_{\delta_R}\delta'_R$$

From Fig. 13.3 $v' = \dfrac{\dot{y}}{V} = -\sin\beta$ and the non-dimensional turn rate $r' = \dot{\psi}\dfrac{L}{V} = \dfrac{L}{R}$ where R is the radius of curvature of the path at that point. The coefficients Y'_v, N'_v, etc., are termed the stability derivatives.

As a typical example

$$Y'_{\delta_R} = \frac{1}{\frac{1}{2}\rho V^2 L^2}\frac{\partial Y}{\partial \delta_R}$$

The directional stability of a ship is related to its motion with no corrective, i.e. rudder, forces applied. In this case the equations become

$$(m' - Y'_{\dot{v}})\dot{v}' = Y'_v v' + (Y'_r - m')r'$$
$$(I' - N'_{\dot{r}})\dot{r}' = N'_v v' + N'_r r'$$

from which it follows that

$$(m' - Y'_{\dot{v}})\left[\frac{(I' - N'_{\dot{r}})\dot{r}' - N'_r r'}{N'_v}\right] = Y'_v\left[\frac{(I' - N'_{\dot{r}})\dot{r}' - N'_r r'}{N'_v}\right] + (Y'_r - m')r'$$

$$(m' - Y'_{\dot{v}})(I' - N'_{\dot{r}})\ddot{r}' - [(m' - Y'_{\dot{v}})N'_r + (I' - N'_{\dot{r}})Y'_v]\dot{r}'$$
$$+ [N'_r Y'_v - (Y'_r - m')N'_v]r' = 0.$$

This equation is of the form

$$\left[a\frac{\mathrm{d}}{\mathrm{d}t^2} + b\frac{\mathrm{d}}{\mathrm{d}t} + c\right]r = 0$$

which has as a general solution of the form

$$r' = r_1 e^{m_1 t} + r_2 e^{m_2 t}$$

where m_1 and m_2 are the roots of the equation

$$am^2 + bm + c = 0$$

$$m = \frac{-b \pm \sqrt{b^2 - 4ac}}{2a}$$

In a stable ship any initial oscillation must decay to zero, which requires both m_1 and m_2 to be negative.

a and b are always positive for ships and the complex solution of the differential equation does not appear to occur. The condition for stability or stability criterion then becomes $c > 0$,

i.e. $N'_r Y'_v - N'_v (Y'_r - m') > 0$

i.e. $\dfrac{N'_r}{Y'_r - m'} > \dfrac{N'_v}{Y'_v}$

Thus the condition for stability reduces to a requirement that the centre of pressure in pure yaw should be ahead of that in pure sway.

Returning to the earlier equations, for a steady turn $\dot{v}$ and $\dot{r}$ are zero giving

$$0 = Y'_v v' + (Y'_r - m')r' + Y'_{\delta_R} \delta'_R$$
$$0 = N'_v v' + N'_r r' + N'_{\delta_R} \delta'_R$$

This leads to a relationship between r' and δ'_R as follows:

$$(N'_r r' + N'_{\delta_R} \delta'_R) Y'_v = (Y'_r - m) N'_v r' + Y'_{\delta_R} \delta_R N'_v.$$

$$[N'_r Y'_v - N'_v (Y'_r - m')]r' = [Y'_{\delta_R} N'_v - Y'_v N'_{\delta_R}]\delta'_R$$

$$\frac{r'}{\delta'_R} = \frac{Y'_{\delta_R} N'_v - Y'_v N'_{\delta_R}}{Y'_v N'_r - N'_v (Y'_r - m')}$$

It will be noted that the denominator is the stability criterion obtained above. This seems reasonable on general grounds. If the denominator in the expression for r'/δ'_R were zero then r'/δ'_R becomes infinite and the ship will turn in a circle with no rudder applied. Thus for a stable ship the denominator would be expected to be non-zero. Also by referring to Fig. 13.2 it will be seen that r'/δ_R must be negative for a stable ship. Following from the sign convention and the geometry of the ship, Y'_{δ_R} is positive, N'_{δ_R} and Y'_v are negative. It will also be seen later that Y effectively acts forward of the centre of gravity so that N'_v is also negative. Thus the denominator must be positive for a stable ship.

An important point in directional control is the so-called *Neutral Point* which is that point, along the length of the ship, at which an applied force, ignoring transient effects, does not cause the ship to deviate from a constant heading. This point is a distance ηL forward of the centre of gravity, where

$$\eta = \frac{N_v'}{Y_v'}$$

Typically, η is about one-third, so that the neutral point is about one-sixth of the length of the ship abaft the bow.

It can be readily checked (Fig. 13.4) that with a force applied at the neutral point the ship is in a state of steady motion with no change of heading but with a steady lateral velocity, i.e. a steady angle of attack.

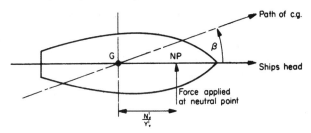

Fig. 13.4 Location of neutral point

When moving at an angle of attack β, lateral velocity $-v$, the non-dimensional hydrodynamic force and moment are vY_v' and vN_v' respectively, i.e. the hydrodynamic force effectively acts at a distance $(N_v'/Y_v')L$ ahead of the c.g. directly opposing the applied force, so that there is no tendency for the ship's head to change. If the applied force is of magnitude F, then the resulting lateral velocity is

$$v = \frac{F}{Y_v}$$

Until the velocity has built up to this required value, there will be a state of imbalance and during this phase there can arise a change of heading from the initial heading.

It follows, that if the force is applied aft of the neutral point and acts towards port the ship will turn to starboard, and if applied in the same sense forward of the neutral point the ship turns to port. Clearly, the greater the distance of application of the force from the neutral point the greater the turning influence, other things being equal. This explains why rudders are more effective when placed aft. If $\eta = \frac{1}{3}$, then the 'leverage' of a stern rudder is five times that of a bow rudder. At the stern also, the rudders gain from the effect of the screw race.

If, in the equation above for r'/δ_R, $N_{\delta_R}' = x'Y_{\delta_R}'$ then

$$\frac{r'}{\delta_R} \quad \text{is proportional to} \quad \frac{N_v'}{Y_v'} - x'$$

That is, for a given rudder angle, the rate of change of heading is greatest when the value of x' is large and negative. This again shows that a rudder is most effective when placed right aft.

THE ACTION OF A RUDDER IN TURNING A SHIP

The laws of dynamics demand that when a body is turning in a circle, it must be acted upon by a force acting towards the centre of the circle of sufficient magnitude to impart to the body the required radial acceleration. In the case of a ship, this force can only arise from the aerodynamic and hydrodynamic forces acting on the hull, superstructure and appendages. It is usual, in studying the turning and manoeuvring of ships, to ignore aerodynamic forces for standard manoeuvres and to consider them only as disturbing forces. That is not to say that aerodynamic forces are unimportant. On the contrary, they may prevent a ship turning into the wind if she has large windage areas forward.

To produce a radial force of the magnitude required, the hull itself must be held at an angle of attack to the flow of water past the ship. The rudder force must be capable of holding the ship at this angle of attack; that is, it must be able to overcome the hydrodynamic moments due to the angle of attack and the rotation of the ship. The forces acting on the ship during a steady turn are illustrated in Fig. 13.5 where F_H is the force on the hull and F_R the rudder force. F_H is the resultant of the hydrodynamic forces on the hull due to the angle of attack and the rotation of the ship as it moves around the circle.

If T is the thrust exerted by the propellers and F_H and F_R act at angles α and γ relative to the middle line plane then, for a steady turn with forces acting as shown in Fig. 13.5, these forces must lead to the radial force $\Delta V^2/R$, i.e.

$$T - \frac{\Delta V^2}{R}\sin\beta = F_H\cos\alpha + F_R\cos\gamma$$

$$-\frac{\Delta V^2}{R}\cos\beta + F_H\sin\alpha = F_R\sin\gamma$$

$$F_H\overline{GE} + F_R\overline{GJ} = 0$$

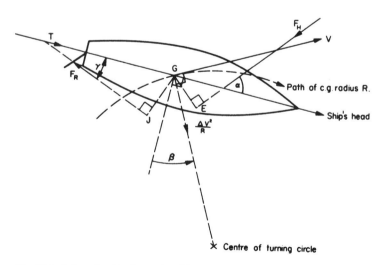

Fig. 13.5 Action of rudder in turning ship

The radial components of the forces on the rudder and the hull, F_R and F_H, must have a resultant causing the radial acceleration.

LIMITATIONS OF THEORY

The student has been introduced to the simple concepts of a linear theory of motion. This is useful in providing an insight into the manoeuvring of ships but many problems are, or appear to be, highly non-linear. This has led to the introduction of higher degrees of derivative to obtain a better representation of the way forces and moments, whose deviations from a steady state condition are other than small, can vary. Such problems concern, for example, steering in a seaway (particularly in a following sea), high-speed large-angle submarine manoeuvring when the body shape may have important effects, athwartships positioning of big ships and drilling vessels. Unfortunately, such approaches are critically dependent upon the validity of the mathematical representation adopted for the fluid forces. One limitation of these analyses is that they assume the forces and moments acting on the model to be determined by the motion obtaining at that instant and are unaffected by its history. This is not true. For instance, it has been shown that when two fins (like a ship and its rudder) are moving in tandem and the first is put to an angle of attack, there is a marked time delay before the second fin experiences a change of force. This approach uses a linear functional mathematical representation which includes a 'memory' effect and shows how the results in the frequency and time domains are related. The approach is limited to linear theory but the inclusion of memory effects provides an explanation for at least some of the effects which arise in large amplitude motions.

Assessment of manoeuvrability

Assessment of manoeuvrability is made difficult by the lack of rigorous analytical methods and of universally accepted standards for manoeuvrability. The hydrodynamic behaviour of a vessel on the interface between sea and air is inherently complex. Whilst reasonable methods exist for initial estimates of resistance and powering and motions in a seaway, the situation is less satisfying as regards manoeuvring. Much reliance is still placed upon model tests and fullscale trials using a number of common manoeuvres which are outlined below.

THE TURNING CIRCLE

Figure 13.6 shows diagrammatically the path of a ship when executing a starboard turn. When the rudder is put over initially, the force acting on the rudder tends to push the ship bodily to port of its original line of advance. As the moment due to the rudder force turns the ship's head, the lateral force on the hull builds up and the ship begins to turn. The parameters at any instant of the turn are defined as:

Drift angle. The drift angle at any point along the length of the ship is defined as the angle between the centre line of the ship and the tangent to the path of the

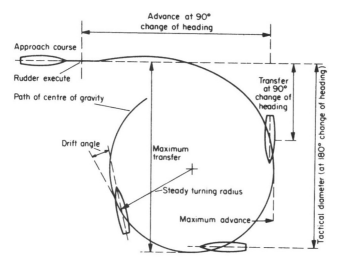

Fig. 13.6 Geometry of turning circle

point concerned. When a drift angle is given for the ship without any specific point being defined, the drift angle at the centre of gravity of the ship is usually intended. Note that the bow of the ship lies within the circle and that the drift angle increases with increasing distance aft of the pivoting point which is defined below.

Advance. The distance travelled by the centre of gravity in a direction parallel to the original course after the instant the rudder is put over. There is a value of advance for any point on the circle, but if a figure is quoted for advance with no other qualification the value corresponding to a 90 degree change of heading is usually intended.

Transfer. The distance travelled by the centre of gravity perpendicular to the original course. The transfer of the ship can be given for any point on the circle, but if a figure is quoted for transfer with no other qualification the value corresponding to a 90 degree change of heading is usually intended.

Tactical diameter. The value of the transfer when the ship's heading has changed by 180 degrees. It should be noted that the tactical diameter is not the maximum value of the transfer.

Diameter of steady turning circle. Following initial application of the rudder there is a period of transient motion, but finally the speed, drift angle and turning diameter reach steady values. This usually occurs after about 90 degrees change of heading but, in some cases, the steady state may not be achieved until after 180 degrees change of heading. The steady turning diameter is usually less than the tactical diameter.

Pivoting point. This point is defined as the foot of the perpendicular from the centre of the turn on to the middle line of the ship extended if necessary. This is not a fixed point, but one which varies with rudder angle and speed. It may be forward of the ship as it would be in Fig. 13.6, but is typically one-third to one-sixth of the length of the ship abaft the bow. It should be noted, that the drift angle is zero at the pivoting point and increases with increasing distance from that point.

The turning circle has been a standard manoeuvre carried out by all ships as an indication of the efficiency of the rudder. Apart from what might be termed the 'geometric parameters' of the turning circle defined above loss of speed on turn and angle of heel experienced are also studied.

Loss of speed on turn

As discussed above, the rudder holds the hull at an angle of attack, i.e. the drift angle, in order to develop the 'lift' necessary to cause the ship to accelerate towards the centre of the turn. As with any other streamlined form, hull lift can be produced only at the expense of increased drag. Unless the engine settings are changed, therefore, the ship will decelerate under the action of this increased drag. Most ships reach a new steady speed by the time the heading has changed 90 degrees but, in some cases, the slowing down process continues until about 180 degrees change of heading.

Angle of heel when turning

When turning steadily, the forces acting on the hull and rudder are F_H and F_R. Denoting the radial components of these forces by lower case subscripts (i.e. denoting these by F_h and F_r respectively) and referring to Fig. 13.7, it is seen that to produce the turn

$$F_h - F_r = \frac{\Delta V^2}{Rg}$$

where $V =$ speed on the turn, $R =$ radius of turn.

$$\text{Moment causing heel} = (F_h - F_r)\overline{KG} + F_r(\overline{KH}) - F_h(\overline{KE})$$
$$= (F_h - F_r)(\overline{KG} - \overline{KE}) + F_r(\overline{KH} - \overline{KE})$$
$$= (F_h - F_r)\overline{GE} - F_r\overline{EH}$$

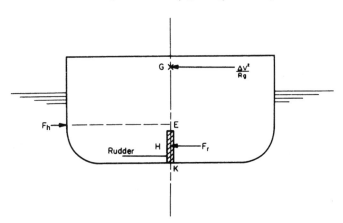

Fig. 13.7 Forces producing heel when turning

For most ships, E, the centre of lateral resistance, and H are very close and this expression is given approximately by

$$\text{Moment causing heel} = (F_h - F_r)\overline{\text{GE}}$$

This moment causes the ship to heel outwards during the steady turn. When the rudder is initially put over, however, F_r acts before F_h has built up to any significant value and during this transient phase the ship may heel inwards. It should also be noted that the effect of F_r during the steady turn is to reduce the angle of heel, so that if the rudder angle is suddenly taken off, the ship will heel to even larger angles. If the rudder angle were to be suddenly reversed even more serious angles of heel could occur.

It will be appreciated that F_h acts at the centre of lateral resistance only if the angle of heel is small. For large heel angles, the position of E is difficult to assess. For small angles of heel

$$\Delta\overline{\text{GM}}\sin\phi = (F_h - F_r)\overline{\text{GE}}$$

$$= \frac{\Delta V^2}{Rg}\overline{\text{GE}}$$

Hence

$$\frac{Rg\sin\phi}{V^2} = \frac{\overline{\text{GE}}}{\overline{\text{GM}}}$$

It must be emphasized, however, that the angle of heel obtained by this type of calculation should only be regarded as approximate. Apart from the difficulty of accurately locating E, some ships, particularly high speed vessels, suffer an apparent loss of stability when underway because of the other forces acting on the ship and appendages due to the flow around the ship when it is turning.

TURNING ABILITY

The turning circle characteristics are not by themselves indicators of initial response to rudder, which may be important when ships are operating in confined waters or in close company. Indeed, some factors which have a major impact on initial response have very little effect on tactical diameter. One indicator that can be used is the heading angle turned through from an initially straight course, per unit rudder angle applied, after the ship has travelled one ship length. Whilst theoretical prediction of tactical diameter is difficult because of non-linearities, linear theory can be used to calculate this initial response and it is possible to derive an expression for it in terms of the stability derivatives.

Multiple regression techniques can be used to deduce approximate empirical formulae for a design's stability derivatives from experimental data. Although they do not accurately account for all the variations in experimental data they are reproduced below as an aid to students.

$$-Y'_{\dot{v}}/\pi(T/L)^2 = 1 + 0.16C_B B/T - 5.1(B/L)^2$$

$$-Y'_{\dot{r}}/\pi(T/L)^2 = 0.67B/L - 0.0033(B/T)^2$$

$$-N'_{\dot{v}}/\pi(T/L)^2 = 1.1B/L - 0.041B/T$$

$$-N'_{r}/\pi(T/L)^2 = 1/12 + 0.017C_B B/T - 0.33B/L$$
$$-Y'_{v}/\pi(T/L)^2 = 1 + 0.40C_B B/T$$
$$-Y'_{r}/\pi(T/L)^2 = -1/2 + 2.2B/L - 0.080B/T$$
$$-N'_{v}/\pi(T/L)^2 = 1/2 + 2.4T/L$$
$$-N'_{r}/\pi(T/L)^2 = 1/4 + 0.039B/T - 0.56B/L$$

THE ZIG-ZAG MANOEUVRE

It can be argued that it is not often that a ship requires to execute more than say a 90 or 180 degree change of heading. On the other hand, it often has to turn through angles of 10, 20 or 30 degrees. It can also be argued that in an emergency, such as realization that a collision is imminent, it is the initial response of a ship to rudder movements that is the critical factor. Unfortunately, the standard circle manoeuvre does not adequately define this initial response and the standard values of transfer and advance for 90 degrees change of heading and tactical diameter are often affected but little by factors which have a significant influence on initial response to rudder. Such a factor is the rate at which the rudder angle is applied. This may be typically 3 degrees per second. Doubling this rate leads to only a marginally smaller tactical diameter but initial rates of turn will be increased significantly.

The zig-zag manoeuvre, sometimes called a Kempf manoeuvre after G. Kempf, is carried out to study more closely the initial response of a ship to rudder movements (see Fig. 13.8). A typical manoeuvre would be as follows. With the ship proceeding at a steady speed on a straight course the rudder is put over to 20 degrees and held until the ship's heading changes by 20 degrees. The rudder angle is then changed to 20 degrees in the opposite sense and so on.

Important parameters of this manoeuvre are:

(a) the time between successive rudder movements;
(b) the *overshoot* angle which is the amount by which the ship's heading exceeds the 20 degree deviation before reducing.

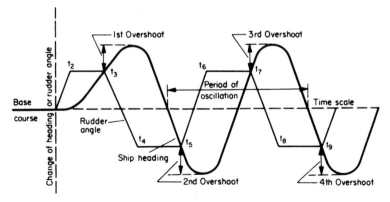

Fig. 13.8 The zig-zag manoeuvre

The manoeuvre is repeated for a range of approach speeds and for different values of the rudder angle and heading deviation.

THE SPIRAL MANOEUVRE

This manoeuvre, sometimes referred to as the Dieudonné Spiral after J. Dieudonné who first suggested it, provides an indication of a ship's directional stability or instability.

To perform this manoeuvre, the rudder is put over to say 15 degrees starboard and the ship is allowed to turn until a steady rate of change of heading is achieved. This rate is noted and the rudder angle is reduced to 10 degrees and the new steady rate of change of heading is measured. Successive rudder angles of 5 °S, 0°, 5 °P, 10 °P, 15 °P, 10 °P, 5 °P, 0°, 5 °S, 10 °S and 15 °S are then used. Thus, the steady rate of change of heading is recorded for each rudder angle when the rudder angle is approached both from above and from below. The results are plotted as in Fig. 13.9, in which case (*a*) represents a stable ship and case (*b*) an unstable ship.

In the case of the stable ship, there is a unique rate of change of heading for each rudder angle but, in the case of an unstable ship, the plot exhibits a form of 'hysteresis' loop. That is to say that for small rudder angles the rate of change of heading depends upon whether the rudder angle is increasing or decreasing. That part of the curve shown dotted in the figure cannot be determined from ship trials or free model tests as it represents an unstable condition.

It is not possible to deduce the degree of instability from the spiral manoeuvre, but the size of the loop is a qualitative guide to this. Of direct practical significance, it should be noted that it cannot be said with certainty that the ship will turn to starboard or port unless the rudder angle applied exceeds δ_S or δ_P, respectively and controlled turns are not possible at low rates of turn.

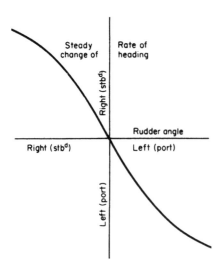

Fig. 13.9(a) Presentation of spiral manoeuvre results (stable ship)

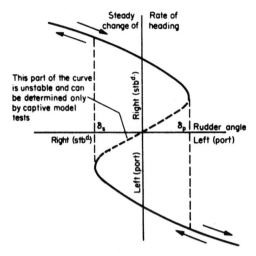

Fig. 13.9(b) Presentation of spiral manoeuvre results (unstable ship)

THE PULL-OUT MANOEUVRE

This manoeuvre is used to determine the directional stability of a ship. The rudder is put over to a predetermined angle and held. When the ship is turning at a steady rate the rudder is returned to amidships and the change of rate of turn with time is noted. If the ship is directionally stable the rate of turn reduces to zero and the ship takes up a new straight path. If the ship is unstable a residual rate of turn will persist. The manoeuvre can be conveniently carried out at the end of each circle trial during ship trials.

It has been found that for a stable ship a plot of the log of rate of turn against time is a straight line after an initial transient period.

It was shown in the section on theory that the differential equation of motion had two roots m_1 and m_2 both of which had to be negative for directional stability. It has been argued that the more negative root will lead to a response which dies

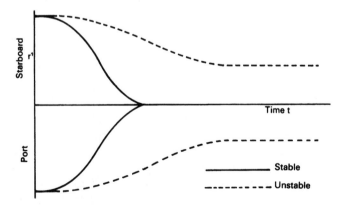

Fig. 13.10 Pull-out manoeuvre, rate of turn on time base

out during the transient phase so that the straight portion of the log rate/time curve gives the root of smaller numerical value. Thus if this root is m_1,

$$r^1 = re^{m_1 t}$$

$$\log r^1 = \log r + m_1 t$$

The area under the curve of turn rate against time gives the total heading change after the rudder is centred. Thus the less the total change the more stable the ship.

STANDARDS FOR MANOEUVRING AND DIRECTIONAL STABILITY

The standards required in any particular design depend upon the service for which the ship is intended but, in any case, they are not easy to define. The problem is made more difficult by the fact that good directional stability and good manoeuvrability are to some extent conflicting requirements, although they are not actually incompatible as has often been suggested. For instance, a large rudder can increase the directional stability and also improve turning performance. Also, in a long fine form increasing draught-to-length ratio can increase stability without detriment to the turning. On the other hand, increasing beam-to-length ratio improves turning but reduces the directional stability. Placing a large skeg aft will improve directional stability at the expense of poorer turning ability.

For a large ocean-going ship, it is usually possible to assume that tugs will be available to assist her when manoeuvring in the confined waters of a harbour. The emphasis in design is therefore usually placed on good directional stability for the long ocean transits. This leads to less wear on the rudder gear, especially if an automatic control system is fitted, and reduces overall average resistance. The highest degree of directional stability is demanded for ships likely to suffer disturbances in their normal service such as supply ships replenishing smaller naval units at sea.

For medium size ships which spend relatively more time in confined waters and which do not normally make use of tugs, greater emphasis has to be placed on response to rudder. Typical of these are the cross channel steamers and anti-submarine frigates.

What are the parameters that are to be used to define the manoeuvring capabilities? They are those parameters measured in the various manoeuvres described in the earlier sections of this chapter. Typical values to be expected are discussed below.

Tactical diameter-to-length ratio. For ships in which tight turning is desirable this may be, say, 3.25 for modern naval ships at high speed, with conventional rudders at 35 degrees. Where even smaller turning circles are required, recourse is usually made to some form of lateral thrust unit.

A *T.D./L* value of 4.5 is suggested as a practicable criterion for merchant types desiring good handling performance. Values of this ratio exceeding 7 are regarded as very poor.

Turning rate. For very manoeuvrable naval ships this may be as high as 3 degrees per second. For merchant types, rates of up to 1.5 degrees per second should be achieved in ships of about 100 m at 16 knots, but generally values of 0.5–1.0 degrees per second are more typical.

Speed on turn. This can be appreciably lower than the approach speed, and typically is only some 60 per cent of the latter.

Initial turning. It has been proposed that the heading change per unit rudder angle in one ship length travelled should be greater than 0.3 generally and greater than 0.2 for large tankers.

Angle of heel. A very important factor in passenger ships and one which may influence the standard of transverse stability incorporated in the design.

Directional stability. Clearly, an important factor in a well balanced design. The inequality presented earlier as the criterion for directional stability can be used as a 'stability index'. Unfortunately, this is not, by itself, very informative. A reasonable design aim is that the spiral manoeuvre should exhibit no 'loop', i.e. the design should be stable even if only marginally so. Using the pull-out manoeuvre it has been suggested that using the criterion of total heading change after the rudder is centred 15–20 degrees represents good stability, 35–40 degrees reasonable stability but that 80–90 degrees indicates marginal stability.

Time to turn through 20 degrees. This provides a measure of the initial response of the ship to the application of rudder. It is suggested that the time to reach 20 degrees might typically vary from 80 to 30 seconds for speeds of 6–20 knots for a 150 m ship. The time will vary approximately linearly with ship length.

Overshoot. The overshoot depends on the rate of turn and a ship that turns well will overshoot more than one that does not turn well. If the overshoot is excessive, it will be difficult for a helmsman to judge when to start reducing rudder to check a turn with the possible danger of damage due to collision with other ships or a jetty. The overshoot angle does not depend upon the ship size and values suggested are 5.5 degrees for 8 knots and 8.5 degrees for 16 knots, the variation being approximately linear with speed.

Rudder forces and torques

RUDDER FORCE

The rudder, being of streamlined cross-section, will be acted upon by a lift and drag force when held at an angle of attack relative to the flow of water. The rudder must be designed to produce maximum lift for minimum drag assuming that the lift behaves in a consistent manner for all likely angles of attack. The lift developed depends upon:

(a) the cross-sectional shape;
(b) the area of the rudder, A_R;
(c) the profile shape of the rudder and, in particular, the aspect ratio of the rudder, i.e. the ratio of the depth of the rudder to its chord length;

(*d*) the square of the velocity of the water past the rudder;
(*e*) the density of the water, ρ;
(*f*) the angle of attack, α.

Hence, the rudder force F_R, can be represented by

$$F_R = \text{Constant} \times \rho A_R V^2 f(\alpha)$$

the value of the constant depending upon the cross-sectional and profile shapes of the rudder. A typical plot for $f(\alpha)$ is as shown in Fig. 13.11.

At first, $f(\alpha)$ increases approximately linearly with angle of attack but then the rate of growth decreases and further increase in α may produce an actual fall in the value of $f(\alpha)$. This phenomenon is known as *stalling*.

Typically, for ships' rudders, stalling occurs at an angle between 35 and 45 degrees. Most ship rudders are limited to 35 degrees to avoid stall, loss of speed and large heel on turn. Stall is related to the flow relative to the rudder; in turning the water flow is no longer aligned with the ship's hull but across the stern, thereby allowing larger rudder angles before stall occurs than are possible when the rudder is first put over. This cross-flow affects also wake and propeller performance.

Many formulae have been suggested for calculating the forces on rudders. One of the older formulae is

$$\text{Force} = 577 A_R V^2 \sin(\delta_R) \text{ newtons}$$

where A_R is measured in m^2 and V in m/s,

V being the velocity of water past the rudder, allowance must be made for the propeller race in augmenting the ship's ahead speed. Typical values assumed are:

Rudder behind propeller, $V = 1.3 \times$ (Ship speed)
Centre-line rudder behind twin screws, $V = 1.2 \times$ (Ship speed).

Haslar used the following formulae for twin rudders behind wing propellers:

Force $= 21.1 A_R V^2 \delta_R$ newtons, for ahead motion, δ_R is measured in degrees.
Force $= 19.1 A_R V^2 \delta_R$ newtons, for astern motion.

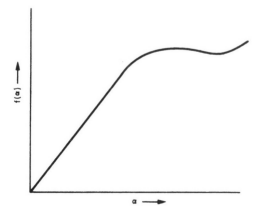

Fig. 13.11 Variation of $f(\alpha)$ with angle of attack

Using the same parameters, Baker and Bottomley have suggested that for middle line rudders behind single screws

$$\text{Force} = 18.0 A_R V^2 \delta_R \text{ newtons}$$

In these formulae V is taken as the true speed of the ship, allowance having been made in the multiplying factors for the propeller race effects.

More comprehensive formulae are given in the literature based on extensive experimental and theoretical work. It is recommended that for naval applications all-movable control surfaces be used with square tips. A good section is the NACA 0015 with a moderately swept quarter chord line. Figure 13.12 illustrates a typical control surface and gives the offsets for the NACA 0015 section, in terms of the chord c and distance x from the nose.

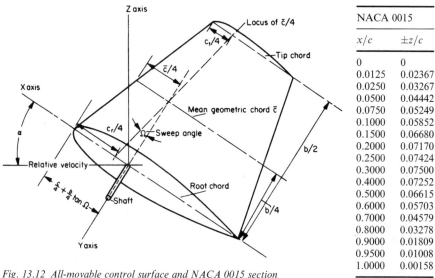

NACA 0015	
x/c	$\pm z/c$
0	0
0.0125	0.02367
0.0250	0.03267
0.0500	0.04442
0.0750	0.05249
0.1000	0.05852
0.1500	0.06680
0.2000	0.07170
0.2500	0.07424
0.3000	0.07500
0.4000	0.07252
0.5000	0.06615
0.6000	0.05703
0.7000	0.04579
0.8000	0.03278
0.9000	0.01809
0.9500	0.01008
1.0000	0.00158

Fig. 13.12 All-movable control surface and NACA 0015 section

The formulae recommended are:

$$C_L = \frac{\text{Lift}}{\frac{1}{2}\rho A V^2} = \left[\frac{a_0 a_e}{\cos\Omega \left(\frac{a_e^2}{\cos^4\Omega} + 4\right)^{\frac{1}{2}} + \frac{57.3 a_0}{\pi}} \right] \alpha + \frac{C_{D_c}}{a_e}\left(\frac{\alpha}{57.3}\right)^2$$

where

a_e = effective aspect ratio = (span)^2/(planform area)

a_0 = section lift curve slope at $\alpha = 0$

   = $0.9(2\pi/57.3)$ per degree for NACA 0015

C_{D_c} = crossflow drag coefficient (Fig. 13.13)

   = 0.80 for square tips and taper ratio = 0.45

and

$$C_D = \frac{\text{Drag}}{\frac{1}{2}\rho A V^2} = C_{d_0} + \frac{C_L^2}{0.9\pi a_e}$$

where

C_{d_0} = minimum section drag coefficient

= 0.0065 for NACA 0015

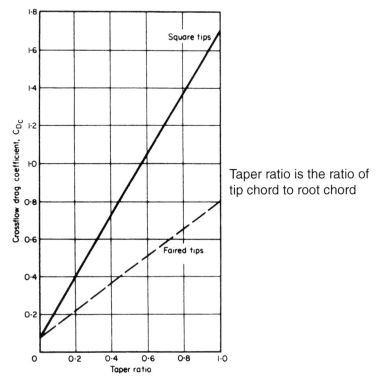

Square tips

Taper ratio is the ratio of tip chord to root chord

Faired tips

Crossflow drag coefficient, C_{Dc}

Taper ratio

Fig. 13.13 Crossflow drag coefficient

CENTRE OF PRESSURE POSITION

It has been seen that it is the rudder force which is important in causing a ship to turn, as the lever of the rudder force from the neutral point is not significantly affected by the position of the centre of pressure on the rudder itself. However, it is necessary to know the torque acting on the rudder to ensure that the steering gear installed in the ship is capable of turning the rudder at all speeds.

For a flat plate, Joessel suggested an empirical formula for the proportion of the breadth of the plate that the centre of pressure is abaft the leading edge and expressed it as:

$$0.195 + 0.305 \sin(\alpha)$$

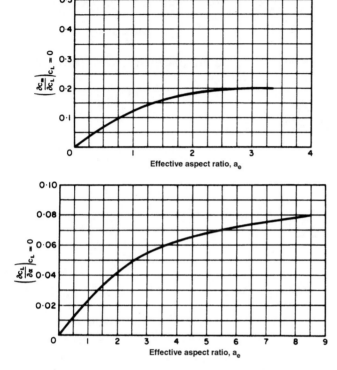

Fig. 13.14 *Variation of chordwise centre of pressure and lift curve slope with aspect ratio*

For rudders, the geometry of the cross-section will have an influence upon the centre of pressure position. Gawn considered that for a rectangular rudder behind a fin or skeg the centre of pressure is 0.35 times the chord length abaft the leading edge. For a rudder in open water this value is reduced to 0.31. For motion astern the rudder is always effectively in clear water and the figure of 0.31 is used in both cases and measured relative to the after edge of the rudder.

One authority recommends a torque (pitching moment) coefficient

$$C_{\mathrm{T}} = \frac{\text{torque}}{\frac{1}{2}\rho A V^2 \bar{c}} = \left[0.25 - \left(\frac{\partial C_{\mathrm{m}}}{\partial C_{\mathrm{L}}} \right)_{C_{\mathrm{L}}=0} \right] \left(\frac{\partial C_{\mathrm{L}}}{\partial \alpha} \right)_{C_{\mathrm{L}}=0} \alpha - \frac{1}{2} \frac{C_{\mathrm{D_c}}}{a_{\mathrm{e}}} \left(\frac{\alpha}{57.3} \right)^2$$

where

$$\bar{c} = \text{mean geometric chord} = \frac{c_{\mathrm{t}} + c_{\mathrm{r}}}{2}$$

$$\left(\frac{\partial C_{\mathrm{m}}}{\partial C_{\mathrm{L}}} \right)_{C_{\mathrm{L}}=0} \quad \text{and} \quad \left(\frac{\partial C_{\mathrm{L}}}{\partial \alpha} \right)_{C_{\mathrm{L}}=0}$$

are defined in Fig. 13.14.

Torque is measured about the quarter-chord point of the mean geometric chord.

The centre of pressure is defined chordwise and spanwise by the following relationships:

Chordwise from leading edge at the mean geometric chord (as percentage of the mean geometric chord),

$$= 0.25 - \frac{C_T}{C_L \cos \alpha + C_D \sin \alpha}$$

Spanwise measured from the plane of the root section (in terms of the semi-span):

$$= \frac{C_L \left(\dfrac{4}{3\pi}\dfrac{b}{2}\right) \cos \alpha + C_D \left(\dfrac{b}{2}\right) \sin \alpha}{\dfrac{b}{2}(C_L \cos \alpha + C_D \sin \alpha)}$$

Typical curves for a NACA 0015 section control surface are reproduced in Fig. 13.15.

In the absence of any better guide, the figures quoted above should be used in estimating rudder forces and torques. However, because of the dependence of both force and centre of pressure on the rudder geometry it is recommended that actual data for a similar rudder be used whenever it is available. In many instances, rudders have sections based on standard aerofoil sections and, in this case, use should be made of the published curves for lift and centre of pressure positions making due allowance for the effect of propellers and the presence of the hull on the velocity of flow over the rudder.

In practice, the picture is complicated by the fact that the flow of water at the stern of a ship is not uniform and may be at an angle to the rudders when set nominally amidships. For this reason, it is quite common practice to carry out model experiments to determine the hydrodynamic force and torque acting on the rudder. As a result of such tests, it may be deemed prudent to set the rudders of a twin rudder ship at an angle to the middle line plane of the ship for their 'amidships' position.

CALCULATION OF FORCE AND TORQUE ON NON-RECTANGULAR RUDDER

It is seldom that a ship rudder is a simple rectangle. For other shapes the rudder profile is divided into a convenient number of strips. The force and centre of pressure are assessed for each strip and the overall force and torque obtained by summating the individual forces and torques.

EXAMPLE 1. Calculate the force and torque on the centre line gnomon rudder shown, Fig. 13.16, for 35 degrees and a ship speed of 20 knots. The ship is fitted with twin screws.

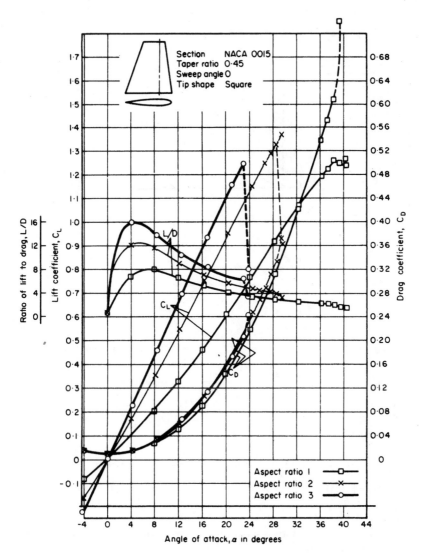

Fig. 13.15 Lift and drag for NACA 0015 sections

Solution: The rudder can conveniently be divided into two rectangular areas A_1 and A_2, A_1 being the smaller. Applying the older formula for force and Gawn's formulae for c.p. position.

Area A_1 is behind a skeg

$$\therefore \quad \text{Force on } A_1 = 557 \times 9 \times (1.2 \times 20 \times 0.51477)^2 \sin 35 = 0.4\,\text{MN}$$

$$\text{c.p. aft of axis} = 0.35 \times 3 = 1.05\,\text{m}$$

$$\text{Moment on } A_1 = 1.05 \times 0.4 = 0.42\,\text{MN m}$$

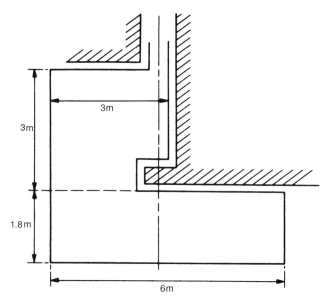

Fig. 13.16

Force on $A_2 = 557 \times 10.8 \times (1.2 \times 20 \times 0.51477)^2 \sin 35 = 0.48\,\text{MN}$

c.p. aft of axis $= 0.31 \times 6.3 = -1.14\,\text{m}$

Moment on $A_2 = -1.14 \times 0.48 = -0.547\,\text{MN}\,\text{m}$

Hence resultant force on rudder $= 0.88\,\text{MN}$

resultant moment $= -0.127\,\text{MN}\,\text{m}$

with c.p. forward of the axis.

EXAMPLE 2. Calculate the force and torque on the spade rudder shown in Fig. 13.17, which is one of two working behind twin propellers. Assume a rudder angle of 35 degrees and a ship speed of 20 knots ahead. If the stock is solid with a section modulus in bending of $0.1\,\text{m}^3$, calculate the maximum stress due to the combined torque and bending moment.

Solution: Assuming that the force on the rudder is given by $21.1\,A_R V^2 \delta R$ newtons, that the c.p. is $0.31 \times$ the chord length aft of the leading edge and that the force acts at the same vertical position as the centroid of area of the rudder.

Area of rudder $= \dfrac{1}{3} \times 1 \times 32.5 = 10.83\,\text{m}^2$

c.p. aft of axis $= 5.37/32.5 = 0.165\,\text{m}$

c.p. below stock $= \dfrac{48.8}{32.5} = 1.502\,\text{m}$

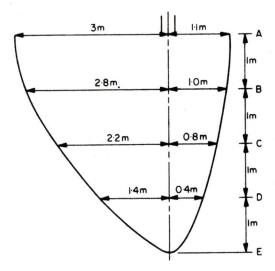

Fig. 13.17

Level	Total chord length (m)	S.M.	F(area)	Lever below stock	F(moment)	c.p. aft of leading edge	c.p. aft of axis	F(torque)
A	4.1	1	4.1	0	0.0	1.27	0.17	0.70
B	3.8	4	15.2	1	15.2	1.18	0.18	2.74
C	3.0	2	6.0	2	12.0	0.93	0.13	0.78
D	1.8	4	7.2	3	21.6	0.56	0.16	1.15
E	0	1	0	4	0.0	0.00	0.00	0.00
			32.5		48.8			5.37

$$\text{Force on rudder} = 21.1 \times 10.83 \times \left(\frac{20 \times 1852}{3600}\right)^2 \times 35 = 847{,}000 \text{ newtons}$$

Bending moment at stock $= 847{,}000 \times 1.502 = 1.272 \text{ MN m}$

Torque on rudder $= 0.165 \times 847{,}000 = 140{,}000 \text{ N m}$

The combined effect of the bending moment M and torque T is equivalent to a bending moment M' given by

$$M' = \frac{1}{2}\left(M + \sqrt{M^2 + T^2}\right)$$

hence $M' = 1.276 \text{ MN m}$

Max stress $= M'/Z = 1.276/0.1 = 12.76 \text{ MN/m}^2$

Experiments and trials

MODEL EXPERIMENTS CONCERNED WITH TURNING AND MANOEUVRING

For accurate prediction of ship behaviour, the model must represent as closely as possible the ship and its operating condition both geometrically and dynamically. It is now customary to use battery powered electric motors to drive the propellers and radio control links for rudder and motor control. The model self-propulsion point is different from that for the ship due to Reynolds' number effects, and the response characteristics of the model and ship propulsion systems differ but the errors arising from these causes are likely to be small. They do, however, underline the importance of obtaining reliable correlation with ship trials.

A number of laboratories now have large tanks in which model turning and manoeuvring tests can be conducted. The facilities offered vary but the following is a brief description of those provided at Haslar (now DERA). The basin is 122 m long and 61 m wide with overhead camera positions for recording photographically the path of the model. The models are typically 5 m long, radio controlled and fitted with gyros for sensing heel. One method used for many years to record the path of the model used lights set up on the model at bow and stern so that when the model is underway they lie in a known datum plane. These lights are photographed using a multiple exposure technique so that the path of the model can be recorded on a single negative. When enlarged to a standard scale, the print has a grid superimposed upon it to enable the co-ordinates of the light positions to be read off and the drift angle deduced. As exposures are taken at fixed time intervals, the speed of the model during the turn can be deduced besides the turning path. The heel angle is recorded within the model. The process is illustrated diagrammatically in Fig. 13.18. A number of alternative tracking methods are available.

The same recording techniques can be used to record the behaviour of a model when carrying out zig-zag or spiral manoeuvres or any other special manoeuvres which require a knowledge of the path of the model. Two points have to be borne in mind, however, if a human operator is used as one element of the control system. One, is that being remote from the model the experimenter has to rely upon instruments to tell them how the model is reacting. They cannot sense the movements of the ship directly through a sense of balance and hence their reactions to a given situation may differ from those in a ship. The second point is that the time they have to react is less. Because Froude's law of comparison applies, the time factor is reduced in proportion to the square root of the scale factor, i.e. if the model is to a scale of $\frac{1}{36}$th full size, permissible reaction times will be reduced to $\frac{1}{6}$th of those applying to the helmsman on the ship. In the same way, the rate at which the rudder is applied must be increased to six times that full scale. It follows that, whenever possible, it is desirable to use an automatic control system or, failing this, a programmed sequence of rudder orders in order to ensure consistency of results.

By using suitable instrumentation, the rudder forces and torques can be measured during any of these manoeuvres. Typically, the rudder stock is strain gauged to record the force and its line of action.

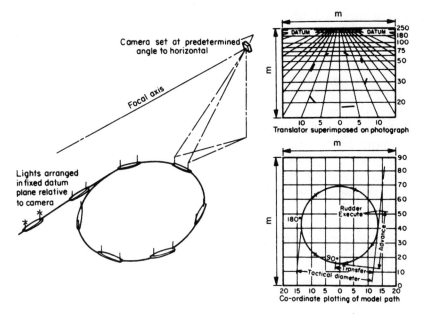

Camera set at predetermined
angle to horizontal

Focal axis

Lights arranged
in fixed datum
plane relative
to camera

m

DATUM   DATUM
250
180
100
75
50

m

30

20

10   5   0   5   10
Translator superimposed on photograph

m

90
80
70
60
50
40
30
20
10
0

Rudder
Execute

180°

90°

Transfer

Tactical diameter

Advance

20 15 10 5 0 5 10 15 20
Co-ordinate plotting of model path

Fig. 13.18

MODEL EXPERIMENTS CONCERNED WITH DIRECTIONAL STABILITY

It has been seen that the spiral manoeuvre can indicate in a qualitative sense whether a design is directionally stable or not and that the pull-out manoeuvre can give an indication of the degree of stability. For a proper study of stability, however, it is necessary to ascertain the derivatives of force and moment as required by the theory outlined earlier. For this constrained model tests are carried out and the assumption made that the forces and moments measured on the model can be scaled directly to full scale.

It is known that there are viscous effects on forces but no suitable correction can be made. Such tests are usually carried out both with and without model propellers fitted and working at model self-propulsion revolutions, as the changed velocity distribution at the stern due to the propeller action is likely to be significant and its effect can be deduced in this way.

The derivatives of forces and moments with respect to transverse velocity (or yaw angle) can be measured in what is termed an oblique tow test in a conventional long ship tank. Measurements are made of the forces and moments required to hold the model at various yaw angles over a range of speeds. Data for the model on a curved path can be obtained using a rotating arm facility with the model at various yaw angles and moving in circular paths of different radii. Speed of advance is controlled by the arm rotational speed which is kept constant for the duration of each run. Submarine models can be run on their side to measure the derivatives with respect to vertical velocity, or trim angle.

Control surface effectiveness can be determined during the oblique tow and rotating arm experiments by measuring the variation in force and moment with control surface angle over a range of yaw angles, path curvatures and speeds of advance. These tests can be carried out over a wide range of parameter values and can thus be used to study situations in which non-linearities exist. They are, however, steady motion tests and as such do not provide any insight into acceleration derivatives.

To measure acceleration derivatives use is made of a planar motion mechanism (PMM) which can yield all the linear derivatives of both velocity and acceleration. The essence of these tests is that the model is force oscillated whilst being towed below the carriage of a conventional ship tank giving rise to sinusoidal yawing and swaying motions. The force and moment records can be analysed into phase components which yield the derivatives associated with the velocity and acceleration components of the motion. It is usual to carry out separate tests in which the model has imparted to it a pure sway and a pure yaw motion (Fig. 13.19).

The rates of turn or curvature that can be applied in PMM tests are limited and the tests do not provide good information on non-linearities and cross-coupling terms. Thus the oblique tow, rotating arm and PMM tests are complementary to one another.

The features of all three tests are incorporated in the Computerized Planar Motion Carriage (CPMC) system used at Hamburg. In this a model can be driven independently in three degrees of freedom and precise transient motions can be generated. Three independent sub-carriages are used to superimpose arbitrary surge, sway and yaw motions on the uniform forward motion of the main towing carriage which does not itself have to accelerate in order to generate transient surge motions in the model.

Besides being used to force oscillate the model the system can be used in a tracking mode to follow closely the movements of a freely manoeuvring model.

Having obtained the hydrodynamic derivatives for a new design they can be substituted in the simple formulae already quoted to demonstrate stability and

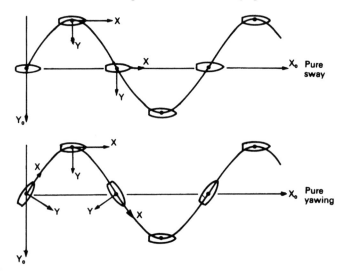

Fig. 13.19 Planar motion mechanism test

position of neutral point. More commonly, the data, including cross-coupling terms, derivatives from control surfaces, etc., are fed into computers which predict turning circles, zig-zag manoeuvres, etc. The validity of this approach is demonstrated in Fig. 13.20.

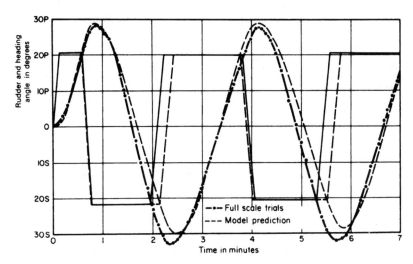

Fig. 13.20 *Comparison of trial and predicted data for zig-zag manoeuvre*

SHIP TRIALS

The credibility of theoretical or model experimental methods for predicting the manoeuvring characteristics of ships depends on establishing reliable correlation with the full-scale ship. Although there is not an exact correspondence of trajectories, there is sufficient correlation between model and ship to believe that the behaviour of the model represents the behaviour of the ship. Several methods can be used for recording the path of a ship at sea including

(a) a log to measure ship speed and a compass to record the ship's head. It is approximate only as most logs are inaccurate when the ship turns;

(b) use of a theodolite or camera overlooking the trial area. Limited by the depth of water available close in shore;

(c) the use of a navigational aid such as the Decca system to record the ship's position at known intervals of time. For accurate results the trials area must be one which is covered by a close grid;

(d) use a satellite navigation system to record the ship's path relative to land. A separate buoy could be tracked to make allowance for water movement;

(e) the use of bearing recorders at each end of the ship to record the bearings of a buoy.

To illustrate the care necessary to ensure a reliable and accurate result a system developed for the last test method is now described. A special buoy is used which moves with the wind and tide in a manner representative of the ship. Automatic recording of bearing angles is used with cameras to correct human errors in tracking the buoy and records are taken of ship's head, shaft r.p.m., rudder angle and heel angle all to a common time base which is synchronized with the bearing records.

The arrangement of the trials equipment in the ship is illustrated in Fig. 13.21.

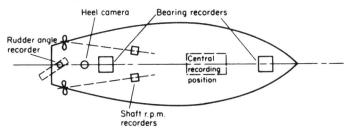

Fig. 13.21 Arrangement of turning trials equipment on board ship

Analysis of turning trials results

Having developed the film records, all of which have a common time base, the bearing of the buoy from each recorder is plotted to a base of ship heading. From this plot, the bearings at every 30 degrees (or any other desired angular spacing) can be obtained. These are then set out relative to a base line representing the distance apart of the two recorders as in Fig. 13.22, where α and β are the two bearings appropriate to 120 degrees change of heading. From the intersection of the two bearing lines a perpendicular $\overline{CD}$ is dropped on to the base line. Then $\overline{CD}$ and $\overline{DG}$ define the position of the buoy relative to the G of the ship and the ship's centre line. Turning now to the right hand plot in Fig. 13.22, radial lines are set out from a fixed point which represents the buoy and the position of the ship for 120 degrees change of heading is set out as indicated.

This process is repeated for each change of heading and the locus of the G position defines the turning path. The drift angle follows as the angle between the tangent to this path and the centre line of the ship. Information on rates of turn is obtained by reference to the time base.

Angle of heel is recorded by photographing the ensign staff against the horizon or using a vertical seeking gyro.

Ship trials involving zig-zag, spiral and pull-out manoeuvres do not require a knowledge of the path of the ship. Records are limited to rudder angle and ship's head to a common time base. The difficulty of recording the spiral manoeuvre for an unstable ship has already been mentioned. Only the two branches of the curve shown in full in Fig. 13.9(*b*) can be defined.

Rudder types and systems

TYPES OF RUDDER

There are many types of rudder fitted to ships throughout the world. Many are of limited application and the claims for a novel type of rudder should be critically examined against the operational use envisaged for the ship. For

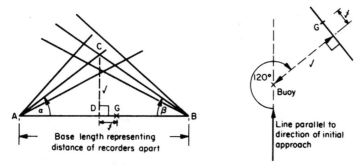

Fig. 13.22 Analysis of turning trial data

instance, some rudders are only of benefit in single screw ships of relatively low speed. It is not possible to cover all the types of rudder in a book such as this and discussion is limited to the four types illustrated in Fig. 13.23.

The choice of rudder type depends upon the shape of the stern, the size of rudder required and the capacity of the steering gear available.

The *balanced spade rudder* is adopted where the ship has a long cut up, the rudder size is not so great as to make the strength of the rudder stock too severe a problem and where it is desired to keep the steering gear as compact as possible.

The *gnomon rudder* is used where the size of rudder requires that it be supported at an additional point to the rudder bearing, but where it is

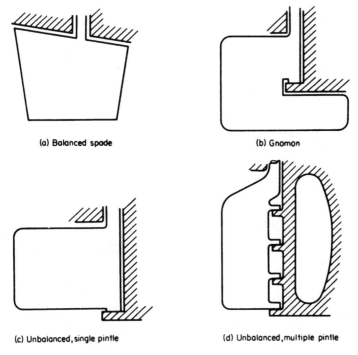

(a) Balanced spade

(b) Gnomon

(c) Unbalanced, single pintle

(d) Unbalanced, multiple pintle

Fig. 13.23 Rudder types

still desired to partially balance the rudder to reduce the size of the steering gear.

Unbalanced rudders are used where the stern shape precludes the fitting of a balanced rudder. The number of pintles fitted is dictated by strength considerations.

BOW RUDDERS AND LATERAL THRUST UNITS

Why should it be necessary to consider bow rudders at all? It has been shown that if a lateral force is applied at the neutral point a ship follows a straight path at an angle of attack, the angle of attack depending on the magnitude of the force. Thus, it would be of considerable use if a navigator could cause a force to be applied at any selected point along the ship's length, i.e. they could control ship's head and path independently. This could be achieved if control surfaces were fitted at both ends of the ship.

Also, if the only rudders fitted are those aft there is a greater danger that damage could render the ship uncontrollable. This is particularly important in wartime when ships are liable to be attacked by weapons homing on the propellers.

Against these considerations it must be remembered that because the neutral point is generally fairly well forward rudders at the bow are relatively much less effective. Neither can they benefit from the effects of the screw race. Unless they are well forward and therefore exposed to damage, the flow conditions over the bow rudder are unlikely to be good. These factors generally outweigh the possible advantages given above and only a few bow rudders as such are fitted to ships in service.

To some extent, these disadvantages can be overcome by fitting a lateral thrust unit at the bow. Typically, such a device is a propeller in a transverse tube. They are particularly useful in ferries when speed in berthing reduces turnround time and enhances economy of operation. Model experiments have shown that the effect of these units may be seriously reduced when the ship has forward speed. The fall in side force can be nearly 50 per cent at 2 knots, 40 per cent occurring between 1 and 2 knots. Placing the unit further aft reduces the effect of forward speed. Contra-rotating propeller systems are recommended for lateral thrust units.

SPECIAL RUDDERS AND MANOEUVRING DEVICES

It has been seen that conventional rudders are of limited use at low speeds. One way of providing a manoeuvring capability at low speed is to deflect the propeller race.

This is achieved in the *Kitchen rudder*, the action of which is illustrated in Fig. 13.24.

The rudder consists essentially of two curved plates shrouding the propeller. For going ahead fast, the two plates are more or less parallel with the propeller race causing little interference. When both plates are turned in plan, they cause the propeller race to be deflected so producing a lateral thrust. When the two plates are turned so as to close in the space behind the propeller, they cause the ahead thrust to be progressively reduced in magnitude and finally to be transformed into an astern thrust albeit a somewhat inefficient one. The same principle is used for jet deflectors in modern high speed aircraft.

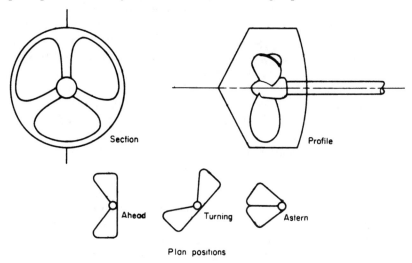

Fig. 13.24 Kitchen rudder action

The Kitchen rudder is used mainly for small power boats. It will be appreciated from the above, that not only can it provide lateral thrust at low ahead speed but that it can also be used to vary the magnitude and/or sense of the propeller thrust. Thus in a boat so fitted the shafts can be left running at constant speed.

An alternative to using deflector plates to deflect the propeller race would be to turn the propeller disc itself. This is the principle of the Pleuger *active rudder* which is a streamlined body actually mounted on a rudder, the body containing an electric motor driving a small propeller. To gain full advantage of such a system, the rudder should be capable of turning through larger angles than the conventional 35 degrees.

The power of the unit varies, with the particular ship application, between about 50 and 300 h.p. With the ship at rest, the system can turn the ship in its own length.

A different principle is applied in the *vertical axis propeller* such as the *Voith-Schneider propeller*, Fig. 13.25(*b*). This consists essentially of a horizontal disc carrying a number of vertical blades of aerofoil shape.

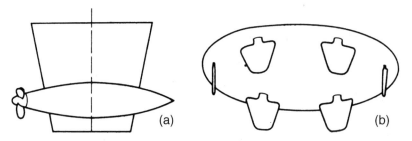

Fig. 13.25 (a) Active rudder, (b) Voith-Schneider propeller

As the horizontal disc is rotated about a vertical axis, a special mechanism feathers the blades in such a way as to provide a thrust in any desired direction. The thrust is caused to act fore and aft for normal propulsion and athwartships for steering. The limitation of this type of propeller is the power which can be transmitted to the disc. In some cases, they are fitted specially for manoeuvring in confined waters as in the case of a number of ferries and other ships operating in canals. They are also used where moderately large turning moments are needed at low speeds as would be the case with some tugs.

Many special rudder forms have been developed over the years. Claims are made for each type of special advantages over more conventional rudder types. Such claims should be carefully examined to ensure that the advantages will be forthcoming for the particular application in mind as, in most cases, this is only so if certain special conditions of speed or ship form apply.

Amongst the special types mention can be made of the following:

(*a*) the *flap rudder* (Fig. 13.26), in which the after portion of the rudder is caused to move to a greater angle than the main portion. Typically, about one-third of the total rudder area is used as a flap and the angle of flap is twice that of the main rudder. The effect of the flap is to cause the camber of rudder section to change with angle giving better lift characteristics. The number of practical applications is not great, partly because of the complication of the linkage system required to actuate the flap;

(*b*) as a special example of the flap rudder, flaps of quite small area at the trailing edge can be moved so as to induce hydrodynamic forces on the main rudder assisting in turning it. Such a rudder is the *Flettner rudder* (Fig. 13.27). The flaps act as a servo system assisting the main steering gear;

(*c*) so-called *balanced reaction rudders* (Fig. 13.28), in which the angle of attack of the rudder sections varies over the depth of the rudder. It attempts to profit from the rotation of the propeller race, and behind propellers working at high slip and low efficiency is claimed to produce a forward thrust;

(*d*) *streamlined rudders* behind a fixed streamlined skeg. This is similar in principle to the flap rudder except that only one part moves. By maintaining a better aerofoil shape at all angles the required lift force is obtained at the expense of less drag and less rudder torque. Such a rudder is the *oertz rudder*.

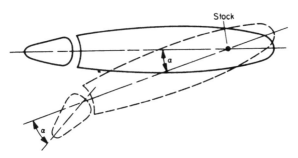

Fig. 13.26 Flap rudder

(*e*) *rotating cylinder rudder*. The fact that a cylinder rotating in a fluid stream develops a lift normal to its axis and the stream flow, has been known for a very long time as the Magnus effect. The principle was considered by NPL (now BMT) to improve ship manoeuvring. Having studied several configurations NPL concluded that the use of a rotating cylinder at the leading edge of the rudder was the most practical. Normal course-keeping was unimpaired and for relatively low cylinder power, attached flow could be maintained for rudder angles up to 90 degrees, i.e. stall which often occurs at 35 degrees could be inhibited. An installation proposed for a 250,000-tonnef tanker had a cylinder one metre in diameter driven at 350 r.p.m. absorbing about 400 kW. It was predicted that the turning circle diameter would be reduced from about 800 m with 35 degrees of rudder to about 100 m at about 75 degrees of rudder with the cylinder in operation. Subsequently sea trials on a 200-tonnef vessel were carried out to confirm the principle and the ship could turn indefinitely almost in its own length.

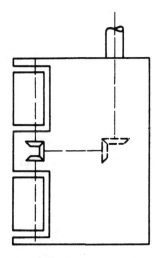

Fig. 13.27 Flettner rudder

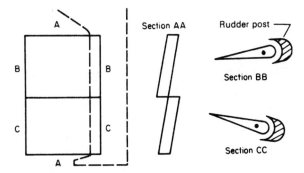

Fig. 13.28 Balanced reaction rudder

DYNAMIC POSITIONING

In ships engaged in underwater activities it may be necessary to hold the ship steady relative to some underwater datum. Typical of this situation are drilling ships and those deploying divers. If the water is shallow then it may be practical to moor the ship. In deeper water use is made of a dynamic positioning system using thrust producing devices forward and aft together with a means of detecting departures from the desired position usually using satellite navigation aids.

AUTOMATIC CONTROL SYSTEMS

Many ships, particularly those on long ocean voyages, travel for long periods of time on a fixed course, the only deviations in course angle being those necessitated by variations in tide, waves or wind. To use trained helmsmen for this type of work is uneconomical and boring for the people concerned. It is in these circumstances that the automatic control system or automatic 'pilot' is most valuable.

Imagine a system which can sense the difference, ψ_e, between the ordered course and the actual course and which can cause the rudder to move to an angle proportional to this error, and in such a way as to turn the ship back towards the desired course, i.e.:

$$\delta_R = \text{Const.} \times \psi_e = a\psi_e, \text{ say}$$

Then, as the ship responds to the rudder the course error will be reduced steadily and, in consequence, the rudder angle will also reduce. Having reached the desired course, the rudder angle will reduce to zero but the ship will still be swinging so that it is bound to 'overshoot'. Thus, by repetition of this process the ship will oscillate about the desired course, the amplitude of the oscillation depending upon the value of the constant of proportionality used in the control equation.

How can the oscillation be avoided or at least reduced? In a ship, a helmsman mentally makes provision for the rate of swing of the ship and applies opposite rudder before the desired course angle is reached to eliminate the swing. By introducing a rate gyro into the control system it also can sense the rate of swing and react accordingly in response to the following control equation

$$\delta_R = a\psi_e + b\left(\frac{d\psi_e}{dt}\right)$$

By careful selection of the values of a and b, the overshoot can be eliminated although, in general, a better compromise is to allow a small overshoot on the first swing but no further oscillation as this usually results in a smaller average error. It would be possible to continue to complicate the control equation by adding higher derivative terms. A ship, however, is rather slow in its response to rudder and the introduction of higher derivatives leads to excessive rudder movement with little effect on the ship. For most applications the control equation given above is perfectly adequate.

The system can be used for course changes. By setting a new course, an 'error' is sensed and the system reacts to bring the ship to a new heading. If desired, the system can be programmed to effect a planned manoeuvre or series of course changes. For an efficient system the designer must take into account the characteristics of the hull, the control surfaces and the actuating system. The general mathematics of control theory will apply as for any other dynamic system. In some cases it may be desirable to accept a directionally unstable hull and create course stability by providing an automatic control system with the appropriate characteristics. This device is not often adopted because of the danger to the ship, should the system fail. Simulators can be used to study the relative performances of manual and automatic controls. The consequences of various modes of failure can be studied in safety using a simulator, including the ability of a human operator to take over in the event of a failure. Part-task simulators are increasingly favoured as training aids. A special example of automatic control systems is that associated with dynamic positioning of drilling or diving ships.

Ship handling

TURNING AT SLOW SPEED OR WHEN STOPPED

It has been seen that the rudder acts in effect as a servo-system in controlling the attitude of the ship's hull so that the hydrodynamic forces on the hull will cause the ship to turn. At low or zero speed, the magnitude of any hydrodynamic force, depending as it does to a first order on V^2, is necessarily small. Since under these conditions the propeller race effect is not large, even the forces on a rudder in the race are small.

The ship must therefore rely upon other means when attempting to manoeuvre under these conditions. A number of possibilities exist:

(a) A twin shaft ship can go ahead on one shaft and astern on the other so producing a couple on the ship causing her to turn. This is a common practice, but leverage of each shaft is relatively small and it can be difficult to match the thrust and pull on the two shafts. Fortunately, some latitude in fore and aft movement is usually permissible.

(b) If leaving a jetty the ship can swing about a stern or head rope. It can make use of such a device as a pivot while going ahead or astern on the propeller.

(c) When coming alongside a jetty at slow speed, use can be made of the so-called 'paddle-wheel' effect. This effect which is due to the non-axial flow through the propeller disc results in a lateral force acting on the stern/propeller/rudder combination in such a way as to cause the stern to swing in the direction it would do if the propeller were running as a wheel on top of a hard surface. In a twin-screw ship the forces are generally in balance. Now, consider a twin-screw ship approaching a jetty as in Fig. 13.29. If both screws are outward turning (that is viewed from aft, the tip of the propellers move outboard at the top of the propeller disc), the port shaft can be set astern. This will have the effect of producing a lateral force at the

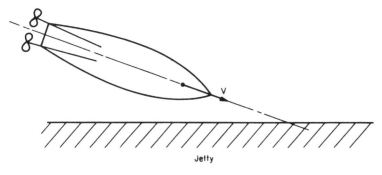

Fig. 13.29 Paddle-wheel effect when coming alongside a jetty

stern acting towards the jetty, besides taking the way off the ship and producing a moment on the shafts tending to bring the ship parallel to the jetty.

(*d*) The screw race can be deflected by a special device such as the Kitchen rudder or, to some extent at least, by twin rudders behind a single propeller. Clearly, unless the race can be deflected through about 90 degrees this system cannot be used without, at the same time, causing the ship to move fore and aft.

(*e*) Use one of the special manoeuvring devices described above.

INTERACTION BETWEEN SHIPS WHEN CLOSE ABOARD

Even in deep water, interaction effects can be significant when two ships are in close proximity. The pressure field created by a ship moving ahead in open water is illustrated in Fig. 13.30, its actual form depending on the ship form.

The pressure field extends for a considerable area around the ship, and any disturbance created in this field necessarily has its reaction on the forces acting on the ship. If the disturbance takes place to one side of the ship, it is to be expected that the ship will, in general, be subject to a lateral force and a yawing moment.

This is borne out by the results reproduced in Fig 13.31 for a ship A of 226 m and 37,500 tonnef overtaking a ship B of 173 m and 24,000 tonnef on a parallel course. From these, it is seen that the ships are initially repelled, the force of repulsion reducing to zero when the bow of A is abreast the amidships of B. The ships are then attracted, the force becoming a maximum soon after the ships are abreast and then reducing and becoming a repulsion as the ships begin to part company.

The largest forces experienced were those of attraction when the two ships were abreast. They amounted to 26.5 tonnef on A and 35.5 tonnef on B at 10 knots speed and 15 m separation. The forces vary approximately as the square of the ship speed and inversely with the separation.

When running abreast, both ships are subject to a bow outward moment but to a bow inward moment when approaching or breaking away.

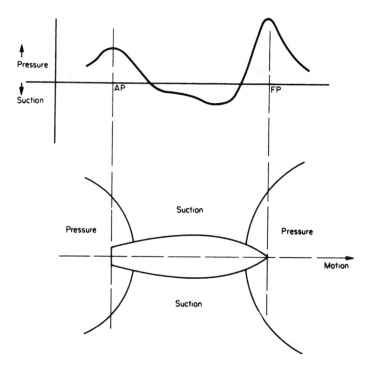

Fig. 13.30 Pressure field for ship in deep water

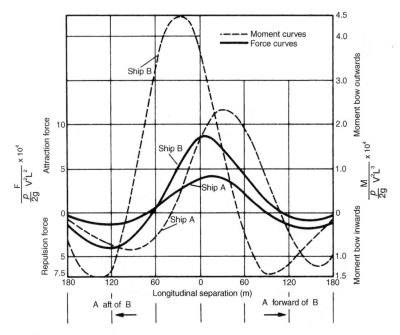

Fig. 13.31 Interaction forces and moments

To be able to maintain the desired course it is necessary to counteract these variations in force and moment, which means that the ship must not only use its rudder but must also run generally at an angle of yaw.

Ship trials show that replenishment at sea operations are perfectly feasible at speeds of up to about 20 knots and such operations are quite commonplace nowadays.

In addition to, or in place of, the disturbance created by a ship in close company, the pressure field of a ship may be upset by a canal bank, pier or by the proximity of the sea bed. In the case of a vertical canal bank or pier, the result will be a lateral force and yawing moment as for the replenishment at sea operation. By analogy, the effect of shallow water is a vertical force and trimming moment resulting in bodily sinkage of the ship and trim by the stern. This effect can lead to grounding on a sandbank which may have been expected to be several feet below the keel.

While naval architects think of these attractions in terms of Bernouilli's equation, mariners recognize the canal effect as 'smelling the ground' and the bodily sinkage in shallow water as 'squat'.

BROACHING

Broaching, or *broaching-to*, describes the loss of directional stability in waves, induced by a large yaw moment exceeding the course keeping ability of the rudders. Orbital motion of water particles in the wave can result in a zero flow past the rudders which become ineffective. This loss can cause the ship to turn beam on to the waves. The vessel might even capsize due to a large roll moment arising from the forward momentum and the large heading angle. The effect is greater because the ship's hydrostatic stability is often reduced by the presence of the waves.

Broaching is likely when the ship is running with, or being slowly overtaken by, the waves. It may be sudden, due to the action of a single wave, or be cumulative where the yaw angle builds up during a succession of waves. Although known well since man put to sea in boats, broaching is a highly non-linear phenomenon and it is only relatively recently that good mathematical simulations have been possible.

When the encounter frequency of the ship with the waves approaches zero the ship can become trapped by the wave. The ship remains in the same position relative to the waves for an appreciable time. It is then said to be *surf riding*. This is a dangerous position and broaching is likely to follow.

The Master can get out of this condition by a change of speed or direction, although the latter may temporarily result in large roll angles.

Stability and control of submarines

The high underwater speed of some submarines makes it necessary to study their dynamic stability and control. The subject assumes great importance to both the commanding officer and the designer because of the very short time available in which to take corrective action in any emergency: many submarines are restricted to a layer of water which is of the order of at most two or three ship lengths deep. To the designer and research worker, this has meant directing attention to the change in the character of the forces governing the motion of the submarine which occurs as the speed is increased. For submarines of orthodox size and shape below about ten knots the hydrostatic forces predominate. In this case, the performance of the submarine in the vertical plane can be assessed from the buoyancy and mass distributions. Above 10 knots, however, the hydrodynamic forces and moments on the hull and control surfaces predominate.

To a certain degree, the treatment of this problem is similar to that of the directional stability of surface ships dealt with earlier. There are differences however between the two, viz.:

(a) the submarine is positively stable in the fore and aft vertical plane in that B lies above G so that having suffered a small disturbance in trim when at rest it will return to its original trim condition;

(b) the limitation in the depth of water available for vertical manoeuvres;

(c) the submarine is unstable for translations in the z direction because the hull is more compressible than water;

(d) it is not possible to maintain a precise equilibrium between weight and buoyancy as fuel and stores are being continuously consumed.

It follows, from (c) and (d) above, that the control surfaces or hydroplanes will have, in general, to exert an upward or downward force on the submarine. Also, if the submarine has to remain on a level keel or, for some reason, the submarine cannot be allowed to trim to enable the stability lever to take account of the trimming moment, the control surfaces must also exert a moment. To be able to exert a force and moment on the submarine which bear no fixed relationship one to another requires two separate sets of hydroplanes. Usually, these are mounted well forward and well aft on the submarine to provide maximum leverage.

Consider a submarine turning in the vertical plane (Fig. 13.32).

Assume that the effective hydroplane angle is δ_H, i.e. the angle representing the combined effects of bow and stern hydroplanes.

In a steady state turn, with all velocities constant, the force in the z direction and the trimming moment are zero. Hence

$$wZ_w + qZ_q + mqV + \delta_H Z_{\delta_H} = 0$$

$$wM_w + qM_q + \delta_H M_{\delta_H} - mg\overline{BG}\theta = 0$$

Fig. 13.32

where subscripts w, q and δ_H denote differentiation with respect to velocity normal to submarine axis, pitching velocity and hydroplane angle respectively. Compare the equations for directional stability of surface ships:

mqV is a centrifugal force term

$mg\overline{BG}\theta$ is a statical stability term

In the moment equation M_w, M_q and M_{δ_H} are all proportional to V^2, whereas $mg\overline{BG}\theta$ is constant at all speeds. Hence, at high speeds, $mg\overline{BG}\theta$ becomes small and can be ignored. As mentioned above, for most submarines it can be ignored at speeds of about 10 knots. By eliminating w between the two equations so simplified

$$\frac{q}{\delta_H} = \frac{M_w Z_{\delta_H} - M_{\delta_H} Z_w}{M_q Z_w - M_w (Z_q + mV)}$$

As with the surface ship problem the necessary condition for stability is that the denominator should be positive, i.e.

$$M_q Z_w - M_w (Z_q + mV) > 0$$

This is commonly known as the *high speed stability criterion*.

If this condition is met and statically the submarine is stable, then it will be stable at all speeds. If it is statically stable, but the above condition is not satisfied, then the submarine will develop a diverging (i.e. unstable) oscillation in its motion at forward speeds above some critical value.

Now by definition $q = d\theta/dt = \dot{\theta}$, so that differentiating the moment equation with respect to time

$$\dot{w} M_w + \dot{q} M_q + \dot{\delta}_H M_{\delta_H} - mg\overline{BG}q = 0$$

But in a steady state condition as postulated $\dot{w} = \dot{q} = \dot{\delta}_H = 0$. Hence $q = 0$ if $\overline{BG}$ is positive as is the practical case. That is, a steady path in a circle is not possible unless $\overline{BG} = 0$. Putting $q = 0$, the equations become

$$w Z_w + \delta_H Z_{\delta_H} = 0$$

$$w M_w + \delta_H M_{\delta_H} - mg\overline{BG}\theta = 0$$

i.e.

$$w = -\delta_H \frac{Z_{\delta_H}}{Z_w}$$

and

$$\theta = \delta_H \left(M_{\delta_H} - \frac{Z_{\delta_H}}{Z_w} M_w \right) \Big/ mg\overline{BG}$$

Now rate of change of depth $= V(\theta - w/V)$ if w is small $= V\theta - w$, i.e.

$$\frac{\text{depth rate}}{\delta_H} = \left(VM_{\delta_H} - VM_w \frac{Z_{\delta_H}}{Z_w} + mg\overline{BG}\,\frac{Z_{\delta_H}}{Z_w} \right) \bigg/ mg\overline{BG}$$

The depth rate is zero if

$$V = -\left(mg\overline{BG}\,\frac{Z_{\delta_H}}{Z_w} \right) \bigg/ \left(M_{\delta_H} - M_w \frac{Z_{\delta_H}}{Z_w} \right)$$

$$= mg\overline{BG} \bigg/ \left(M_w - M_{\delta_H} \frac{Z_w}{Z_{\delta_H}} \right)$$

From the equation for θ, if the hydroplanes are so situated that

$$\frac{M_{\delta_H}}{Z_{\delta_H}} = \frac{M_w}{Z_w}$$

then θ is zero. The depth rate will be $\delta_H Z_{\delta_H}/Z_w$ which is not zero. The ratio M_w/Z_w defines the position of the *neutral point*. This corresponds to the similar point used in directional stability and is usually forward of the centre of gravity. A force at the neutral point causes a depth change but no change in the angle of pitch.

The equation for depth rate can be rewritten as

$$\text{depth rate}/\delta_H = \frac{Z_{\delta_H}}{Z_w}\left(1 - \frac{V}{V_c} \right)$$

where

$$V_c = mg\overline{BG} \bigg/ \left(M_w - \frac{M_{\delta_H}}{Z_{\delta_H}} \cdot Z_w \right)$$

or

$$\text{depth rate}/\delta_H = \frac{Z_{\delta_H} V}{mg\overline{BG}}\left\{ \frac{M_{\delta_H}}{Z_{\delta_H}} - x_c \right\}$$

where

$$x_c = \frac{M_w}{Z_w} - \frac{mg\overline{BG}}{VZ_w}$$

The first of these two expressions shows that $(\text{depth rate})/\delta_H$ is negative, zero or positive as V is greater than, equal to or less than V_c respectively. V_c is known as the *critical speed* or *reversal speed*, since at that speed the planes give zero depth change and cause reverse effects as the speed increases or decreases from this speed. Near the critical speed the value of $(1 - (V/V_c))$ is small— hence the hydroplanes' small effect in depth changing.

It will be seen that θ is not affected in this way since

$$\frac{\theta}{\delta_{\mathrm{H}}} = -\frac{Z_{\delta_{\mathrm{H}}}}{Z_w}\frac{1}{V_{\mathrm{c}}} = -\frac{Z'_{\delta_{\mathrm{H}}}}{Z'_w}\frac{V}{V_{\mathrm{c}}}$$

The magnitude of $\theta/\delta_{\mathrm{H}}$ changes with V but not its sign. If stern hydroplanes are considered, a positive hydroplane angle produces a negative pitch angle (bow down), but depth change is downwards above the critical speed and upwards below the critical speed.

The second expression for depth change illustrates another aspect of the same phenomenon. x_{c} denotes a position $mg\overline{BG}/VZ_w$ abaft the neutral point,

$$\frac{mg\overline{BG}}{VZ_w} = \frac{mg\overline{BG}}{\frac{1}{2}\rho L^2 Z'_w V^2}$$

hence x_{c} is abaft the neutral point by a distance which is small at high speed and large at low speed. The critical situation is given by $x_{\mathrm{c}} = M_{\delta_{\mathrm{H}}}/Z_{\delta_{\mathrm{H}}}$, i.e. centre of pressure of the hydroplanes. The position defined by x_{c} is termed the *critical point*. Figure 13.33 illustrates the neutral and critical point positions. Figure 13.34 shows a typical plot of x_{c}/L against Froude number. The critical speed can be obtained by noting the Froude number appropriate to the hydroplane position, e.g. in the figure

$$\frac{V_{\mathrm{c}}}{\sqrt{gL}} = 0.05$$

i.e.

$$V_{\mathrm{c}} = 3\,\text{knots if } L = 100\,\text{m}$$

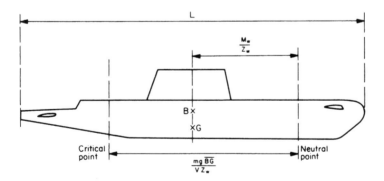

Fig. 13.33 Neutral and critical points

EXPERIMENTS AND TRIALS

As in the case of the directional stability of surface ships, the derivatives needed in studying submarine performance can be obtained in conventional ship tanks using planar motion mechanisms and in rotating arm facilities. The model is

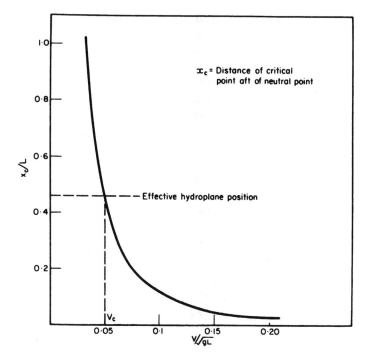

Fig. 13.34 Variation of critical point with speed

run upright and on its side with and without propellers, hydroplanes and stabilizer fins to enable the separate effects of these appendages to be studied. Data so obtained are used to predict stability and fed into digital or analogue computers. The computer can then predict the manoeuvres the submarine will perform in response to certain control surface movements. These can be used to compare with full-scale data obtained from trials. A computer can be associated with a tilting and rotating cabin, creating a simulator for realistic training of operators and for studying the value of different display and control systems.

Design assessment

MODIFYING DYNAMIC STABILITY CHARACTERISTICS

In common with most features of ship design, it is likely that the designer will wish to modify the dynamic stability standards as defined by the initial model tests. How then can the desired standards be most effectively produced?

In most cases, the basic hull form will be determined by resistance, propulsion and seakeeping considerations. The designer can most conveniently modify the appendages to change the dynamic stability. The procedure is similar for submarines and surface ships but is illustrated below for the former.

Assuming that the hydroplanes are correctly sized, the designer concentrates on the stabilizer fins (skeg for the lateral plane). If the contributions of these fins to Z'_w and M'_w, as determined from the model results with and without fins, are $\delta Z'_w$ and $\delta M'_w$ the effective distance of the fins from the centre of gravity is X_s, say, where:

$$\frac{X_s}{L} = -\frac{\delta M'_w}{\delta Z'_w}, \quad \text{i.e. } \delta M'_w = -\frac{X_s}{L}\delta Z'_w$$

The negative sign arises because the fins are aft.

The effect of the fins on the curvature derivatives can be deduced similarly or, if not available from direct model tests, it can be argued that the rotation causes an effective change of incidence at the fin, such that:

$$\delta Z'_q = \frac{X_s}{L}\delta Z'_w$$

and

$$\delta M'_q = \left(\frac{X_s}{L}\right)^2 \delta Z'_w$$

If the derivatives, as originally determined, give rise to an unstable motion, the required increase in fin area can be deduced using the above relationships and assuming that $\delta Z'_w$ is proportional to the fin area.

EXAMPLE 3. The stability derivatives found for a certain submarine, complete with all appendages are:

$$Z'_w = -0.02, \quad Z'_q = -0.01$$
$$M'_w = 0.012, \quad M'_q = -0.005$$
$$m' = 0.024$$

The corresponding figures for Z'_w and M'_w without fins are 0 and 0.022. Calculate the percentage increase in fin area required to make the submarine just stable assuming m' is effectively unaltered.

Solution: The stability criterion in non-dimensional form is

$$M'_q Z'_w - M'_w(Z'_q + m') > 0$$

Substituting the original data gives -0.000068 so that the submarine is unstable.

If the fin area is increased by p per cent then the derivatives become

$$Z'_w = -0.02 + p(-0.02)$$
$$M'_w = 0.012 + p(-0.01), \text{ in this case } X_s = -\tfrac{1}{2}L$$
$$Z'_q = -0.01 + p(-0.01)$$
$$M'_q = -0.005 + p(-0.005)$$

Substituting these values in the left-hand side of the stability criterion and equating to zero gives the value of p which will make the submarine just stable.

Carrying out this calculation gives $p = 14.8$ per cent, and the modified derivatives become

$$Z'_w = -0.023, \quad Z'_q = -0.0115$$
$$M'_w = 0.0105, \quad M'_q = -0.00575$$

EFFICIENCY OF CONTROL SURFACES

Ideally, the operator of any ship should define the standard of manoeuvrability required in terms of the standard manoeuvres already discussed. The designer could then calculate, or measure by model tests, the various stability derivatives and the forces and moments generated by movements of the control surfaces, i.e. rudders and hydroplanes. By feeding this information to a computer a prediction can be made of the ship performance, compared with the stated requirements and the design modified as necessary. By changing skeg or fin and modifying the areas of control surfaces, the desired response may be achieved.

As a simpler method of comparing ships, the effectiveness of control surfaces can be gauged by comparing the forces and moments they can generate with the forces and moments produced on the hull by movements in the appropriate plane. Strictly, the force and moment on the hull should be the combination of those due to lateral velocity and rotation, but for most purposes they can be compared separately; for example the rudder force and moment can be compared with the force and moment due to lateral velocity to provide a measure of the ability of the rudder to hold the hull at a given angle of attack and thus cause the ship to turn. The ability of the rudder to start rotating the ship can be judged by comparing the moment due to rudder with the rotational inertia of the ship. The ability of hydroplanes to cope with a lack of balance between weight and buoyancy is demonstrated by comparing the force they can generate with the displacement of the submarine. It is important that all parameters be measured in a consistent fashion and that the suitability of the figures obtained be compared with previous designs.

Effect of design parameters on manoeuvring

The following remarks are of a general nature because it is not possible to predict how changes to individual design parameters will affect precisely the manoeuvring of a ship.

Speed. For surface vessels increased speed leads to increased turning diameter for a given rudder angle although the rate of turn normally increases. For submerged bodies, turning diameters are sensibly constant over the speed range.

Trim. Generally stern trim improves directional stability and increases turning diameter. The effect is roughly linear over practical speed ranges.

Draught. Somewhat surprisingly limited tests indicate that decrease in draught results in increased turning rate and stability. This suggests that the rudder becomes a more dominant factor both as a stabilizing fin and as a turning device.

Longitudinal moment of inertia. Changes in inertia leave the steady turning rate unchanged. A larger inertia increases angular momentum and leads to larger overshoot.

Metacentric height. Quite large changes in metacentric height show no significant effects on turning rate or stability.

Length/beam ratio. Generally speaking the greater this ratio the more stable the ship and the larger the turning circle.

Problems

1. A rudder placed immediately behind a middle line propeller is rectangular in shape, 3 m wide and 2 m deep. It is pivoted at its leading edge. Estimate the torque on the rudder head when it is placed at 35 degrees, the ship's speed being 15 knots.
2. A ship turns in a radius of 300 m at a speed of 20 knots under the action of a rudder force of 100 tonnef. If the draught of the vessel is 5 m, $\overline{KG}$ is 6 m and $\overline{GM}$ is 2 m find the approximate angle of heel during the steady turn.
3. A rudder is shaped as shown. If it is on the middle line in a single screw ship, how far abaft the leading edge is the centre of pressure?

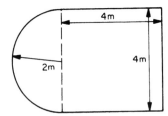

Fig. 13.35

4. A rudder has a profile as sketched in Fig. 13.36.
   Calculate the force on this rudder when operating behind a single centre-line screw at a ship speed of 20 knots ahead with the rudder at 35 degrees. Use the formula due to Baker and Bottomley.

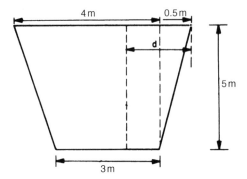

Fig. 13.36

What is the value of *d* in order that the torque is zero in this condition assuming the rudder is effectively in open water.

5. Calculate the force and torque on the spade rudder shown, which is one of two working behind twin propellers. Assume a rudder angle of 35 degrees and a ship speed of 18 knots ahead.

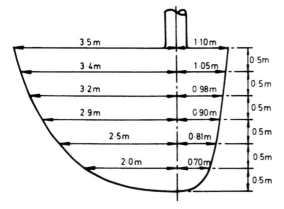

Fig. 13.37

6. The rudder sketched in Fig. 13.38 has sections similar to NACA 0015. Calculate the force and torque on the rudder for 20 knots ahead speed, with the rudder at 35 degrees, assuming no breakdown of flow occurs.

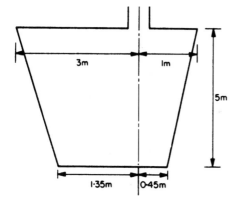

Fig. 13.38

7. Describe the action of the rudder on a ship
   (*a*) when it is first put over,
   (*b*) when the ship is turning steadily.

   Sketch a typical turning circle, giving the path of the c.g. of the ship from the point when the helm is first put over up to the point when the ship has turned through 360°. Show the position of the ship (by its centre line) at 90°, 180°, 270° and 360° turn. Show on your diagram what is meant by Advance,

Transfer, Tactical Diameter and Drift Angle. Which way would you expect a submarine to heel when turning? Give reasons.

8. The balanced rudder, shown in Fig. 13.39, has a maximum turning angle of 35° and is fitted directly behind a single propeller.

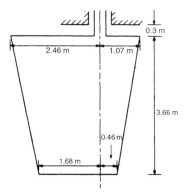

Fig. 13.39

What torque and bending moment are applied to the rudder stock at the lower end of the sleeve bearing, when the rudder is put over at a ship's speed of 26 knots?

In the force equation, $P = KAV^2\delta_R$ (P in newtons, A in m², V in knots), take the constant $K = 0.041$. Also, if the length of an elemental strip of the rudder surface, drawn at right-angles to the centre line of the stock, is 1 then assume the centre of pressure of the strip to be 0.321 from the leading edge.

9. Using Fig. 13.12, plot an NACA 0015 section with a chord length of 5 m. Calculate the area of the section, the distance of the centroid from the nose and the section modulus about each principal axis, assuming a solid section.

Take the x/c values at intervals of 0.1.

10. A twin-screw vessel has a rectangular spade rudder 1.5 m wide and 2 m deep. The axis is 0.5 m from the leading edge. If friction at the rudder stock bearings and in the stearing gear may be taken as 5000 Nm, estimate the range of possible angles which the rudder can take up if the steering gear is damaged while the ship is underway at 30 knots. Distance of centre of pressure abaft leading edge may be taken as chord $(0.195 + 0.305 \sin \theta)$.

11. Details of two ship designs A and B are given below.

	Ship A	Ship B
Length on WL, L (m)	215	252.5
Beam, B (m)	24	26.75
Draught, T (m)	7.625	8.0
Area of rudder, A_R (m²)		50

Design B achieved a tactical diameter of 4.3 ship lengths at 28 knots. Assuming that tactical diameter = const. $\times L^3 T / BA_R$, calculate the rudder areas necessary to give tactical diameters of 3, 3.5 and 4.0 ship lengths in design A at the appropriate speed.

12. Calculate the approximate heeling moment acting on a ship of 60 MN displacement assuming

$$\text{Force on rudder} = 21 A_R V^2 \delta_R \text{ newtons, } A_R \text{ in m}^2, V \text{ in m/s, } \delta_R \text{ in degrees}$$

$$\text{Length of ship} = 150 \, \text{m}$$

$$A_R = \text{rudder area} = 50 \, \text{m}^2$$

$$V = \text{ship speed} = 18 \text{ knots on turn}$$

$$\delta_R = \text{rudder angle} = 35 \text{ degrees}$$
$$\text{Draught} = 8 \, \text{m}$$
$$\overline{KG} = 10 \, \text{m}$$

Height of centroid of rudder above keel $= 5 \, \text{m}$

$$\frac{TD}{L} = 3.6.$$

13. Two designs possess the following values of derivatives

	Y_v'	N_v'	Y_r'	N_r'	m'
Design A	-0.36	-0.07	0.06	-0.07	0.12
Design B	-0.26	-0.10	0.01	-0.03	0.10

Comment on the directional stability of the two designs.
Assuming both designs are 100 m long how far are the neutral points forward of the centres of gravity?

14. The directional stability derivatives for a surface ship 177 m long are:

$$Y_v' = -0.0116, \quad N_r' = -0.00166$$
$$N_v' = -0.00264, \quad m' = 0.00798$$
$$Y_r' = -0.00298$$

Y_v' and N_v' without a 9.29 m² skeg were -0.0050 and 0 respectively.
   Show that the ship, with skeg, is stable and calculate the distance of the neutral point forward of the c.g. and the effective distance of the skeg aft of the c.g. What increase in skeg area is necessary to increase the stability index by 20 per cent?

15. A submarine 100 m long has the following non-dimensional derivatives:

$$Z_w' = -0.030, \quad M_q' = -0.008$$
$$M_w' = 0.012, \quad m' = 0.030$$
$$Z_q' = -0.015$$

Calculate the distance of the neutral point forward of the c.g. Is the submarine stable?
   If $\overline{BG} = 0.5 \, \text{m}$ and displacement is 4000 tonnef, calculate the critical speed for the after hydroplanes which are 45 m aft of the c.g.

14 Major ship design features

So far in this book we have considered the behaviour of the total ship and how that can be manipulated. Such safe and satisfactory behaviour of the ship as an entity is under the total control of the naval architect. Many professions contribute to the elements of a ship. The naval architect while not directly controlling each of them, has the responsibility of integrating them into the whole design and will usually be the Project Manager. The design must be balanced, each element demanding no more and no less than is a proper share of the total, contributing just enough to the overall performance. Standards of behaviour must be adequate but not more and facilities provided must be no more than is necessary for satisfactory functioning of the vessel. Such a balancing act among so many disparate elements is not easy. The naval architect must be in a position to judge to what extent the demands upon the ship by the individual specialist should be met. This requires enough knowledge of the specialisms to be able to discuss and cajole and, if necessary, to reject some of the demands.

Some of the specialisms are addressed in this chapter before we are able to move on to the ship design process itself. They are all bound up closely with the ship but none more so than the choice of the propulsion machinery.

Machinery

Propulsion machinery for ships used to be tailor-made to conform to the size and predominant speeds of the ship, such as top and cruising speeds. Today, even for warships, it is a question of selecting standard units and combining them in a satisfactory manner. Most merchant ships are expected to proceed at their economical speed for their entire lives so that their propulsion machinery may be optimized in a relatively straightforward manner. So dominant are economic considerations that we should begin with an examination of where the energy contained in the fuel goes. This is shown in Fig. 14.1 for a diesel-driven

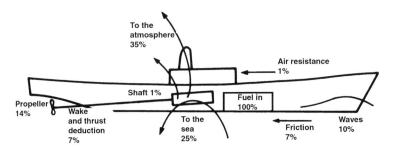

Fig. 14.1 Where the energy goes

frigate at high speed; while exact figures will differ for other ships they will be of the order of size shown. Of the total energy available in the fuel about 60 per cent is lost immediately to the sea via the condenser water and to the atmosphere via the exhaust gases. Less than 20 per cent is actually spent in overcoming the friction of the sea and in creating waves. Yet the diesel engine is the most efficient of the propulsion machinery options open to designers at present. Thermal efficiencies at maximum power of various systems are roughly:

diesels	43%
gas turbine	35%
steam turbine	20%

Relative costs of the basic fuel are currently as shown in Table 14.1 compared to a light diesel oil figure of unity. They do not remain static for long and the economic choice of machinery does require some wise foresight. Note that medium and high-speed diesels and marine gas turbines use exactly the same fuel which is standardized for NATO warships. Because coal is bulky as Table 14.1 shows and is so far associated with relatively low system efficiencies, despite the introduction of fluidized bed boilers, it has not yet found favour with many owners. A diesel engine run on heavy oil is often a preferred fit because of the relative cost advantage and as well as the large slow-running diesels, some medium-speed diesels can successfully use heavy oil. Oil producing countries are well aware of the competition from coal and a trend back to coal remains a persistent possibility. There has also been a significant trend towards high speed diesels, particularly in small vessels.

Table 14.1

	Relative cost/tonne	Relative cost/kJ	MJ/kg	Stowage m^3/tonne	Stowage m^3/MJ
Light diesel oil	1.00	1.00	45	1.2	27
Heavy fuel oil	0.62	0.64	43	1.05	24
Coal	0.15	0.27	25	1.5	60

Choice, however, is not dependent entirely upon running costs. The choice of main machinery for all ships is made after an examination of many aspects:

(*a*) demands upon the ship in terms of mass and volume;
(*b*) overall economy in terms of procurement, installation, running and logistics costs;
(*c*) range of speeds likely to be needed;
(*d*) availability, reliability and maintainability;
(*e*) signature suppression, e.g. quietness, stealth and noxious efflux;
(*f*) vulnerability, duplication and unitization;
(*g*) engineering crew and automation;
(*h*) vibration induced in the ship.

For most merchant ships the space and mass demands upon the ship of a diesel engine installation are acceptable and they show also good overall economy and high reliability. Nor are great flexibility in speed or reduced signatures often requirements. Motor ships are therefore very common. There is a choice in the type of diesel engine. The huge diesels developed for the large tankers of the 1960s are now capable of delivering 40 MW or more at such low rotary speed that they do not need a gearbox. They are very big and heavy and unsuitable for anything but the largest ships. Medium- and high-speed diesels have also benefited from recent developments and are available in a wide range of powers. Double and selective supercharging have increased the output per tonne of diesel machinery and also ameliorated such problems as coking up at fractional power outputs. Figure 14.2 shows a very rough order of installed power necessary for merchant ships of various displacements and speeds. (It is not intended to supplant the need for proper assessment.)

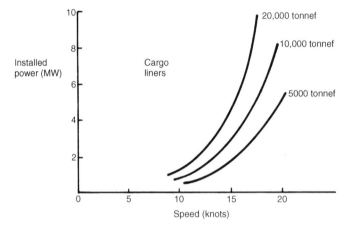

Fig. 14.2 Approximate installed power for cargo liners

Power requirements for small warships are indicated very roughly in Fig. 14.3. For these, the other factors above are often more important. It is usual for a warship for example to proceed for much of its life at an economical speed but to be capable of bursts of high speeds. Furthermore, military operations some-times demand extreme quietness underwater for anti-submarine warfare and other signature treatments that influence the choice of main machinery. For these reasons it is usual, save in the smallest and simplest warships to arrange standard power units to combine in various ways so that different elements may be selected for each operating condition that has to be met.

Standard units are the diesel engine, steam turbine, gas turbine and electric motor which may be used alternatively or in combination. Acronyms have been developed to describe these succinctly, e.g. CODOG meaning combined diesel or gas turbine and CODLAG meaning combined diesel electric and gas turbine. The switchover point from one grouping to another is important. It depends primarily upon the power delivered and the economy of running measured by the specific fuel consumption. Curves such as those shown in Fig. 14.4 are developed for many combinations which, with an assumption concerning the

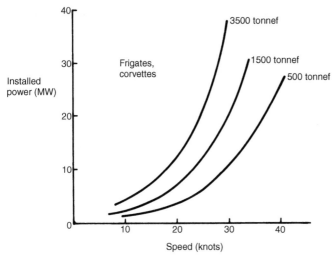

Fig. 14.3 Approximate installed power for warships

likely operating pattern, enable overall economy to be assessed. There are other considerations too such as quiet operating speed, engine wear, diesel coking and gearing. Sudden changes in efficiency or specific fuel consumption with speed are undesirable and the flat characteristics of modern diesel engines and large marine gas turbines have much eased the problems of combining units.

Many engines run with a maximum efficiency at high rotary speed so that a gearbox is necessary to reduce speeds to values acceptable to the propulsor. This is not the place to discuss gearbox design, which is an important study for marine engineers. However, there are step changes in gearbox design driven by maximum tooth loading and other factors which importantly affect the choice of machinery unit combinations. The adoption of a third gear train or epicyclic

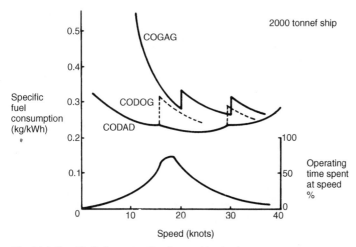

Fig. 14.4 Specific fuel consumption for combined units

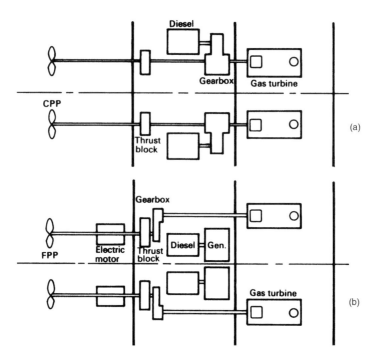

Fig. 14.5 (a) CODOG, (b) CODLAG

gearing may boost cost and make demands upon the ship that significantly affect the considerations concerning machinery choice. Reversing of the propulsion may be effected by the gearing or by the pitch reversal of the propulsor blade, by clutching in an auxiliary drive or by reversal of the prime mover itself.

Two of many machinery configurations are shown in Fig. 14.5.

Choice of numbers of shafts and the type of propulsors are matters of compromise. A single shaft and a large diameter slow helical propeller are propulsively the most efficient arrangement. Twin shafts in lower wake conditions and with necessarily rather smaller diameter propellers may lose 0.05 on propulsive coefficient comparatively and are slightly more expensive. However, they do allow more total power to be transmitted, they provide better standby propulsive power in case of failure and they can provide a turning moment on the ship when there is no way on the ship or when steering failure occurs. So far as the propulsor is concerned, nothing is more efficient than a well-designed helical propeller, approaching the theoretical maximum of the actuator disc (Fig. 14.6). Alternative propulsion devices need to be considered however for other reasons. Vertical axis propellers give remarkably responsive steering for such vessels as ferries and river craft and dispense with rudders; shrouded propellers may be quiet; water jets can be especially compact in small craft; paddle wheels provide good manoeuvrability; controllable pitch propellers, while a few per cent less efficient, do give a rapid means of reversing thrust and an ability to select the correct pitch for each condition of operation.

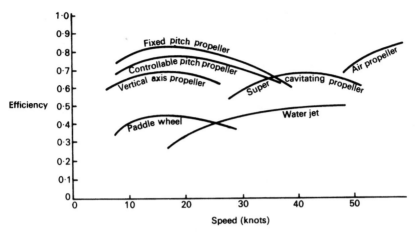

Fig. 14.6 Approximate efficiency of propulsion devices

In concluding this small introduction to the considerations affecting machinery choice we might cast our minds forward. With world reserves of oil running out and oil becoming increasingly uneconomical in the coming years, thoughts are already turning to alternative fuels. Wind is again being seriously studied both for main propulsion and as a means of economizing. Oil from shale rock, while currently uneconomical, will become increasingly attractive. Solar panels do not yet look very promising because of the enormous areas required to provide enough energy for a ship. Coal, on the other hand, will remain plentiful for many years yet and can be conveyed in pulverized form, pneumatically, in slurry form or even converted into oil. Nuclear propulsion has already been safely used in hundreds of vessels, mainly submarines. In merchant ships it has been successfully used although barely economically; it does give rise to social problems. While nuclear reactors are indeed exceedingly safe and reliable, the consequences of serious malfunction do give rise to understandable anxieties.

For small fast ships the diesel looks like remaining a favourite choice for a long time yet. The reasons are very clear. Despite being relatively light for their power output, gas turbines suffer three significant disadvantages. Their specific fuel consumption is as much as double that of a comparable diesel for powers less than about 10 MW. Their power output is markedly sensitive to the temperature of the ambient air drawn in. Finally they need quantities of air three or four times those needed by diesels of similar power, so that designers of small ships find themselves constrained by large deck openings.

AIR INDEPENDENT PROPULSION (AIP)

Diesel electric submarines suffer from limited underwater endurance, particularly at high speed. While they operate their diesels when snorkelling they are vulnerable to detection at such times. The problem was overcome by applying nuclear power to produce the 'true submarine' but this solution is expensive.

It also presents problems in the disposal of nuclear waste products and eventually the boats themselves.

In the maritime field the need for a cheaper way to obtain long underwater endurance led to research into AIP systems such as the fuel cell. In other fields, for example the automotive, environmental concerns drove research for powering cars and other forms of land transport.

In this section we consider AIP systems, other than nuclear, for ship propulsion, the emphasis being on submarine applications. Fuel and an oxidant are required to generate electricity by either a heat engine (e.g. closed cycle diesel, Stirling engine, closed cycle gas turbine), or an electro-chemical cell (e.g. lead acid battery, fuel cells). Different fuels can be used. There are obvious attractions in using diesel fuel but methanol and hydrogen are also contenders. Diesel is most easily stored; methanol requires a more complex stowage and a reformer is needed to extract the hydrogen in pure form; hydrogen, for a given amount of oxygen, generates about a third more energy than diesel but it introduces safety problems if stored in liquid form; metal hydrides are compact and a safer means of storing the hydrogen.

It would be impractical to carry all the air required for normal combustion within the submarine even using cryogenics. Instead oxygen is carried in liquid form under pressure at low temperature. Including the tanks needed to provide the extra buoyancy to compensate for a heavier system, a typical AIP system requires about 15 times the displacement of the corresponding diesel oil storage of a conventional submarine.

The choice of AIP system will depend upon the power and endurance, that is the size and mission profile of the submarine. There is no single best buy. The designer must consider the impact on the total submarine system. Contenders are:

1. fuel cells with metal hydride or reformed methanol for hydrogen storage. Chemical energy is converted directly into electrical energy. Efficiencies are about 50 per cent. The Proton Exchange Membrane Fuel Cell (PEMFC) is felt to have good potential and a production model of a 300 kW PEMFC plant has been produced for a German submarine design. This PEMFC system consists of a stack of single cells separated by a bipolar plate. Between each plate is a proton exchange membrane coated with a platinum-based electro- catalyst;
2. the closed cycle diesel. The engine runs on a synthetic atmosphere in which the exhaust gases are treated by absorbing carbon dioxide and adding oxygen, and then recycled to the engine inlet. Efficiency is about 30 per cent;
3. the Stirling engine which converts fuel into heat and then into mechanical work. Efficiency is similar to the closed cycle diesel;
4. a steam turbine fed with steam generated by burning fuel with oxygen.

Efficiency figures refer to the plant itself. The high efficiency of the fuel cell arises because it converts the hydrogen and oxygen directly into electricity. When that plant is integrated with other design features to produce the total submarine, the differences in the overall effectiveness of the different systems is very much reduced.

Fuel cells are also being considered for surface ships to provide ship service power to reduce on-board fuel consumption and meet future, increasingly strict, emission standards. Other advantages are that they have no moving parts, are reliable and easy to maintain, and may be friendly to the environment.

ELECTRICAL GENERATION

Depending upon the type of main propulsion machinery, one or more types of prime mover will be used for electrical generation, e.g. in a steam ship, steam turbo-alternators are fitted with back-up plant powered by diesels or gas turbines. The number of plants depends upon the total capacity required. This capacity for a ship might at first sight be thought to be the 'total connected load', i.e. the sum of all electrical demands. It is not so simple as this because, first, there are several conditions in which the ship requires different electrical equipment working and, secondly, in any one such condition there is a diversity of equipment in use at any moment. The diversity factor is largely a matter of experience. Conditions for which electrical demands are calculated vary with types of ship but may include the following:

(*a*) normal cruising, summer and winter (in a cargo ship, 40–50 per cent of the normal cruising load is due to machinery auxiliaries);
(*b*) harbour, loading or unloading;
(*c*) action, all weapons in use;
(*d*) salvage, ship damaged and auxiliaries working to save the ship;
(*e*) growth during the ship's life (typically 20 per cent is allowed in merchant ships and warships),

There has been a rapid growth in the generator capacity of ships since 1939— a frigate design, for example, which might have needed 1 MW in 1950, needed 3 MW by 1960 and 50 MW by 2000, including 40 MW for electrical main propulsion on two shafts. Generating capacity must be provided to meet suitably any of these loads. Machines should be loaded in these conditions near maximum efficiency. Other considerations enter into the problem too—the needs of maintenance, availability of steam in harbour, break down, growth during the ship's life, capacity when damaged.

From all of these considerations, the numbers, sizes and types of generators are decided. The greatest flexibility would be provided by a large number of small capacity machines, but this is not the most efficient way. Too many generators could be very heavy and involve complex control systems. In addition to the main generators usually a salvage generator is sited remote from the primary generators and above the likely damaged waterline if possible.

Where the electric motor is part of a propulsion system some special problems arise. Switching such large power demands causes surges which need control equipment so that other parts of the ship are not deprived. Often they require their own generators and switchgear. They may, however, be incorporated into the hotel load and may even be connected into buffering arrangements using batteries and rectifier units. When it has been fully developed, superconducting electrical machinery may become attractive although it must be remembered that such a system must include also gas compressors and refrigeration machinery.

Systems

Apart from the electrical systems, over fifty different systems may be found in a major warship for conveying fluids of various sorts around the ship. Even simple ships may have a dozen systems for ventilation, fire fighting, drainage, sewage, domestic fresh water, fuel oil, compressed air, lubricating oil, etc. Because their proper blending into the ship affects very many spaces, it is desirable that the naval architect should have complete control of them, except those local systems forming part of a machinery or weapons installation, even if certain of them are provided by a subcontractor. The naval architect must be completely familiar with the design of all such fluid systems and be capable of performing the design.

ELECTRICAL DISTRIBUTION SYSTEM

The generation of electrical power has already been discussed briefly. Classification Societies allow considerable flexibility in the method of distribution of this power, but the basic design aims are maximum reliability, continuity of supply, ease of operation and maintenance and adaptability to load variation. All this must be achieved with minimum weight, size and cost. The actual system adopted, depends very much upon the powers involved. Prior to 1939, installations were small, e.g. 70 kW in a typical cargo ship. Few passenger liners had as much as 2 MW. Most installations were d.c. and it was not until after 1950 that shipowners began to require a change to a.c. generation and distribution. This change was influenced very much by the savings in weight and the reduction in maintenance effort accruing which became more important as powers increased. Typical figures are:

10,000 tonf dwt. dry cargo ship	1 MW
Tankers	1.5–5 MW
Container ship (3.3 kV)	8 MW

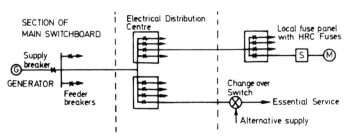

Fig. 14.7 Typical distribution system

All major warship installations in the Royal Navy are a.c. Distribution is achieved by feeding from a small number of breakers grouped on the switchboard associated with each generator to electrical distribution centres throughout the ship.

Distribution from the EDC is by moulded case circuit breakers of 250 and 100 amp capacity. Finally power is supplied to small circuits of less than 30 A by high rupturing capacity fuses. Because of their relatively high starting currents, motors in excess of $4\frac{1}{2}$ kW are supplied through MCBs. Important services such as steering are provided with alternative independent supplies by well-separated cable routes from two generators feeding a change-over switch. In case of damage, a system of emergency cables is provided for rigging through the ship to connect important services to generators which are still running.

Passenger ships must have independent emergency lighting. Typical voltages of distribution are:

Merchant ships: d.c. 220 V (power and lighting)
110 V for some small ships
a.c. 440 V at 60 Hz or 380 V at 50 Hz
3.3 kV at 50 or 60 Hz generation in a few ships
115 V or 230 V at 60 Hz (lighting)

Warships: a.c. 440 V at 60 Hz 3-phase
115 V at 60 Hz for lighting and domestic single phase circuits.

Most warships within NATO and some merchant ships adopt the insulated neutral earth system in order to preserve continuity of supply under fault conditions. Neutral earthing however does permit the economy of single-pole switching and fusing. Many shipowners have preferred a single solid bus bar system with all connected generators operating in parallel. This system gives maximum flexibility with minimum operating staff. The maximum installed capacities are limited by the circuit breaker designs, e.g. assuming breakers of 100 kA interrupting capacity, the system is limited to 1800 kW at 240 V d.c. and 3000 kW at 440 V a.c. Other disadvantages are that a fault at the main switchboard may cause total loss of power, and maintenance can be carried out only when the ship is shut down. The alternative is the split bus bar system which gives greater security of supply and enables maintenance to be carried out by closing down one bus bar section.

Alternating current is converted to direct current by transformer rectifier units or to a.c. of a different frequency by static frequency converters. Both of these distort the sinusoidal waveforms so that spurious signals may be created within the electrical equipment throughout the ship. Design of the electrical distribution system therefore involves complicated assessments of the electromagnetic compatibilities.

PIPING SYSTEMS

Design of any liquid piping system begins by plotting on ship plans the demands for the fluid and by joining the demand points by an economical

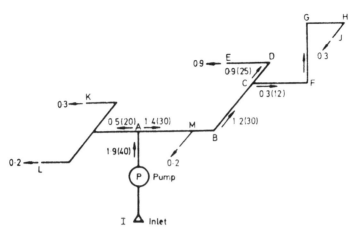

Fig. 14.8 *Typical simple open ended network*

piping layout compatible with the architecture of the ship. (A discussion of the type of piping network to be adopted occurs later but it needs to be chosen at this stage.) Pipe sizes are then allocated on a trial basis by permitting velocities generally about 1.5–2 m/s at which level experience has shown that erosion and noise are not excessive. It may later be found desirable to allow some stretches to run at speeds up to 3 m/s. From this network, must be estimated the pressure

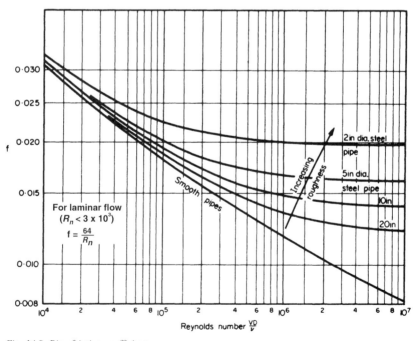

Fig. 14.9 *Pipe friction coefficient*

required from the pump to create this flow. Figure 14.8 shows in perspective a very simple open-ended network that might result. By working back from the presumed simultaneous demands, the flows in the different parts of the system can be found by simple addition and pipe sizes allocated. The pump must then be capable of drawing water from I and delivering it to the remotest point against (*a*) the resistance of the system and (*b*) against gravity. Resistance of the system is due to friction in the pipes and resistance (or losses) due to bends, junctions, valves, expansions, contractions, filters and nozzles. Fortunately, all of these fittings losses can be expressed in similar form

$$\text{fitting loss} = K_\rho \frac{V^2}{2}$$

where K is a factor defined in Table 14.2, ρ is the mass density of the fluid, V is the velocity of flow.

Table 14.2
Losses in fittings

Fitting	$K = \dfrac{\text{loss}}{\rho V^2 / 2g}$	Equiv. length $\dfrac{L}{D} = \dfrac{K}{f}$
Gate valve	0.2 for $D = 25\,\text{mm}$	11
	0.1 for $D = 300\,\text{mm}$	9
90° Angle valve or 60° oblique valve	3.0 for $D \geq 50\,\text{mm}$	190
	4.5 for $D = 12\,\text{mm}$	215
Globe valve	6.0 for $D \geq 50\,\text{mm}$	375
	9.0 for $D = 12\,\text{mm}$	430
Oblique valve 45°	2.5 for $D \geq 50\,\text{mm}$	160
	3.5 for $D = 12\,\text{mm}$	150
Diaphragm valve	1.5	70
Plug or straight through cock	0.4	20
Sudden contraction	On outlet velocity	
	0.4 for $A_1/A_2 = 10$	—
	0 for $A_1/A_2 = 1$	
Sudden expansion	On inlet velocity	
	0.8 for $A_1/A_2 = 0.1$	—
	0.2 for $A_1/A_2 = 0.5$	
Inlet, smoothed entry	0	0
90° bend, radius R	0.3 for $R > 2D$	15
Outlet	1.0	50
Equal tee,		
flow past	0.3	15
flow round	1.2	60
flow from branch	1.8	90
Elbow, 90°	0.6	30
45°	0.1	12
Strainer	0.8	40

Note: Interpolate linearly. Do not extrapolate

Now, the frictional loss in a circular pipe is

$$\text{frictional loss} = f\frac{L}{D}\rho\frac{V^2}{2}$$

where f is a factor dependent on Reynolds' number and pipe roughness as given in Fig. 14.9; L is pipe length, and D is pipe bore.

If a pipe is assumed smooth, the fitting loss can conveniently be expressed in the same form as the frictional loss by calling it a number of diameters of equivalent length of pipe—a gate valve, for example (Table 14.2), is equivalent to the frictional loss due to about eleven diameters of pipe length. Ship pipes are relatively smooth with the possible exception of small diameter steel pipes and this is a common and convenient artifice; while there are clearly approximations involved, it is accurate enough for most ship systems.

The total head, therefore, required of the pump is

$$H = \sum f\frac{L}{D}\rho\frac{V^2}{2} + \sum (K_1 + K_2 + \cdots)\rho\frac{V^2}{2} + \rho h$$

where h is the vertical separation of inlet and outlet. What is important to realize is that if the pump delivers along the most resistful path, all other paths will be satisfactory. This worst path, called the index path or circuit may not always be obvious from inspection and several paths will have to be examined to find the most resistful one. Having determined this, other paths may have to be made equally resistful by the insertion of obstructions such as orifice plates to avoid too high a delivery and pressure. Let us illustrate these points by an example.

EXAMPLE 1. Calculate the performance required of the salt water pump for the system shown in Fig. 14.8 which meets the demands shown in litres per second. Estimated lengths and fittings, not shown in Fig. 14.8, are given in the table below. The inlet to the pump, I, is at a pressure of 50 kN/m^2 and the vertical heights separating I and E and I and J are respectively 15 m and 20 m. Delivery is required at a pressure of 150 kN/m^2.

	Length, m	90° Elbows	45° Oblique valves
IA	10	0	0
AM	45	2	2
MB	6	1	1
BC	12	4	1
CD	20	2	2
DE	12	0	1
CF	3	1	0
FG	6	1	1
GH	6	2	0
HJ	6	0	1

Solution: It is not obvious from inspection whether IAE or IAJ is the index path, although IAK or L are clearly not. It will be assumed that the pipes are smooth. Tabular form is most convenient for this calculation to find $\Sigma\,[f(L/D) + K_1 + K_2 + \cdots]V^2$ which must then be multiplied by $\rho/2$. Take v to be $1.05 \times 10^{-6}\,\mathrm{m^2/s}$.

Even without gravity head, the index path is clearly IAJ. For this path the total losses are

$$\frac{1}{2} \times 632.2\,\frac{\mathrm{m^2}}{\mathrm{s^2}} \times 1025\,\frac{\mathrm{kg}}{\mathrm{m^3}} = 324{,}000\,\mathrm{N/m^2}$$

The pump must also deliver against gravity head which is

outlet pressure + gravity head − inlet pressure

$$150{,}000\,\mathrm{N/m^2} + 20\,\mathrm{m} \times 1025\,\frac{\mathrm{kg}}{\mathrm{m^3}} \times 9.807\,\frac{\mathrm{m}}{\mathrm{s^2}} - 50{,}000\,\mathrm{N/m^2} = 301\,\mathrm{kN/m^2}$$

Therefore, total delivery required by the pump = 1.9 litres per second at a pressure differential of $625\,\mathrm{kN/m^2}$.

This example illustrates the principles of pipe system design. For a given fluid, it is possible to devise charts showing frictional loss plotted against pipe velocity, diameter and quantity which speed up the calculation. The example

	Length (m)	L/D	Equiv. fitt. L/D	Total L/D	V (m/s)	$R_n = \dfrac{VD}{v}$	f	$f\dfrac{L}{D}V^2$
IA	10	250	} 60	310	1.5			14.3
Tee								
AM	45	1500	370 } 15	1885	2.0	5.7×10^4	0.0205	154.6
Branch								
MB	6	200	185	385	1.7 }	4.9×10^4	0.021	23.4
BC	12	400	275	675	1.7 }			41.0
Total (i)								233.3
Tee			15 }					
CD	20	800	370 }	1185	1.8 }	4.3×10^4	0.0215	82.5
DE	12	480	155 }	685	1.8 }			47.7
Outlet			50 }					
Total (ii)								130.2
Tee			60					
CF	3 }		30					
FG	6 }	1750	180 }	2280	2.7	3.1×10^4	0.024	398.9
GH	6 }		60					
HJ	6 }		150					
Outlet			50					
Total (iii)								398.9
(i) + (ii)								363.5
(i) + (iii)								632.2

shows also the relatively large effects of valve losses and the effects of pinching
the pipe diameter. If the final 21 m of piping, for example, were to be of 18 mm
diameter instead of 12 mm, the pressure required of the pump would be reduced
by 138 kN/m^2 and this would be an obvious next step in designing this particular system (when IAL would become the critical path).

Several factors affect the choice of the type of system for a particular purpose.
By the nature of the demand, some systems must be closed and, once the system
is primed, gravity does not influence pump characteristics (except for impeller
cavitation), so that the pump delivers solely against system resistance—a hot
water heating system is typical. Whether open ended, like firemain and domestic
systems, or closed, the designer must consider how important is system reliability. If a simple distribution, as in the example, is adopted, pump failure will
cause a cessation of supply which may be acceptable in some systems. Such a
system is called a tree system. More often, a standby supply will be needed and
this is achieved by cross connecting two adjacent tree systems, the whole ship
being served by several tree systems each of which normally operates independently. (Fig. 14.10.) Emergency supply by one pump to two cross connected tree
systems will, of course, reduce the pressure at the demand points and, unless
properly designed, may result in totally inadequate supply. This case must
therefore be the subject of calculation during the design stage.

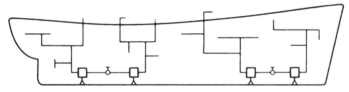

Fig. 14.10 Cross connected tree systems

Systems of importance where no interruption of supply can be tolerated are
designed as ring mains with a number of pumps running in parallel. This type
of system permits pumps to be rested or to break down or pipes to be damaged
with a high chance of maintaining supply to equipment. Chilled water systems
in warships, where deprivation of cooling water to some equipment for even a
few seconds would damage performance, are typical. There are many different
index paths in this case depending upon which pumps are operating or whether
cross connections are opened to isolate a damaged section; moreover, the
positions of null points where there is no flow in the system are not obvious.
There will be a number of trial flow patterns to be tried and here again is a
fruitful area for computer programming. A typical large ship chilled water ring
main system is shown simplified in Fig. 14.11. There are many variations
possible to this scheme and different safety and emergency devices can be
incorporated; for example, automatic pressure actuated starting for alternative
pumps. In a complex system, it is necessary to regulate the flow to each facility
concerned by means of constant flow devices in order that the system may be
balanced. Such devices meter the flow to the required amount for a wide range
of pressure differential.

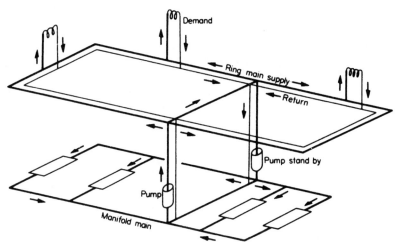

Fig. 14.11 Ring main chilled water system

In this short description of the design of piping systems, it has not been possible to cover all design features, for which separate textbooks are available. The matching of the pump characteristic normally designed by a sub-contractor, to the system demands must be considered with care to ensure stable operation. Positioning of the pump in the ship is also important to avoid a high suction demand on the pump. System priming must be considered. The need for a reservoir or gravity tank to even out the fluctuations in demand is also important. Filtering and flushing of the system must be considered and bleed valves at local high points provided to remove trapped air.

AIR CONDITIONING AND VENTILATION

It is commonly believed that ventilation in a ship is required to enable people to breathe. In fact, an atmosphere fit for breathing can be achieved on very small quantities of fresh air—a small fraction of that needed for ventilation. The major purposes of ventilation are:

(*a*) to remove heat generated in the ship;
(*b*) to supply oxygen for supporting burning;
(*c*) to remove odours.

For most compartments outside machinery spaces, the need to remove heat predominates. Looked at from this point of view, it is clear why normal ventilation often fails to provide comfort. Air drawn in from outside must leave hotter than it entered; a hot, muggy day outside will produce hotter, muggier conditions inside. Heat created within a compartment will be collected by the ventilation air which will be exhausted, hotter, to the atmosphere—one of the two natural sinks for heat available to a ship.

Air conditioning uses the other major natural heat sink, the sea. Heat produced within the ship is ultimately exhausted to the sea and an air con-

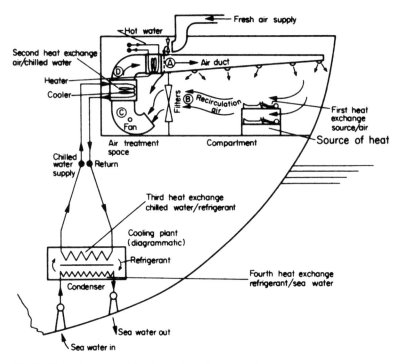

Fig. 14.12 Typical air conditioning system, diagrammatic

ditioning system must efficiently permit this transfer, as will presently be described (Fig. 14.12). While ventilation does remove heat and may create complete comfort for machinery, it cannot effect proper control of the three principal factors which affect human comfort, namely:

(*a*) air temperature;
(*b*) air humidity;
(*c*) air purity.

It is these three factors which air conditioning controls, to varying degree.

A typical air conditioning system is illustrated diagrammatically in Fig. 14.12.

The source of heat in the compartment first loses some heat to the air blown over it; this air is passed through a fan, together with a modicum of fresh air and over the surfaces of a coil heat exchanger rendered cold by chilled water, before returning to the compartment. The chilled water carries this heat to refrigeration machinery which conveys it to the sea in its condenser. Heat produced by the source must therefore be efficiently transferred from one medium to another three or four times until it reaches the sea. There are three principal components of the process; in reverse order they are:

(*a*) Refrigeration machinery for cooling and calorifiers for heating. Design of this is a specialized task which can conveniently be isolated from the design of the whole system, provided that the tasks required of it are adequately defined by the system designer (i.e. pressure–quantity relationship, temperature ranges, etc.).

(b) Water systems, chilled water and hot water. These are designed in the manner already described previously for piping systems. There may well be requirements for dual chilled water supply for important equipment and devices providing a constant flow are often essential.

(c) Air system. This system, involving two heat exchanges, is the crux of air conditioning and is the subject of the calculations described presently. In so doing, it is necessary to assume on the part of the reader, a familiarity with some elementary physics; some definitions of air measurement are given in Chapter 9. Basically, the air system is a recirculatory one with a small quantity of fresh air make up, sufficient only to keep bacteria levels and odours down. Physiologists recommend between 0.15 and 0.30 m³/min, although less will limit bacteria.

All ambient air contains water. The amount carried can be measured by two thermometers, one of which is kept wet and the relationships between wet and dry bulb readings have been related by a chart known as the *psychrometric chart*. This chart relates wet and dry bulb temperatures, latent and total heats, percentage relative humidity and specific volume of air in any condition. Such a chart is shown in Fig. 14.13.

What condition of air is comfortable to personnel? Such is the accommodating nature of the human body that there is a good deal of latitude, but experiments have shown that the areas shaded in Fig. 14.13 for summer and winter are the most suitable. Any air conditioning system should therefore aim to produce communal compartment conditions in the middle of this shaded area and to enable individual cabin occupants to select conditions over such a range.

A further measure of human comfort is provided by the *effective temperature* (ET) *scale* (it is not a temperature), which is a variable ratio of wet/dry temperatures and also air velocity, found experimentally to accord a feeling of comfort. Above 25 ET discomfort increases and it is this limit, known also as the *threshold of comfort*, to which warship systems are designed to operate in extreme tropical ambients. This limit, although the extreme design condition, is infrequently met in warships which operate, like merchant ships, within the comfort zones.

The first step in the process of air conditioning design is the determination of the sources of heat. A given compartment gains heat by conduction through deck and bulkheads, from electrical equipment, hot pipes, lighting, the ventilation fan itself and from personnel. Each of these is a source of sensible heat and

Table 14.3
Personnel heat (Watts per person)

People in:	Sensible heat	Latent heat
Normal mess	45	135
Recreational space	45	163

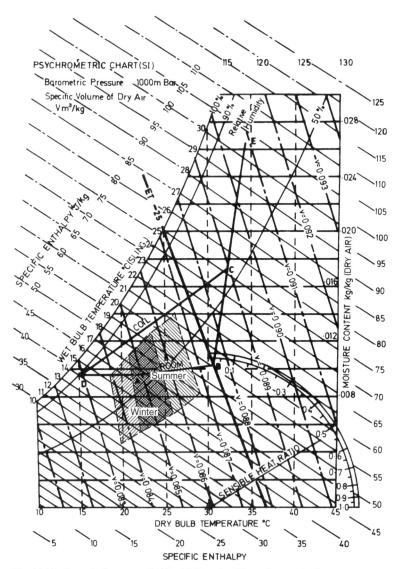

Fig. 14.13 Air cycle for system in Fig. 14.12, using SI psychrometric chart

personnel, in addition, are a source of latent heat of evaporation. Heat given off by personnel depends on how heavily they are working; average figures are given in Table 14.3.

Gains through boundaries follow the law

Heat gain $= U \times$ surface area $\times$ temperature difference

U is the rate of heat transferred per unit area per degree and is derived from the thermal conductivity constant k for the material of the boundary. Values for some materials are given in Table 14.4.

To obtain U, the insulating effect of the still air next to the boundary, known as *surface film resistance* R, must be added in. For air outside the ship R is about 0.039 in metric units, for air inside manned compartments it is 0.110 and for a closed air space 0.18 and for water is 0.0009, for steam 0.0001 and for Freon 0.0009.

Table 14.4
Thermal conductivity constants

Material	k(W/mK)
Standard glass fibre marine board	0.043
Cork slab	0.042
Block asbestos	0.055
Plywood, fireproofed	0.152
Teak, across grain	0.139
Cork linoleum	0.072
Aluminium	120
Steel	46
Glass (in shadow)	1.01

$$U = \frac{1}{R + \sum(x/k)}$$

where x is the insulation thickness. Furthermore, due to imperfections in fitting, the k values for insulants given in Table 14.4 are, from experience, doubled (except steel). Thus for a surface of 25 mm steel + 25 mm glass fibre + 6 mm plywood inside manned compartments.

$$U = \frac{1}{0.110 + 0.5(0.0011 + 0.5814 + 0.0395)} = 2.375 \text{ W/m}^2\text{K}$$

Temperature difference is normally that of compartment temperatures but where surfaces are exposed to the sun, as in the tropics, they are assumed to acquire temperatures of 60 °C if horizontal and 49 °C if vertical. Machinery space deckheads are assumed to be at 82 °C and galleys and auxiliary machinery rooms at 49 °C.

The whole heat gain calculation is performed on standard forms which are derived

(a) total heat

(b) sensible heat ratio $= \dfrac{\text{sensible heat}}{\text{total heat}}$

A typical heat load calculation might appear as in Table 14.5. t_0, t_i and t_d are temperatures outside, inside and the difference.

Having performed the heat gain calculation, it is necessary to return to the psychrometric chart. The cycle of operation of the air in the compartment is: heat and moisture gained by recirculation air which is returned, augmented by

Table 14.5
Heat gain calculation

Source	Dimensions (m)	Type	t_0 °C	t_i °C	t_d K	Area A, m²	U W/m²K	Heat gain UAK (W)
Port bulkhead	10 × 3 25 mm	MFMB	49	29	20	30	2.38	1428
Stbd. bulkhead	10 × 3 25 mm	MFMB	34	29	5	30	2.38	357
After bulkhead	8 × 3 25 mm	MFMB	29	29	0	24	2.49	0
Fwd. bulkhead	8 × 3 30 mm	MFMB	49	29	20	24	2.18	1046
Crown	10 × 8 6 mm	lino	29	29	0	80	6.58	0
Deck	10 × 8 50 mm	MFMB	49	29	20	80	1.45	2320
								5151
Body heat, sensible	28 men × 45							1260
Lights	2000 W							2000
Equipment	5.85 kW							5850
Space heaters	2 kW							2000
Fan	1.6 h.p. × 745.7							1193
TOTAL, SENSIBLE								17,454
Body heat, latent	28 men × 135							3780
TOTAL HEAT								21,234

$$\text{SENSIBLE HEAT RATIO} = \frac{17,454}{21,234} = 0.822$$

some fresh air, to the cooling coil; at the coil it is cooled below its dew point to remove heat and moisture and, with possibly some after warming, returned to the compartment. This cycle must be constructed on the psychrometric chart as illustrated in Fig. 14.13 for the positions illustrated in Fig. 14.12.

After leaving the after warmer at A, the air picks up sensible and latent heat along AB, constructed to a slope representing the sensible heat ratio obtained from the heat gain calculation and of length given by the total heat gain. At B, the air leaves the room near the edge of the comfort zone along the 25 ET line. BC is obtained by proportioning BE in the ratio of recirculation/fresh air, E representing the condition of the intake air. (This, will generally have been pre-treated at the point where it enters the ship in order to avoid excessively hot or cold trunks between that point and the compartment.) At C, the air mixture enters the cooler. CD represents the extraction by the coil to give a point D horizontally from A since the heat added by the after warm DA is all sensible. D needs to be compatible with the available chilled water temperature and, for an efficient cooler, on the 90 per cent relative humidity line. CD then represents the total heat and sensible heat ratio of the coil (or coil slope) which permits a suitable coil to be selected from a standard range. There is a certain amount of trial and error about this process, but it is not difficult to reach a satisfactory cycle with B in a comfort zone or at least below the 25 ET line. Often, this can be achieved without after warm. The quantity of air recirculated is now obtained from the sensible heat pick up in the room over the dry bulb temperatures represented by A and B. If this, as in the heat gain example, is 9 °C, with the specific heat of air 1009 J/kgK,

$$\text{recirculation air returned at B} = \frac{17,454}{9 \times 1009}$$
$$= 1.92 \text{ kg of dry air per second}$$

This must now be proportioned up with the fresh air to give the total air to be handled by fan and cooler. Alternatively, this may be obtained directly from the sensible heat and dry bulb readings of C and D.

This process is fundamental to good air conditioning design and represents the nub of system calculation. There are, however, many other calculations which the space here available does not permit to be described in full:

(a) Heater sizing. It is unlikely that the capacity of the heater will be determined by after warm necessary in the tropical cycle. Calculations similar to those for heat gain are performed for heat loss during the winter cycle, often on the same form. Because humidity is not involved the calculation is straight-forward.

(b) Trunk sizing calculation. Frictional losses in air ducting are similar in form to those in piping, although there is a greater variety of fittings. Pressures are an order lower and expressed normally in millibars. Velocity of air flow in the tortuous ducting expected in congested spaces is about 10 m/second. Higher speeds produce unacceptable noise except in long straight runs where speeds as high as 30 m/second permitting much smaller trunking, can be adopted. Such speeds are not uncommon in the space available in passenger liners where twin ducts, one cool and one hot, are also sometimes fitted, so permitting each passenger to adjust the temperature of the cabin.

(c) Air quantity required for the heating cycle is determined usually by the cooling cycle which is the more demanding. If not, it is found in the same way as for normal ventilation, viz.:

$$\text{Air quantity} = \frac{\text{Total sensible heat rate} \times \text{specific volume of air}}{\text{Temperature rise} \times \text{specific heat of air}}$$

If the sensible heat gain H is in Watts and the temperature rise is 3°C the air quantity Q in m^3/s is given by

$$Q = \frac{H \times 0.88}{3 \times 1009} = \frac{H}{3500} \text{ approximately}$$

Finally, the purity of the air in an air conditioning system is controlled to a degree dependent on its application. Filters are fitted in most systems to trap dust, fluff and soot. Public rooms are often fitted with tobacco smoke filters.

In warships, some filters may also have to be fitted to remove radioactive and other dangerous particles from incoming fresh air. In a submarine, submerged for long periods, the purity of the air must be controlled more closely by removing carbon dioxide and hydrogen and by generating oxygen.

FUEL SYSTEMS

There are several reasons why the type of fuelling system needs to be decided early in the design. Principally, the weight and disposition of fuel have an important effect on stability and trim of the ship and the demands on space need early consideration. Being low in the ship, a large quantity of fuel has a stabilizing effect. As is explained later, special consideration must be given to stability when fuel is used up. Delivery of a satisfactory quantity of fuel oil from service tank to boiler or engine, while important, is a simple matter, effected by a simple local system. An important influence on the ship design, the whole system is considerably more extensive in order to achieve some or all of the following:

(*a*) to store sufficient fuel to enable the ship to achieve its required range in rough weather;
(*b*) to accept high pressure re-fuelling without spillage or structural damage to the ship;
(*c*) to provide to the service tank fuel of sufficient quality;
(*d*) to ballast empty fuel tanks with sea water and to discharge such water overboard without polluting coastal waters;
(*e*) to accept fuel at positions consistent with the supply ship supply points.

Stowage of oil is usually effected in double bottoms and in deep tanks right forward and right aft in cargo ships. An adequate margin of quantity—often 20 per cent—should be provided over that needed for the planned route before refuelling in port or at sea, to allow for rough weather and emergencies.

Warships alongside oilers provide easy targets and need therefore to accept fuel at high speed and pressure. Also, it is desirable that escorts should spend a minimum of time off the screen. Bunkering may be a critical part of the turn round time for a merchant ship. When replenishing at sea through a 15 cm hose pressures may be as high as 1 MPa and rates 400 to 600 tonnef per hour. A typical system is shown in Fig. 14.14.

To avoid an excessively high beam-to-draught ratio, and yet retain adequate transverse stability, it is becoming common in warships to adopt sea water replacement of fuel. Submarines employ sea water displacement of diesel oil. Oil tankers may ballast their cargo tanks with sea water for the return journey to the oil ports to ensure seaworthiness. This use of sea water in fuel tanks creates two basic problems:

(*a*) fuel must be provided to the service tank in an uncontaminated state;
(*b*) oily ballast must not be discharged in many areas of the world unless containing less than 100 parts per million of oil. This requirement by IMO to reduce beach pollution, has been ratified by most seafaring countries and incorporated into their maritime laws. In the UK it is the Merchant Shipping (Dangerous goods and marine pollutants) Regulations.

Oil and water separate naturally for the most part and (*a*) is achieved by providing deep settling tanks which permit this to occur in time. The tanks are not completely emptied to avoid using any emulsified mixture at the oil/water interface, which is removed by the stripping system when the tank is

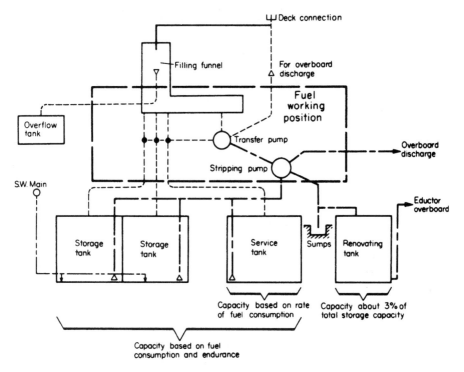

Fig. 14.14 Diagrammatic fuel system

being rested. Similar natural separation occurs in water ballasted fuel or cargo tanks where the oil collects on the surface and clings to the structure. After a period of settling, most of the ballast water can be pumped overboard and the residue pumped by the stripping system to a fuel renovating tank. There it is sprayed by a chemical additive or de-mulsifier which assists separation and the water can again be discharged. Oil centrifuges can be fitted to speed separation.

Delivery from the service tank to engines sensitive to water contamination may be achieved as shown in Fig. 14.15 incorporating centrifuge, prefilter and separator. Also shown is an emergency fuel tank which supplies fuel by

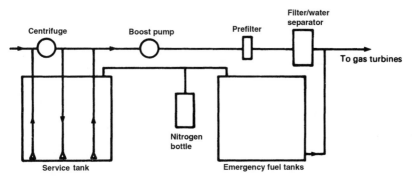

Fig. 14.15 Service and emergency fuel supplies (no valves shown)

pressurizing with nitrogen from a bottle. With such an arrangement gravity plays no part so that emergency and ready-use tanks can be kept low in the ship. Spillage during action damage is consequently less dangerous.

There are four systems associated with fuelling and ballasting, some using common piping. They are fuel filling which delivers oil to tanks fitted with air escapes and sounding tubes; fuel transfer transferring fuel from low in the storage tanks to service tanks; sullage stripping to remove residues, etc.; de-ballasting.

MARINE POLLUTION

After many years of endeavour following some disastrous spills into the seas, IMO has succeeded in producing a complex set of regulations for ships known as MARPOL 73/78. These became internationally accepted in 1983. They now deeply affect the design and construction of ships and their use to control pollution of the seas. Discharge of oily water, sewage, waste and chemical dumping are now all closely controlled and a world exchange of information via satellites about breaches of the codes of behaviour is possible.

It is essential to consult the latest regulations concerning what may or may not be permitted. At the time of writing the broad limitations were:

(a) Raw sewage cannot be discharged at less than 12 NM (nautical miles) from the nearest land.
(b) Macerated and disinfected sewage cannot be discharged at less than 4 NM.
(c) At less than 4 NM and in harbour discharge is only permitted from approved sewage treatment plants.
(d) For discharge of sewage from holding tanks a ship must be moving at 4 knots at least and be further than 12 NM from the nearest land.
(e) No plastic may be dumped at sea.
(f) Dunnage must not be dumped less than 25 NM from the nearest land.

Consequences upon both ships and ports are considerable. Crude oil washing of the heavy oil deposits in bulk carrier oil tanks and separate exclusive water ballast tanks are now common. Steam cleaning of oil tanks is slowly being discontinued. Levels of pollution of all effluents from ships are required to be very low and, in some cases are barely attainable. At the ports, new facilities are being built up to receive different effluents and deal with them more easily than is possible in a ship. Thus, ships are needing to be made more capacious to incorporate segregated ballast and hold tanks for discharge in port.

Sewage can be dealt with in several ways. Heat processes can be used to produce a dry flammable product which is then burnt. Chemical treatment tends to leave a hard deposit which is difficult to remove and dispose of. Bacteria are most often used to break down the solid materials but the bacteria must be fed and special steps have to be taken if the throughput of the system falls below about a quarter of the designed capacity. The problem is further exacerbated by the big variation in loading throughout a 24 hour period. Systems utilize a vacuum collection system or rely upon gravity. Whichever system is used safeguards must be provided to prevent unpleasant, and potentially lethal, fumes penetrating to other parts of the ship. Flap valves in a vacuum sewage system have been known to stick open and allow hydrogen sulphide to flow back into living spaces with fatal consequences. To meet demands for discharge close to land the final effluent must

(*a*) have a suspended solids content not more than 50 mg/l above that of the flushing water;
(*b*) not discolour the surrounding water;
(*c*) have a coliform bacteria count not exceeding 250 per 100 ml;
(*d*) have a five-day biochemical oxygen demand of not more than 50 mg/l;
(*e*) be disinfected, but with residual chlorine level not greater than 10 mg/l.

The safest way to deal with sewage is to hold it until the ship enters a port with satisfactory reception facilities. Such hold tanks are not small even if much of the liquid is first removed in settling tanks and they do need to be of adequate dimensions and in sensibly accessible places. Sewage mains in the ship, if kept under a slight suction head, do not need to rely so much on gravity and are not so sensitive to offensive leakage. Wastes from bathrooms, laundries and scuppers tend now to be kept separate from sewage because overboard discharge of washing water remains acceptable.

In a warship the average daily arisings of garbage are 0.9 kg/person/day food waste and 1.4 kg/person/day other garbage. This is dealt with by a selection of

(*a*) incinerators: modern equipment can burn all types of garbage, both wet and dry, including plastics and waste oil; the residual ash is sterile.
(*b*) pulpers: mainly food waste but can handle paper and cardboard.
(*c*) shredders: primarily for cans, bottles, wood packing, thick card and miscellaneous metal items up to 20 litre drums.
(*d*) compactors: capable of compacting all waste products into securely sealed leak-proof cardboard containers.

CATHODIC PROTECTION

Electrochemical corrosion occurs when two dissimilar metals are present in an electrolytic medium. Sea water is an efficient electrolyte. Different parts of the

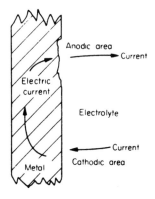

Fig. 14.16 Electrolytic corrosion

same metal made dissimilar, by treatment, or a metal and its oxide are sufficiently dissimilar to create such corrosion as shown in Fig. 14.16. An anodic
area, such as iron oxide, is eaten away creating more rust, while an electric
current is created, leaving the metal at the anodic area and entering it at the
cathodic area, where no corrosion occurs. Painting, if perfect, increases electrical resistance and retards the process but at any imperfection in the paint,
deep pitting may be caused by concentrating the electrolytic effects.

Average values of the electrical potential for a number of different metals
against the same standard in sea water at 25°C are given in the electro-chemical
scale of Table 14.6. Where the difference between two potentials exceeds about
0.25 volts, appreciable corrosion of the metal with the higher potential (the
anode) will occur, if the junction is moist.

Table 14.6
Electro-chemical table

Material	Potential, volts
Magnesium alloy sheet	−1.58
Zinc base die casting	−1.09
Galvanized iron	−1.06
Aluminium alloy (14% Zn) casting	−0.91
Aluminium alloy (5% Mg)	−0.82
Cadmium plating	−0.78
Aluminium alloy extrusion	−0.72
Mild steel	−0.70
Cast iron, grey	−0.70
Duralumin (Al/Cu) alloy	−0.60
Chromium plating on mild steel	−0.53
Brass	−0.30
Stainless steel, austenitic	−0.25
Copper	−0.25
Gunmetal	−0.24
Aluminium bronze	−0.23
Phosphor bronze	−0.22
Millscale (Fe_3O_4)	−0.18
Monel (nickel alloy)	−0.16
Nickel plating	−0.14
Silver plating	−0.01
Platinum	+0.20
Graphite	+0.30

Another metal, higher up the electro-chemical scale, placed nearby in the
electrolyte, transforms the whole of the first metal of Fig. 14.16 into a cathode if
the effect is sufficient to overwhelm the local action and no corrosion occurs
there (Fig. 14.17). All corrosion occurs at the new metal which is the sacrificial
anode. Such protection is commonly applied to small static objects such as
buoys, and individual piles and materials used for the anodes include very pure
zinc, magnesium or aluminium. The basic principle is to swamp the local
corrosion currents by imposing an opposing current from an external source.

A more effective system is that using an impressed current (Fig. 14.18). The
potential of all areas of metal must be depressed to a value more negative than

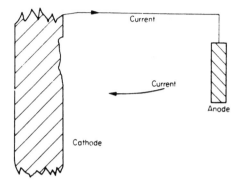

Fig. 14.17 Sacrificial anode

any naturally anodic area. This can be measured against a standard reference electrode in the sea water. The current densities (mA/m^2) required are about 32 for painted steel, 110 for unpainted steel, 150 for non-ferrous metal and 540 for propellers. They vary with factors such as ship speed, condition of paintwork, salinity and temperature of sea water. The system can be used for ships building or laid up, as well as those in service. It can be used in large liquid cargo tanks. Automatic control units are needed to give this adjustment relative to a half cell datum fixed to the hull. A suitable permanent anode is provided by a plate of platinum-covered titanium fixed to an area of the hull covered with an epoxy resin insulant designed to spread the protective effects. Sacrificial impressed current anodes of trailing aluminium wire are not uncommon.

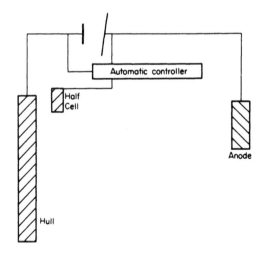

Fig. 14.18 Impressed current cathodic protection

There are several reasons why a decision needs to be made on whether cathodic protection is to be adopted before the design of the ship commences. Space and power demands need to be known but primarily, corrosion margins have to be decided. Classification societies permit a reduction of about 10 per cent in section structural modulus of ships protected by an approved process of cathodic protection. Certainly a substantial reduction in corrosion margin is justified.

As well as the external surfaces of the hull and the internal surfaces of liquid cargo tanks, cathodic protection is applicable to piping and machinery systems.

Some adverse effects of cathodic protection have been reported. Alkaline by-products of the sacrificial anodes can, over a long period, degrade timbers. Similar action can occur through any stray electrical currents. Prevention of this problem can be effected by protection of the timbers by impregnation and by judicious positioning of the minimum number of zincs.

Equipment

Layout of the weatherdeck is an important aspect of ship design and needs to be performed early. Equipment that needs to be on the weatherdeck requires some preliminary examination because interferences are inevitable. Space and geometrical demands will occur for moving equipment, arcs of fire and angles of view, free movement of personnel and for satisfactory wind flow to assure safe landing of aircraft. There will also be direct conflict for the best siting of ladderways, boats and davits, cranes, replenishment at sea gear, hatch coamings, anchors and cables, capstans, ventilation inlets and outlets, fairleads and hawser routes and, of course, aerials, armament, masts, funnels and superstructure. As always in ship design, compromise among the competing demands has to be achieved through understanding discussion with the seamen who will run the ship. The design of the bridge itself is now subject to considerable study by ergonomists as well as experienced pilots and seamen.

CARGO HANDLING

Not so long ago all cargo was loaded and unloaded by union purchase and large gangs of dockworkers manually organizing matters. Union purchase is the use of two derricks in concert to hoist cargo and transfer it simultaneously. It was an expensive and time consuming process keeping ships tied up alongside for long periods when they might have been at sea transporting their goods.

To effect greater economy in cargo handling, it was first arranged methodically in packages, in pallets or in containers of standard sizes which could be split up ashore and consigned to road or rail transportation without delaying the ship. By achieving a higher utilization of the ship and fewer handling costs, the whole transport system efficiency was improved. Container terminals had to be specially built and great gantry cranes were installed to transfer containers rapidly and efficiently from ship to shore. Container ships which ply between smaller ports not so equipped must rely upon dockside cranes or their own derricks and cranes to load and unload containers or their palletized cargo.

When trade is slack, overhead costs at container ports continue; cranes must be maintained and handling personnel paid to be on standby. On the other hand, when the terminal is full, queueing for facilities occurs. This has caused some owners to reconsider their total reliance upon container terminals and to install gantries and cranes on each of their ships. There is a general trend towards such self-unloaders for many types of cargo. Slim centreline cranes or gantries spanning the whole ship with cantilever extensions plumbing the dockside are now common on container ships.

The ultimate integration of sea and land transportation is, perhaps, the Roll-on Roll-off ferry whereby lorries, often carrying standard containers, are driven on and off ships with large garage decks which mate with ramps at their terminals. These are discussed further in Chapter 16.

Bulk carriers have also benefited from developments in rapid cargo handling. Many cargoes are suitable for movement within the hold by gravity through gates on to conveyor belts to a point where the cargo is raised by bucket elevator or rotating helices to deck level and then to a horizontal conveyor belt, usually contained within a large tube to discharge points overboard. There are many variations. Residues in holds may still have to be cleared by bull-dozers or grabs but the process can be very fast.

What is clear is that the processes of handling cargo must be considered at the earliest stages of the design. It is intimately related to the type of trade, to ports of call, to company policy and to the economics of trade.

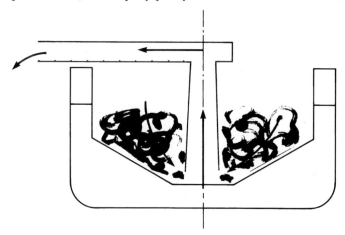

Fig. 14.19 Overboard discharge of bulk cargo

REPLENISHMENT OF PROVISIONS

For a passenger ship or warship carrying 2000 people, the problem of trans-porting twenty tonnes of victuals a day is large. Ideally, the designer will group storerooms and cold and cool rooms in a block, served by a lift or conveyor to the preparing spaces and to the weather deck, close to the point where, in port, the provisions are brought on board. Additional mechanical handling equip-ment may be provided for horizontal transfer of provisions.

The rig for replenishment at sea comprises, basically, a kingpost or high point on each ship between which is stretched a jackstay or highwire which is

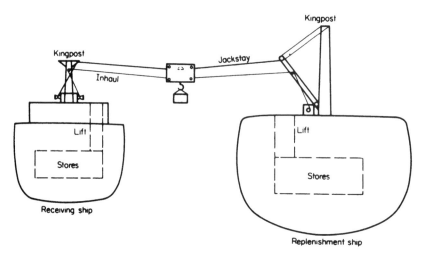

Fig. 14.20 Rig for replenishment at sea

kept at a constant tension by a self rendering winch. A traveller on pulleys is then run along the jackstay pulled either on a closed loop by one ship or by each ship in turn on an inhaul. To land the load on the deck, tension on the jackstay is slackened off. In fair weather a heavy jackstay can transfer 40 loads (each up to 2 tonnef) per hour with the ships 30 m apart. In heavy weather 25 loads with ships 45 m apart are more typical figures. Liquids are transferred by hose. Rates are limited by hose size and the filling systems on the receiving ship. Realistic rates are 450 tonnef/hour through a 5 cm hose or 150 tonnef/hour by 7.5 cm hose.

On the market today there are proprietary replenishment at sea rigs which introduce, at some expense, more automation and control. Jackstay and inhaul are connected at the kingpost to a jockey which is able to slide vertically up and down the kingpost, capturing the incoming loads and placing them with care on the deck or on to a fork-lift truck. These may permit more delicate loads, like guided weapons, to be transferred in higher sea states.

LIFE SAVING APPLIANCES

Cargo ships are required now to carry sufficient totally enclosed motor driven lifeboats for 100 per cent of the crew each side of the ship. Inflatable liferafts sufficient for all of the crew must be carried in addition. If the lifeboats are free-fall they must be fitted at the stern, sufficient for all of the crew and inflatable liferafts for all of the crew must also be carried each side of the ship. In addition, there are requirements for rescue boats and special demands on oil tankers, chemical carriers and gas ships.

Passenger ships on long or short international voyages must carry partially enclosed motor lifeboats for 50 per cent of all personnel on board each side of the ship. Inflatable liferafts for 25 per cent of personnel must be carried in addition. Alternatively, they must carry lifeboats sufficient for 37.5 per cent and liferafts sufficient for 12.5 per cent each side. Inflatable liferafts sufficient for

25 per cent must also be carried. These figures also vary for two-compartment ships on short voyages and for small ships.

An important development for some passenger ships, notably large capacity ferries, has been the fitting of complete marine escape systems. Similar in principle to aircraft escape, they comprise long chutes to sea level where they mate with large liferafts. They are not cheap and, if fitted, must be exercised regularly at some expense and inconvenience.

The demands of evacuation upon the architecture of the ship are substantial and should be considered as a complete system within the overall safety case for the ship. Boats and davits are but one element of an evacuation system that involves consideration of communications, alarms, drills, mustering, free passage for alarmed people, the probable environment external to the ship and within the ship at the time.

Creating a fighting ship

GENERAL

The ship weapon-system must be conceived and developed as an integrated whole. The ship must have a low susceptibility to detection by the enemy and be capable of sustaining some degree of damage whilst remaining a viable fighting unit. Even with modern detection systems a considerable degree of stealth can be conferred on a ship by reducing its radar reflectivity by avoiding flat vertical surfaces and its various signatures: acoustic, magnetic, infra-red and pressure. This makes it harder to detect and classify in the first instance, makes it more difficult for weapons to home in on it and makes its own decoys and sensors more effective.

Helicopters and other aircraft can dramatically increase the area a ship can monitor and over which it can interact with an enemy. There are other ways of extending the Command's knowledge and control of its environment. Bathythermographs can probe the depth to establish thermal layers, important in hunting submarines; sensors can be towed at a distance from the ship, e.g. variable depth sonars and towed arrays; remotely operated vehicles can hunt for mines and destroy them by setting charges near to them.

WEAPONS AND FIGHTING CAPABILITIES

Any weapon system contains four main elements: surveillance; guidance, tracking and illumination; data handling and mechanical handling (Fig. 14.21).

The surveillance system awakens the ship to a potential threat. It may be a long range radar high in the ship, a long range search sonar below the keel or information from an aircraft or other external source. Pride of place must be reserved for this system, the eyes and ears of the ship. Information on the threat is passed to the tracking device which locks on to the echo and provides range and bearing information and guidance to the weapon, the target sometimes being illuminated by electromagnetic waves or pulses to assist the process. The tracking and guidance device may be narrow beam radar or sonar, visual or television sight.

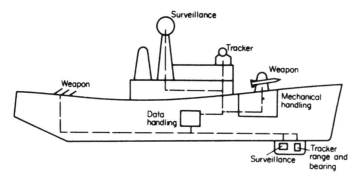

Fig. 14.21 Typical weapon system

The information received from sensors and other sources is processed by computer and passed to the Command and the mechanical handling system, launcher and weapon, the launcher following its directions until the weapon is fired. Once fired the weapon may receive further guidance to direct it to the target. It may ride a radar beam, receive commands from the ship or satellite or use an in-built homing system.

Within the ship the data is transmitted using a common data highway or bus which circulates information packaged electronically by a process called multiplexing. Modern warships can look significantly different to their predecessors. Revolving radars are being replaced with static electronic scanning systems; frequency agility is used to avoid jamming; silo launchers are replacing rotating mounts.

The overall ability and effectiveness of a warship is best considered in terms of its fighting capabilities. These can include detection and classification of enemy targets; destruction of enemy submarines; preventing enemy air-launched missiles hitting own ship; and so on. These must be couched in operational terms by the Naval Staff. Thus it may be necessary to detect enemy aircraft presenting a minimum radar cross-section out to a stated range in certain ambient conditions with a given probability of success.

It may be possible to provide some capabilities in more than one way in which case one or more systems may be fitted. Thus preventing enemy air-launched missiles striking one's own ship can be achieved by destroying the missile in flight (a hard kill solution), by jamming its homing system, by seducing it with a decoy or by taking rapid avoiding action. Clearly the decisions the Command must take are complex and must be made rapidly.

A layered defence is adopted so that some sensor/weapon systems will be long range, for example surveillance radars which may operate out to ranges of several hundred miles. Others will be designed to deal with targets which evade the longer range systems. Finally short range systems are primarily for self defence. The combination of capabilities specified for a given design will depend upon its intended role. A primarily anti-submarine vessel would expect to be able to detect and destroy submarines at some distance from the ship but also have a self defence capability against air attack.

INTEGRATION OF SHIP, SENSORS AND WEAPONS

The designer, working closely with the naval staff and weapon engineers must integrate the various equipments into the design so as to maximize the fighting capabilities in the undamaged state. Maximum areas of view or fire will be provided; surveillance radars will be placed high up; missile launchers well clear of obstructing superstructure (arcs of fire must allow for deviations in missile path, possible aerodynamic interference effects and relative movements due to ship motion); systems may be double headed to enable two targets to be engaged simultaneously; interference between different elements will be minimized, for example the effects of telecommunications on radar reception, of self noise on sonar performance or of electro-magnetic radiation on missile control and firing systems.

To minimize degradation of capabilities after damage the designer will reduce the area of the ship over which a hit can immobilize a system; duplicate important items, locate critical areas deep in the ship or provide local protection.

Whilst much of this is common sense, a methodical approach is needed. For this the designer can produce dependency diagrams showing which physical elements contribute to each capability. Then, within the context of the ship environment the effectiveness of each element and hence the overall capability can be assessed. The effect on level of capability due to certain design changes can then be studied, e.g. modifying upper deck layout to improve arcs of fire, introducing a second sensor to give all round vision, duplicating vital elements to make the system more robust. This produces trade-offs for discussion with the naval staff which will be expressed in terms of overall ship performance in likely operational scenarios and in statistical terms based on the probabilities of the ship meeting certain combinations of enemy forces and weather conditions and with more or less friendly forces in support.

Parts of modern weapon systems are delicate and must be protected from shock and vibration (Chapter 9). Alignment of parts of a system may be critical and the natural elasticity of the ship's structure must be allowed for. If a modular approach is used then modules can be developed, tested and repaired in controlled factory conditions. Usually now each weapon system has its own computer rather than relying on a central ship's computer which can make the ship vulnerable to a single hit. These concepts can also ease the problems of up-dating weapon fits during refit or fitting weapons into ships taken up from trade.

An influence diagram illustrating the influences governing the shipfitting of a weapon system, is shown in Fig. 14.22.

Accommodation

Demands on a warship for crew accommodation are so great that they comprise a major design feature of the ship and the size of the ship is profoundly affected by its complement. Part of the growth in size of warships since the second world war is traceable to:

(*a*) an increase in standards of accommodation;
(*b*) an increase in complements.

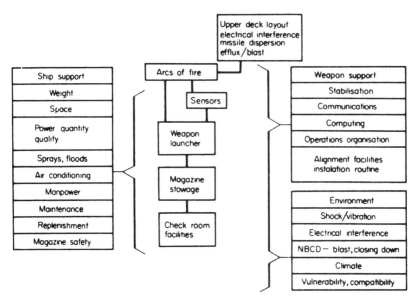

Fig. 14.22 Shipfitting influences on weapon systems

The first results from the social changes ashore, which have also resulted in women forming part of a warship's complement, while the second is due to the greater complexity of ships and a higher maintenance load. Each person requires not only space in a cabin and recreation space but also a small proportion of galley, heads, provision rooms, air conditioning, etc. All these enlarge the ship, requiring bigger machinery, more fuel and so on. Standard weight allowances in warships are fourteen people to the tonnef alone and seven people to the tonnef with their immediate effects, but the overall effect of an addition to complement is much more than this for the reasons stated.

Spaces in a warship coming under the general heading of accommodation are cabins, sea cabins, wardroom, anteroom, messdecks, galleys, serveries, vegetable preparing spaces, refrigerated spaces, aircrew refreshment bar, bakery, pantries, drying rooms, dining halls, sculleries, incinerators, bathrooms, heads, laundries, cinema, sound reproduction room, television studio, chapel, schoolroom, library, recreation spaces, canteen, bookstall, barber's shop, ice cream bar, tobacco kiosk, sick bay, dental clinics and prisons. Not all ships have all these spaces. Layout of the ship to effect a suitable juxtaposition of all of these and other spaces is a considerable task and requires a familiarity with all features of life on board ship. The crew does not like noise or smell, they do not like sleeping where they eat, they are happier if they can see daylight, they like opportunities for privacy as well as communal activities, they need to be reasonably close to adequate bathrooms and toilets, their sleep should be uninterrupted by others changing watch and their food should be well served,

Table 14.7
Initial accommodation space estimates for warships

	Officers	Warrant and Chief Petty Officers	Petty Officers	Junior Rates
Cabins	4.0 to 8.5	4.2 (WO) 2.4 (CPO)	2.1	1.6
Wardroom/anteroom	2.3 to 2.5			
Recreation area		0.8	0.8	0.55
Dining area		0.5	0.5	0.4
Bathrooms, etc.				
Showers per person	1 to 8	1 to 15	1 to 20	1 to 25
Washbasins	1 to 3	1 to 5	1 to 8	1 to 10
WCs	1 to 6	1 to 10	1 to 15	1 to 15
Urinals	1 to 18	1 to 30	1 to 45	1 to 45
Laundry	0.1 to 0.3 ⎫			
Galley	0.1 to 0.4 ⎬ depending on complement			
Canteen	0.04 to 0.07 ⎭			

Notes:
1. All figures in square metres per person. Separate allowance to be made for access passageways, hatches, etc.
2. Senior officers have larger cabins with their own bathrooms.
3. Warrant officers, chief petty officers and petty officers are in single, double and quadruple cabins respectively.
4. Junior rates are in six berth cabins in new construction and in 18–42 person messes (multiples of 6) in older ships.

diverse and dependable. The naval architect has done well if all these are achieved without compromising the ship's fighting ability.

In the early stages of design, space must be allocated to each feature of the ship, in order to estimate the ship's size. As a first guide for warships, the allowances of Table 14.7 are useful. The figures given are nett floor areas inside stiffeners or linings. Gross figures, 20 per cent larger may suffice initially. All space allocation depends much on the shape available, and rough layouts of selected spaces should be made to confirm the values used. In arranging the layout of compartments, dining halls must be arranged adjacent to serveries with a suitable flow of traffic for self service without cross flow or congestion; cold and dry store rooms should be readily accessible to the preparing spaces for daily supplies, lifts being provided where possible.

Because the tasks are quite different, the magnitude of the accommodation problem in a merchant ship (except passenger ships) is much smaller and it is doubtful whether the size of a cargo ship is appreciably affected by its complement. Large aircraft carriers may have complements of 3–4000, a 5000 tonnef guided missile destroyer may carry 500 people while a 200,000 tonnef deadweight oil tanker may carry less than 30. Accommodation standards in merchant ships, as a result, are relatively much higher, minimum standards being enforced by the laws of the country of registry.

For merchant ships, crew's accommodation is generally grouped as follows:

(a) Deck and engineer officers. In single or double cabins. Bathroom with one bath or shower and one washbasin for every six persons. Separate smoke room and dining saloon;

(b) Petty officers. Cabins and washing facilities as for officers. Separate mess-room. Messrooms based on $1\,m^2$ per man;

(c) Engine room hands. Separate sleeping and dining accommodation. Bathrooms as for officers. The ILC recommended minimum floor area per person in sleeping rooms as:
$3.75\,m^2$ in ships 1000–3000 tonnef
$4.25\,m^2$ in ships of 3000–10,000 tonnef
$4.75\,m^2$ in ships 10,000 tonnef or over.
Where two ratings share one room the figures are reduced by $1\,m^2$ per person.

(d) Deck hands. As for engine room hands. Bathrooms as for officers. Crew's smoke room shared with (c).

Clothes washing facilities are required with drying and ironing facilities. The whole accommodation should be sited above the summer loadline, be provided with natural light, ventilation, artificial light and heating. Passengers are to be totally segregated from the crew. The rules are variable depending on size and type of ship and reference should be made to the appropriate legislation.

Measurement

There are two measurements of a merchant ship's earning capacity which are of fundamental importance to its design and operation. These are *deadweight* and *tonnage*. Deadweight is related to the weight of cargo and tonnage is related to the volume of cargo. *Deadmass* is now increasingly used.

The deadweight of a ship is the difference between the load displacement up to the minimum permitted freeboard and the light displacement. The light-weight comprises hull weight and machinery. Deadweight is therefore the weight of all cargo, oil bunkers, fresh and feed water, stores, crew and effects. The weight of the cargo alone is called the *cargo deadweight*. A ready reckoner of deadweight against draught for a particular ship is often supplied to the ship's master in the form of a *deadweight scale*. This shows diagrammatically for salt and fresh water, relative to the load line markings, the draught in metres and feet, the deadweight and displacement in tonnef, the TPC and MCT one cm. It provides a ready means for the master to estimate the change of draughts, or permissible load, when loading in water of density different to that of sea water. The extra permissible draught in fresh water is indicated on the load line markings.

In many ships the data is provided in tabular form or as a computer program.

A coefficient used in the early estimation of dimensions and the study of economics is the ratio of deadweight to deep displacement; this is called the

Table 14.8
Typical values of deadweight coefficient

Ship	Deadweight coefficient $= \dfrac{\text{Deadweight}}{\text{Deep } \Delta}$
50 m, Coaster	0.62
80 m, Shelter deck and one deck	0.70
100 m, Three island and two decks	0.74
150 m, Refrig. shelter deck and two decks	0.58
200 m, Container ship	0.60
250 m, Bulk carrier	0.82
300 m, Oil tanker	0.86

deadweight ratio or *deadweight coefficient* and normally refers to summer load draught. The deadweight coefficient for the type ship will be a guide for the new design; typical values are given in Table 14.8 and in Chapter 15, although, since there is considerable variation in apparently similar ships, these should be treated with caution.

The volume of a ship is expressed in tons of 100 ft^3 (2.83 m^3) and is referred to as its *tonnage*. On its tonnage are based the charges for berthing the ship, docking, passage through canals and locks, and for many other facilities. It is often used as a coarse measure of a ship's size and is also confused, by the layman, with displacement.

Records of measurement of a ship's size in this manner can be traced back in the United Kingdom to the thirteenth century for the carriage of wine. A standard size of barrel, called a 'tun', was decreed in the fifteenth century for this purpose and taxes and harbour dues were based upon a ship's 'tunnage'. Over the centuries, various rules for assessing this tonnage, as it became, were devised, the most influential being the Moorson system of 1853.

Tonnage regulations have not always led to safe design. In 1773, the formula on which tonnage was assessed was $0.515(L - 0.6B)B^2$, where L is the length and B the breadth of the ship. Because this did not include draught, owners required small beam and large draught to reduce tonnage; this, however, created poor stability and many ships were lost as a result. More recently, in order to exempt the 'tween decks and still permit them to carry cargo, 'the shelter deck was 'opened' by the provision of a tonnage hatch in the upper deck and openings in main transverse bulkheads resulting in minimal safety standards; the regulations were, in fact, formed to permit this exemption and were responsible for a huge number of shelter deck ships constructed in this artificial manner which, nevertheless, became very efficient and safe cargo ships later.

By the middle of the twentieth century, many different methods of assessment of tonnage existed; among the important tonnage regulations were the International, British, United States, Suez and Panama. Efforts by the International Maritime Organization (IMO) were directed in the 1960s towards

producing an internationally agreed system of tonnage measurement, prefer-
ably by the application of simple formulae and this proved successful in 1969
when an international conference adopted the formulae. Assessment of Suez
and Panama tonnages were not changed at that time however.

There are two tonnages of primary interest no matter which authority is
measuring or registering the ship. These are *gross tonnage* and *register tonnage*
(*net*). The regulations governing their measurement are complicated but are
summarized for the British Tonnage Regulations below. Other regulations
differ only in detail.

The *tonnage deck* is the upper deck in vessels having one deck and the second
deck in all other cases. Spaces above are 'tween decks and superstructures.

The *underdeck tonnage* is the total volume in tons of $2.83\,\mathrm{m}^3$ of the ship below
the tonnage deck to the inside of frames, underside of deck plating and above
the inner bottom. This is obtained by detailed calculations, in a manner somewhat similar to that for displacement calculations and varies in detail for
individual regulations; tonnage measured according to the tonnage regulations
may, therefore, be slightly different from the actual volume of the spaces below
the tonnage deck calculated for, say, cargo capacity.

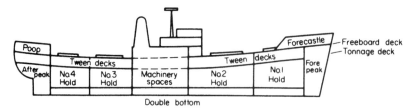

Fig. 14.23 General cargo ship

The *overdeck tonnage* is the volume to the inside of frames and deck plating
of the 'tween decks, poop, bridge, forecastle, deckhouses and erections above
the tonnage deck less the exempted spaces. Spaces *exempted* include dry cargo
space (unless in a break in the deck) and certain closed-in spaces associated with
machinery, safety equipment, navigation, galleys, washrooms, water ballast
and workshops. *The gross tonnage* is the sum of the underdeck and overdeck
tonnages plus volumes of hatchways (less one-half per cent of gross tonnage
computed without hatchways), plus light and air spaces which are included at
the owner's request in the measurement of the machinery spaces.

The register tonnage is the gross tonnage less the following deductions:

(i)   Master's accommodation, crew's accommodation and provision store-
      rooms (limited to a maximum of 15 per cent of the total tonnage of
      Master's and crew's accommodation) but not fresh water.
(ii) Spaces below deck allocated exclusively to steering, navigation, safety
      equipment and sails (for ships propelled exclusively by sail, up to a max-
      imum of $2\frac{1}{2}$ per cent of gross tonnage).
(iii) Spaces below deck allocated, with certain provisos, to workshops, pumps,
      donkey engine and boiler.

(iv) Spaces below deck used exclusively for water ballast up to a maximum of 19 per cent of the gross tonnage, including exempted spaces such as the double bottom and capacity below line of floors containing water ballast, oil fuel, etc.

(v) Spaces below the upper deck occupied by propelling machinery. In ships propelled by screws, if the volume of this space is between 13 and 20 per cent of the gross tonnage, the deduction is 32 per cent of the gross tonnage; if the space occupied is less than 13 per cent of the gross tonnage, the deduction is in direct proportion to 32 per cent; if the volume of the propelling machinery is in excess of 20 per cent of the gross tonnage, the propelling power deduction is $1\frac{3}{4}$ times the tonnage of the propelling space.

Happily, this complicated assessment has now disappeared for new ships. The new formula tonnage regulations came into force in 1982 and are applicable to all new and converted ships and, at the owner's request, to existing ships. The formulae now applied to ascertain the gross and net tonnages are as follows:

Gross tonnage, $GT = K_1 V$

Net tonnage, $NT = K_2 V_c \left(\dfrac{4d}{3D}\right)^2 + K_3 \left(N_1 + \dfrac{N_2}{10}\right)$

where $K_1 = 0.2 + 0.02 \ \log_{10} V$

$\qquad V = $ total volume of all enclosed spaces, m^3

$\qquad K_2 = 0.2 + 0.02 \ \log_{10} V_c$

$\qquad V_c = $ total volume of all cargo spaces, m^3

$\qquad d = $ moulded draught amidships, m

$\qquad D = $ moulded depth amidships, m

$\qquad K_3 = 1.25 \left(1 + \dfrac{GT}{10,000}\right)$

$\qquad N_1 = $ number of passengers (in cabins with not more than 8 berths)

$\qquad N_2 = $ number of passengers not included in N_1

note that: $\left(\dfrac{4d}{3D}\right)^2$ must not be taken as greater than unity

$\qquad K_2 V_c \left(\dfrac{4d}{3D}\right)^2$ must not be taken as greater than $0.25 \, GT$

$\qquad NT$ must not be taken as less than $0.30 \, GT$

$\qquad$ if $N_1 + N_2$ is less than 13, N_1 and N_2 shall be taken to be zero

$\qquad d$ is the assigned summer loadline draught or, for passenger ships, the deepest subdivision loadline draught

$\qquad V$ and V_c are calculated for metal ships to the inside of shell plating and are to include appendages

While these new regulations are considerably simpler than the old ones, they still require a great deal of precise definition of geometry and phrases such as 'enclosed spaces' and 'excluded spaces'. National regulations should be consulted for the definitions.

The measurements of ships described above are statutory, i.e. required by the law of the country of registry. Two measurements not required by law but of use, where pertinent, to the owner and designer are *grain capacity* and *bale capacity*. Grain capacity is the cargo volume in cubic metres out to the bottom and deck plating, excluding space filled by frames and other structure. Bale capacity is the cargo volume to the inside of the frames or sparrings on frames and beams.

Problems

1. Discuss the bases on which machinery may be chosen for a merchant ship. How do the arguments differ for a warship? What types of main propulsion machinery are available? Discuss the main properties of each.
2. Write a short description of the problems of

   (*a*) nuclear main propulsion;
   (*b*) all gas turbine propulsion;
   (*c*) diesel-electric propulsion.

3. What factors affect the size and type of electrical generation in a warship? Describe briefly how electrical distribution may be achieved in merchant ships and warships.
4. Compare the properties of tree systems and ring main systems for fluid distribution in a ship. How are the calculations performed for each?
5. What does air conditioning seek to achieve that normal ventilation does not? Describe the series of heat exchanges in a ship's air conditioning system and how each is efficiently secured.
6. How does the choice of fuel system affect a ship design? Describe a system for rapid fuelling.
7. How do the laws on oil and sewage pollution affect the design of a ship?
8. Describe the principles of cathodic protection and how they are put into effect.
9. What is a container ship? Discuss the various ways of mechanical handling of cargo into and out of general cargo carriers.
10. Describe the various elements of a ship–weapon system and how they interact. What problems face the naval architect in siting these elements and what effects has the ship on the design of the weapon system?
11. A pump is required to deliver 400 tonnes of fresh water an hour at a pressure of $150\,kN/m^2$ along 50 m of straight horizontal steel piping containing two diaphragm valves and a strainer. Estimate the pump delivery pressure required for (*a*) 15 cm diam. and (*b*) 20 cm diam. piping. Ambient temperature is 20°C.
12. A lubricating pump is sited in one side of a square network of 12 mm bore smooth piping circulating $0.25\,m^3$ of oil per minute in closed circuit. Each side of the square is 6 m long and there are two 45 degree oblique valves

and two strainers in the complete circuit. If the loss in the equipment being supplied is negligible, calculate the pressure differential at the pump and the corresponding power required. The kinematic viscosity and specific gravity of the oil are respectively $5.1 \times 10^{-4}\,\mathrm{m^2/s}$ and 0.90. What would be the figures if the piping were increased to 36 mm?

13. The following table represents an open ended main salt water service in a merchant ship. If a delivery at a pressure of 0.55 MPa is required at the remote end of the system and there is positive pressure of 0.10 MPa at the drowned pump suction, estimate the performance required of the pump. If the overall efficiency of the pump is 0.72, what steady power is required of the electrical supply to the pump motor?

Leg	Length (m)	Bore (mm)	Fittings	Delivery (litres/min)
PA	15	102	1 strainer 3 easy bends 1 globe valve	182 at A
AB	18	76	1 90° angle valve 5 easy bends	136 at B
BC	12	51	2 90° angle valves 4 easy bends	182 at C
CD	49	51	1 globe valve 7 easy bends 1 plug cock	546 at D

14. A package protecting a guided weapon is 4.88 m long and 1.22×1.22 m in section and is constructed of 1.6 mm thick aluminium. It is taken from an air conditioned magazine where the dry bulb temperature is 29 °C to the upper deck where it is 43 °C in the shade. What is the rate of heat gain through the box? What would it be with 19 mm glass fibre all round?

15. The refrigerated hold of a cargo ship is 15 m × 15 m in plan and 8 m high above the tank top which is 1 m above the keel. The draught of the ship is 5 m in the Red Sea. On the after side of the hold there is a machinery space whose temperature is 50 °C and on the forward side a hold at 30 °C. Sea temperature is 25 °C, the deck head is at 35 °C and the sides 30 °C on the cold side and 45 °C on the sunny side.

All surfaces are lined with 1 cm of plywood and 15 cm of cork slab. All steelwork is 1 cm thick.

Calculate the capacity required for refrigeration machinery to maintain a temperature of −10 °C in the hold.

16. The package of question 14 is loaded on deck off Singapore in an ambient of 31/25.5 °C before being sealed. How much can the package be cooled before condensation occurs inside? If taken into a magazine at 20 °C, how much moisture will collect in the package? The missile occupies 20 per cent of the space.

17. Recirculation air is required to leave a room at 26.7/20 °C and be mixed with an equal quantity of fresh air at 32/29 °C before being passed to a coil

having a slope of 0.42. The sensible heat ratio of the room is 0.65. If the air leaves the cooler at 90 per cent relative humidity, estimate how much heat per kg of air must be supplied by the after warmer.

18. Heat gain calculations for a messdeck for sixty men show the total heat to be 44,000 Watts of which 40 per cent is latent. Fresh air is drawn in at the rate of $0.005 \, m^3/\text{sec}$ per man and is at 34/31 °C. Conditions in the room should not exceed 25 ET. Avoiding the need for after warm, construct a suitable psychrometric cycle, stating the performance required of the coil and the air quantity needed. Air should leave the coil at 90 per cent relative humidity.

19. A frigate design is to have a complement of fourteen officers, fifty chief petty officers, fifty-four petty officers and 206 junior ratings. Make a first estimate of the space which would be needed for living spaces, toilet facilities, laundry, galley and sick bay.

20. A warship has a length of 110 m and a beam of 13 m. The only decks suitable for accommodation are No. 2 deck and the forward half of No. 3 deck, each of which is expected to have a waterplane coefficient of 0.70. Estimate what proportion of these spaces should be given over to the accommodation of the previous question, assuming that the laundry is sited on No. 4 deck and that sewage is discharged directly overboard. What would be this proportion if a sewage system enabled all heads and bathrooms to be sited on No. 4 deck?

21. A cargo ship has a moulded depth and assigned freeboard of 20 m and 4 m respectively. Its gross volume is $10^5 \, m^3$ and cargo volume is 78 per cent of this.

    Calculate the formulae gross and net tonnages. What would the net tonnage become (*a*) if freeboard could be reduced to 3 m or (*b*) if 15 cabin passengers were carried?

15 Ship design

Design is a creative iterative process serving a bounded objective.

This definition of design brings out the four essential features of engineering design which serve well as the divisions of this chapter. Objective there must be, even if like Michaelangelo's it is simply 'To please God'. Naval architects require the discipline of a well-defined objective, if their creation is to answer properly the owner's need. Moreover, it has to be bounded; that is to say the limits to which the designer may go need stating. It has become common to address a system of which the ship is but a part; a science known these days as systems engineering. Having defined what it is the designer is to address and the limits within which to work, the creative activity can start. This is normally a circular process, a first shot, corrected and re-created often many times until it satisfies the objective. It is iterative, as will later be clear.

Students must not be dismayed that they are yet denied the chance to apply their hard-won skills to this satisfying aspect of their profession. Such skills must be applied in a constructive and disciplined manner if they are not to be wasted. Students will find that a disciplined framework will not only enrich their natural creative ability but provide an opportunity for the computer to assist them. Data bases, full of successful history can provide them with a firm start to the process of converging upon the best solutions.

It is first necessary to be clear about the role of the naval architect throughout the design process, because it changes with time. The earliest stages are generally a debate with the owner, proposing various ways in which the owner's wishes could be fulfilled, matching the operations envisaged to the investment that would be necessary to perform them. In the case of maritime trading this would be suggesting many different possible transportation systems and assessing their profitability and chance of success. In the case of military operations it would consist of proposing many different ways of achieving offensive or defensive operations and the cost and effectiveness of each solution. These early studies comprise concept design or conceptual design, sometimes called pre-feasibility. They consist of a lively debate among all those who have a contribution to make, designers, operators, economists and many others all directing their creative thoughts towards an objective declared by the owner who may be a trader or a military commander representing a government. The result of this debate is in outline a few promising ideas which lead into a phase of development usually known as feasibility study.

This second phase of the work is directed more clearly at the engineering and the management. Its aim is to identify the whole system and to quantify its profitability, its material elements and the risks attendant upon its development.

It attempts to establish the viability of the best of the promising concepts. At the completion of this phase the owner chooses the way in which to proceed. This is an important decision because it commits significant funds in the succeeding phase.

That phase is the full design worked out in every necessary detail so that material may be ordered and construction may begin. The design team is not necessarily the same as the one which completed the first two phases and the nature of the work has changed significantly. It is now directed towards the definition of the ship—and other parts of the transportation system—for contract and production. Indeed at the latter end of the full design phase the means of supporting the ship during its life are produced by the designers.

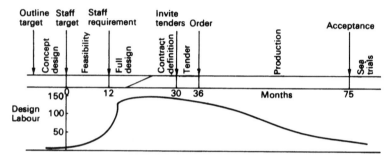

Fig. 15.1 Phases of warship design

Phases of design of a ship are shown in Fig. 15.1. While they may appear under different names (in the USA, concept and feasibility are reversed) their character is the same throughout the world both for merchant ships and warships. The figure also shows the times likely for each phase of a major warship design (excluding the time taken by government approval to proceed) and the design staff required for a frigate. For a destroyer the effort required to define a building contract properly now amounts to about 50,000 man-days compared with 5000 in 1960, some of this dramatic increase being due to complexity.

Objectives

It is not sufficient to say that a ship must be designed to meet the requirements of the owner until the earlier phases of design have defined what they should best be. The fundamental aim of an owner or potential owner of a merchant ship is, of course, to make a profit on the investment. For example, a tramp steamer may be ordered to ply for trade along a chosen route, picking up and delivering goods that other people want moved. It may be necessary to borrow money to buy the ship, pay taxes, pay the crew, meet bunkering costs and to arrange the flow of cash to maximize profit. There will be constraints of regulation, social pressures and competition affecting the choices so that there will be a limit to profit which a budget will reveal. In more complex cases, investors will wish to consider the total distribution of goods in a fleet of ships, the means of handling, the choice of taxes and regulations to which they wish to be subject, the pattern of trade and their cash flows. The transportation system to be designed in that case needs to address an objective of optimum profitability in the predicted trading conditions and environment over many years. Characteristics of the ship and all other parts of the transportation system emerge from the economics of the venture which now requires deeper study.

ECONOMICS

As a first step a potential owner decides on the sort of trading in which to indulge. From market research comes an estimate of the amount of goods that would have to be transported annually on the chosen routes and how quickly they would have to be delivered. This annual transport capacity might be met by a few large ships or many small ships and it is necessary to carry out various economic examinations to decide not only the best size of ship and its speed but the profitability over a period of time. The most important element of this calculation is the freight rate, which is the rate payable in the free markets of the world for the transport of units of specific goods. It is the subject of some competition, it fluctuates in accordance with market pressures and rates are published regularly. Sometimes such rates are determined by conferences; although the Conference System does smack of price fixing, it does give a degree of stability to trading.

Freight rate determines the flow of cash into a company. There may also be inward cash flow through Government subsidies and loans. Outward cash flow is caused by:

(a) down payment and instalments on building costs;
(b) loan repayments and interest;
(c) running costs for bunkering, crew, port charges, etc.;
(d) maintenance and repair;
(e) profit;
(f) corporation and other taxes.

Note that depreciation is not a cash flow although it may have to be calculated for tax purposes.

Now it is necessary to compare on the same basis the profitability of the various possible schemes of trading. This basis is the compound interest that would be earned by an investment P_0 at an interest rate r after a period of years i:

$$P_i = P_0(1 + r)^i$$

Inverting this, the present sum P_0 which would produce P_i in i years' time is

$$P_0 = P_i(1 + r)^{-i}$$

Over a period of n years, the annual cash flow A_i in each year i has a net present value

$$\text{NPV} = \sum_{i=0}^{i=n} A_i(1 + r)^{-i}$$

636 Basic ship theory

This is the basis of comparison; the higher the positive net present value, the better the investment. Arguments concerning the correct discount rate r are difficult but are based upon either the assessment of alternative opportunities when it is called the opportunity cost rate or upon a personal preference for speedy profit when it is known as the time preference rate.

There are several important variations upon net present value preferred by some economists. Yield or Required Rate of Return (RRR) is that discount rate r which gives a zero NPV. Required Freight Rate (RFR) or Shadow Price is the minimum cargo rate which the shipowner has to charge the customer just to break even.

EXAMPLE 1. A ship is estimated to have a capital cost of £1M. A useful life of 20 years is predicted with a scrap value of £50,000. It is expected that the ship will earn £0.5M for each full year's operations but that due to special survey requirements this will be reduced by £60,000 and £100,000 in the 12th and 16th years. The running costs in each year, after allowance for tax, are estimated to be as in column 3 of the table below. Calculate the net present value assuming a discount rate of 7 per cent.

Solution

Year (i)	Cash flows after tax			Discount factor $(1 + 0.07)^{-i}$	Discounted cash flows $A_i(1 + 0.07)^{-i}$
	(+)ve	(−)ve	Nett $= A_i$		
1	500,000	300,000	200,000	0.93458	187,000
2	500,000	50,000	450,000	0.87344	393,000
3	500,000	305,000	195,000	0.81630	159,000
4	500,000	310,000	190,000	0.76290	145,000
5	500,000	320,000	180,000	0.71299	128,000
6	500,000	315,000	185,000	0.66634	123,000
7	500,000	320,000	180,000	0.62275	112,000
8	500,000	325,000	175,000	0.58201	102,000
9	500,000	330,000	170,000	0.54393	92,000
10	500,000	320,000	180,000	0.50835	92,000
11	500,000	330,000	170,000	0.47509	81,000
12	440,000	335,000	105,000	0.44401	47,000
13	500,000	315,000	185,000	0.41496	77,000
14	500,000	330,000	170,000	0.38782	66,000
15	500,000	335,000	165,000	0.36245	60,000
16	400,000	350,000	50,000	0.33874	17,000
17	500,000	320,000	180,000	0.31657	57,000
18	500,000	345,000	155,000	0.29586	46,000
19	500,000	360,000	140,000	0.27651	39,000
20	550,000	370,000	130,000	0.25842	34,000

Total £2,057,000
Less initial cost £1,000,000

NPV £1,057,000

Note: Variations in negative cash flows reflect costs of surveys, varying earnings and allowances on capital cost. This latter leads to the very low figure for year 2, it being assumed that tax allowances suffer an effective delay of 1 year.

While this has demonstrated the basic elements of economic assessment, there is rather more to it. At the end of these economic exercises the owner will know the cargo carrying capacity required of the ship and probably the speed. This is the start required by the naval architect and marine engineer in their design activities. It will also indicate characteristics of other parts of the system such as the cargo handling and support facilities. They constitute subsidiary elements of the overall objective to which we shall return presently in discussing boundaries.

It is clear that the effective assessment of the economic aspects of ship design depend vitally upon a good prediction of procurement costs. Technical cost estimating draws upon the history of similar activities and procurement of equipments in the elements which have been found conducive to extrapolation into the future. Labour, materials and overheads are the three basic elements, subdivided into shipyard and manufacturing industry. Shipyard costs are subdivided into steelwork, outfitting, pipework, cabling and many hundreds of other elements, each of which is found to be governed by different variables, e.g. steelwork costs by weight, main cabling by power, paint by $L(B + D)$. These algorithms by which technical cost estimating is performed are usually kept covert but occasionally an interesting paper on this subject appears.

COST EFFECTIVENESS

The basic objective of a warship designer is to provide an advantage to a government in military action against a potential enemy. There are three elements to the value or effectiveness of a warship—or other military artefacts:

(*a*) capability;
(*b*) availability;
(*c*) military worth.

Capability is a measure of the offensive ability of the ship. There are very many parts to such an ability, e.g. speed, detection, range of sensors, accuracy of delivery of a missile, crew efficiency, signature suppression, reaction time to a threat. Moreover, the measures will vary with the environment in which the ship is working, e.g. the level of electromagnetic countermeasures, the deployment of an enemy, the climate and geography, all put together into various scenarios of operation. The assessment is the task of the operational analyst, who attempts to produce, for each scenario and for each postulated military mission, the probability that the ship will succeed. The measure of capability can therefore be expressed as a matrix of probabilities based upon many different assumptions. When the moment for action arrives, the necessary devices must of course be ready to use. This second element to the effectiveness of a warship, availability, can also be quantified in terms of probability as is discussed in the next section. Availability depends on the intrinsic likelihood that a device will work when called upon to do so, called the reliability; and the likelihood that, should it break down, it can be repaired in an acceptable timescale, called maintainability. The measures again depend upon the mission and the environment, so that availability can be constructed as a matrix of probabilities.

The third element, military worth, is an assessment of the military advantage over a potential enemy which the possession of the warship confers. Not only does this depend upon the postulated scenarios and the specific missions but upon the intelligence of an enemy's future abilities. While theoretically also a matrix, quantification is exceedingly difficult and often a matter of judgement.

Thus system effectiveness is a conjunction of three matrices, often very large and complicated.

$$\text{S.E.} = \mathbf{C\ A\ W}$$

The primary constraint upon maximizing effectiveness is cost. A useful measure to which warship designers may apply optimization techniques is therefore

$$\text{Cost effectiveness} = \frac{\text{S.E.}}{\text{Cost}} = \frac{\mathbf{C\ A\ W}}{\text{F}}$$

In short this represents an expression of value for money. This must not be regarded as a simple fraction even though it might, in exceedingly simple cases, be possible to quantify. It is nevertheless a useful discipline to remind designers of limitations to their desire to increase effectiveness. Incidentally, cost should not be discounted unless value or military worth is also discounted because the value of an artefact which is not available for use because there is no money to repair it is zero. Value, or military worth, is also likely to degrade with time as a potential enemy's technology advances.

This measure is applicable also to mercantile operations even though availability is often very high. In that case, capability is a measure of transport capacity while military worth is replaced by utility. Utility is then a measure of the correctness of the predicted assessment of market forces or even the gamble that the owner is prepared to accept. Only later will it be known whether decisions had been correct! Utility theory is an important aspect of the study of economics.

It is clear that each element of these objectives may be traded off against another. Endurance of a warship, for example, may be increased at the expense of weapons payload; reliability of a weapon system may be increased by redundant equipments but at a higher cost; the chance of encountering a defeating situation may be low enough to accept so that a weapon system may be omitted to the benefit of other aspects of the ship; increased vulnerability may be acceptable to reduce initial costs. These trade-offs or compromises are many and complex and occupy a great deal of effort in the early design stages. The requirements for a ship consequently have to evolve until the process of design development of the ship itself is worthwhile.

Much the same occurs with merchant ships where the owner's requirements evolve steadily with the trade-off exercises. There the trade-offs would be related to such issues as choice of flag, variability of terminal facilities, bunkering positions, insurance rates.

Boundaries

The bald statements of objectives so far considered are inadequate. A designer does not have a free hand to change the world and must be constrained within boundaries that have to be defined by the owner or the government at the outset. Some of the boundaries and constraints will be quickly settled, others will depend on profitability. We will consider them very briefly under three headings:

economic, ethical and social
geographical, organizational and industrial
time and system

ECONOMIC, ETHICAL AND SOCIAL BOUNDARIES

Boundaries of the economic system as discussed earlier have a profound effect upon the ship and its supporting system. Flags of convenience and an imposed condition or preference for the carriage of goods only by national flag ships are important considerations which affect tax, construction standards, subsidies and loans. The wisdom of easily obtained credit by linking a shipbuilding industry to a merchant bank has been observed by several nations. Easy credit, i.e. low interest rates and long repayment periods, are often made available for political or social reasons by governments especially in times of recession to keep their industries alive. Lower insurance premiums are sometimes now available for ships designed to a higher degree of safety.

Of course, designers' activities must be bounded by the law, maritime law, health and safety law, consumer protection law and civil damages law. They must adhere to professional standards of conduct prescribed by their peers and in default may expect retribution. Even within the law, however, there remain choices that will be determined by ethical standards for which conduct may be judged by society at large.

The economic boundaries available will be much determined by the client and relationships with the finance houses. Registration in some flags of convenience nations might attract low fees and minimal corporation tax but engender standards which place at risk crew, innocent third parties and the environment through pollution, inadequate survey or corner cutting. Poor quality classification societies may appear financially attractive. The letter of the law might be served rather than its spirit so encouraging rule-cheating. These are ethical issues which deserve deep consideration by owners, insurers and ship designers at all levels. Naval architects will be wise to record carefully the discussions that they have with clients or superiors in their employment and the standards of safety that they have agreed to observe. Enquiry into a subsequent accident may absolve nobody who has made a contribution to the cause and employers cannot indemnify their employees against criminal charges.

An owner's attitude to social questions, conditions on board, care of family, conditions of employment and many others demonstrate company image and affect crew efficiency in both mercantile and military vessels. They must be chosen with deliberation.

GEOGRAPHICAL, ORGANIZATIONAL AND INDUSTRIAL BOUNDARIES

Not all ships are designed for a specific route. If they are, then the constraints upon dimensions imposed by geographical boundaries will be clear. Length, beam and draught constraints are imposed by canal passage and upper works overhang problems apply particularly to the Panama Canal for such ships as aircraft carriers. Bars at harbour entrances, depths of water at berths and in confined waterways, heights of quays, cranes and travellers all may affect the limits to dimensions. RoRo and container ships are especially affected by port facilities while warship length may be constrained by base port docks.

Organizational and industrial boundaries that will require definition may follow from company. policy on its links with other companies and whether it intends to charter, whether it hires crew as it requires or keeps a nucleus in-house. Training and recruitment may be an important factor affecting the design of ships and their supporting organization and may persuade owners to standardize equipment and machinery. Bunkering places and methods, victualling, financing, repairing, agencies and communication policies should all be known to the designer. The most important boundaries however are those concerned with the transportation system and with time.

TIME AND SYSTEM BOUNDARIES

A ship designed to a minimum procurement cost may not appeal to many potential owners. Designed to minimum ownership costs it may be a very different ship. It is thus important to know the period of time over which the profitability of ownership is to be assessed.

Through-life costs over 25 years of a frigate and annual costs of a RoRo cargo liner are shown in pie chart form in Fig. 15.2. It will be seen that a frigate costing £300M to build may cost another £500M to keep for 25 years. This may encourage a designer to invest, for example, in a labour-saving device in order to save one member of the crew for 25 years which will allow the ship to be a shade smaller, saving fuel and onboard facilities. These life cycle cost trade-offs are important once the intended life of the ship is known. For example,

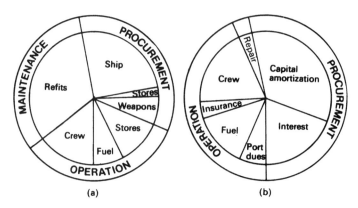

Fig. 15.2 (a) Frigate life costs, (b) RoRo annual costs

adoption of non-corrosive piping systems, fastidious paint preparation, cathodic protection, etc., may save many times the initial investment on maintenance costs. There is no point in thus economizing, however, if the cost of maintenance do not fall upon the buyers of the ship because they intend to sell early. A designer must, therefore, know the boundaries of the owner's intentions and, if relevant, the character of the whole of the support system.

Ownership costs associated with warships exceed the annual expenditure on new construction many-fold. Support of the Fleet is organized on a worldwide basis to provide fuel, food, clothing, stores, ammunition, spares, repair and regular maintenance. There are standardization policies, standards of fuel quality, problems of transporting ammunition and missiles, availability of tools and facilities, coding and documentation and many other factors that must be known to designers if they are to devise a ship which is to be compatible with the existing organization.

Merchant ships face similar but rather less complicated problems. One of their most important interfaces is with the port facilities and it is this interface which has caused a revolution in merchant ship design in the last few years.

Table 15.1
Annual cost of cargo liners UK–Far East

	Conventional	Container	RoRo
Crew	4	2	5
Insurance	4	4	5
Fuel	40	50	43
Port charges	2	4	4
Cargo handling	20	7	6
Capital (ship)	30	25	30
Capital (units)	0	8	7
	100%	100%	100%

Table 15.1 compares the running costs of a conventional cargo liner on a long voyage with similar capacity container and RoRo ships. Among the striking features of this table is the dramatic reduction in cargo handling charges when the cargo is packaged into units. This is an excellent example of extending the boundaries of economic interest beyond the ship, embracing the port handling facilities and, even, the road transport system. It is for this reason that we have been referring to transportation system design instead of simply to ship design. The economy of such systems is significantly different and the ship itself is changed as a result. Craneage on board may not be necessary at all and such is the automated container handling at the container depots that the crew may be minimized and port turn-round time halved.

Creativity

We have now to create a design which will satisfy the owner. Earlier chapters largely dealt with what might be called the *attributes* of a ship—its stability, strength, seaworthiness and so on. Now it is necessary to consider the ship as a whole, possessing these attributes to an agreed standard and able to meet the needs of its owner. As far as that owner is concerned the ship must be capable of doing various things. That is to say it must possess a number of *capabilities* within agreed operational scenarios. Contributing to these capabilities will be a range of systems and sub-systems, some devoted to one capability and others contributing to several. Yet others will support most or all of the capabilities.

To illustrate this:

- In a single-shafted ship the shaft supports the ship's capability to move. In a multi-shaft ship it will also support its ability to manoeuvre.
- A radar system will support an ability to navigate safely. In warships it will also support a number of fighting capabilities such as detecting, tracking and destroying enemy aircraft or missiles.
- An air conditioning system will be necessary to support most of a ship's capabilities.

One way, then, of regarding a ship for design purposes is as a series of systems. In this context the term system is given a broad meaning in order that all elements are contained in one system or another. Thus the main hull of the ship is a system supporting the ability of the ship to float. Most systems will include equipments, piping, cabling and some supporting structure.

For each capability, or sub-capability, a diagram can be produced showing how individual ship elements contribute to that capability. Such diagrams are called *dependency diagrams* (see Fig. 15.18).

Some elements will be in series and others in parallel and the effect of losing any one element (by failure, accident or enemy action) can be assessed. For example:

- The loss of one main propulsion diesel out of a total of four, will mean that only 75% power is available.
- The loss of a propeller in a single shaft ship means the loss of all propulsive power.
- The loss of one steering motor, where two are in parallel, will not affect the immediate performance but leaves the ship vulnerable to the loss of the second motor.

As is discussed elsewhere, dependency diagrams are fundamental to a systematic evaluation of ship availability. In this context they are usually termed *availability diagrams*. They are also used for vulnerability studies when they are called *vulnerability diagrams*.

Iteration in design

Any rational creative endeavour is iterative. It begins with a guess; this is tested against the criteria imposed by the objective, like maximum economic yield; and, found wanting, the guess is modified so that it can be re-tested and so on.

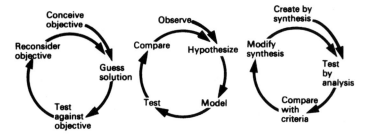

Fig. 15.3

In other circles this is called the scientific method; observe, hypothesize, model, test, compare with observations, modify hypothesis, etc.

So far as a ship is concerned, the initial guess is made easier by experience which may be in two forms:

(*a*) a basis ship which may be manipulated into a new ship;
(*b*) basic parameters derived from large numbers of previous ships.

The criteria against which the newly created ships are tested are derived not only from the prime objective but from the boundaries within which the design is to be derived such as national safety regulations, international law and reliability. Thus there will be standards which must be reached in respect of stability, structural behaviour, pollution control, seakeeping, habitability, manoeuvring, internal environment and hazard protection as well as speed and payload.

Table 15.2
Major design stages

Concept design	Choose prime parameter
	Make first estimate of size, cost, speed
	Re-assess objective
	Re-estimate
	Check other prime parameters
	Repeat as necessary
Feasibility design	Assess payload characteristics
	Develop dimensions
	Assess power requirements
	Define propulsion machinery options
	Sketch first layouts
	Consider vulnerability, safety
	Embark on spiral convergence, repeating
	Establish validity of design
Full design	Establish design management
	Develop each element of the ship, checking behaviour iteratively:
	geometry
	stability and flotation
	architecture
	dynamics
	propulsion
	structure

DESIGN PHASES

The amount of work involved in the design process will depend upon the size and complexity of the ship—a liquefied gas carrier will be more complicated than a simple tanker—and how closely the new ship follows the pattern of a previous, successful design. The division of work between the owner's naval architects and those of the main contractor will vary. This does not change the sum total of the technical work but can impact upon the management of the process.

Typically the naval architect's involvement can be split into seven stages although with different people at different stages. Terminology will vary from organization to organization but the basic ideas are the same.

First stage. Usually called the concept or feasibility stage.

In this the customer's requirements are established together with the criteria for customer acceptance of the total system. The overall system can then be broken down into a number of sub-systems, or functions, which can be further sub-divided to create a description of all systems.

The designer then considers various ways of meeting the needs of each element of the design and hence, the feasibility of meeting the overall requirement. Preferred options can be costed and a broad solution agreed with the customer.

Second stage. The feasibility or functional design.

The systems and sub-systems selected from the first stage are designed and the interfaces established. It is important to define accurately all the interfaces so as to avoid duplication or omissions. This is a time of progressive refinement using a wide range of inputs—design reviews, analyses, specialist advice, modelling, information from equipment suppliers and feed back of experience from sea.

Third stage. The full design.

In this stage the detailed layout of compartments and the sizing and routing of all cabling, trunking and piping is undertaken. Again it is a matter of iteration, compromise and final acceptance. The information obtained in this phase enables a list of all the material needed to build the ship can be defined.

Fourth stage.

During this phase the information—drawings, computer inputs and so on—to support manufacture and assembly is produced. This must be associated with a build plan to ensure that material and information are available in a timely fashion to support the planned building sequence.

Fifth stage. The build stage.

This is the stage that covers the manufacture and assembly of the ship and all its associated systems and equipment.

Sixth stage. Testing and commissioning.

As sections of the ship are completed they must be tested to ensure the requirements have been met, including those of safety. Early tests will indicate the air or water tightness of structure. Later will come tests to show that systems have been installed correctly. Finally the overall performance will be established by trials in the basin or at sea.

Seventh stage. In-service.

The owner will be concerned with support of the ship throughout its life. There will be initial periods which will be covered by warranty. The owner may also make the original contractor responsible for extended periods of operation. This encourages the production of a reliable product.

The first three stages are the true design phases and they are summarized in Table 15.2. Each is iterative in nature.

PRIME PARAMETERS

There are only three prime parameters by which the first estimate of size of a ship may be made. They are volume or mass or linear dimension. One of these will dominate the choice of size of ship to carry the required payload. Volume-limited ships (or capacity carriers) are those which when full are not down to their minimum freeboard. A mass-limited ship is one where there will be unused volume when the mass required is carried. Dimensionally-limited ships are those whose minimum dimensions are determined by the dimensions of their payload. It is useful to anticipate what sort of prime parameter is relevant because it is possible then to converge more quickly upon the solution.

Typical of mass-limited ships are ore carriers and the armoured battleships of old; passenger ships and many light warships are typical of volume-limited ships. Some warships, container ships and river boats are typical of ships which are wholly or partially constrained by particular dimensions; the standard container must fit into the dimensions of container ships while river boats often have a maximum draught limitation. Most modern warships are volume-limited but may be constrained or even dominated by weather deck layout (see Fig. 15.6).

So important are these three prime parameters that we must study each in some detail.

Volume

The volumes given over to various functions in a frigate and in a cargo ship are shown in Fig. 15.4. While there are variations depending mainly on endurance, these values are surprisingly constant for many countries of origin.

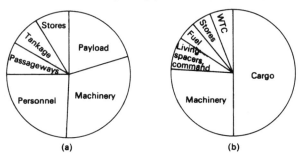

Fig. 15.4 Division by volume: (a) Frigate, (b) Cargo liner

Most destroyers and frigates devote about 20 per cent of their enclosed volume to armament and about 5 per cent to tankage. As a starting point a pseudo block coefficient is often useful; this is defined as

$$C'_B = \frac{V_T}{\text{L.B.D.}} = \frac{\text{totally enclosed volume}}{\text{length o.a.} \times \text{beam W.L.} \times \text{depth}}$$

which for frigates and destroyers is about 0.9. Thus, by knowing the armament to be carried, its estimated volume may be multiplied by 5 to give a first shot at the totally enclosed volume from which, using the ratios later discussed an estimate of dimensions may be derived. Displacement can be derived from the enclosed volume by a 'density' defined as the total enclosed volume divided by the light displacement. Figures for this density vary from $4.5\,\text{m}^3/\text{tonne}$ for rather densely packed designs to $5.5\,\text{m}^3/\text{tonne}$ for the more spacious.

Cargo ships of some 20,000 tonnef deadweight would normally expect to devote at least 50 per cent of their volume to cargo and 25 per cent to machinery. The breakdown of volume can be made quite easily however for a wide range of ships by rules of thumb. Not all of the moulded internal cargo volume is available for cargo because of the shape of the cargo. Percentages available are roughly:

grain 98
bale 88
insulated 72

Of course, most naval architects prefer to work from a basic reliable proven ship design if there is one available so that they can converge more quickly than is possible from the parameter approach. However the new ship is derived, these first crude steps will need refinement and variational analysis as later described.

Mass

Whether or not volume is more dominant than mass, an early estimate of displacement is necessary. Typical values of the deadweight ratio or, more correctly, deadmass ratio are:

	Deadmass/Σ
Passenger liner	0.35
Container ship	0.60
Liquid gas carrier	0.62
General cargo liner	0.67
Ore carrier	0.82
Large tanker	0.86

Historical data may be used to give estimates of each weight group of various merchant ships.

Some typical percentages of the deep displacement for various elements in merchant ships are given in Table 15.3.

Table 15.3
Typical mass group percentages

Group	16 kt Cargo ship L = 150 m	22 kt Passenger ship L = 240 m	16 kt Oil tanker L = 200 m
Net steel	21	36	17
Outfit	5	16.5	1.6
Hull systems	1.5	6	1.6
Propulsion machinery	5	5.9	2.3
Light mass	32.5	64.4	22.5
Crew and passengers	0.2	0.6	1.6
Fuel	11	14	2.6
Fresh water	0.3	13	1.3
Dry cargo	35	8	—
Liquid cargo	21	—	72
Deadmass	67.5	35.6	77.5
Deep displacement	100	100	100
Deadmass ratio	0.675	0.356	0.775

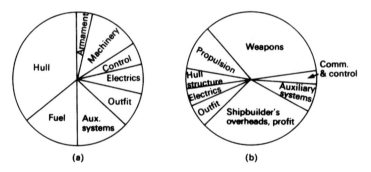

Fig. 15.5 *Division of frigate (a) by mass, (b) by procurement cost*

The percentages of deep displacement mass given over to various functions in a frigate are shown again in pie chart form in Fig. 15.5(*a*). This figure contrasts with Fig. 15.4. Payload, accounting for 20 per cent of the internal volume needs less than 10 per cent of the displacement; structure accounting for 36 per cent of the mass needs little volume. An even greater contrast occurs in the percentages of ship cost. While the figures do vary quite a lot and depend much on the particular accounting conventions adopted, Fig. 15.5(*b*) does give some idea of the relative costs of elements of a frigate. The hull, for example, now demands less than 10 per cent of the cost even with its slice of overheads while the weapons payload, including the helicopter but without ammunition rates very high, as might be expected.

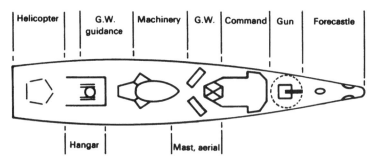

Fig. 15.6 Frigate weatherdeck layout

Linear dimensions

Obvious limitations to linear dimensions occur through such needs as crossing the bar over a harbour or dock entrance, passing under a bridge, docking in a dock of a particular length, operating in shallow waters, berthing at a confined quay, movement down a canal and unloading beneath gantry cranes. Rather less obvious are the minimum dimensions needed for a collection of functions. The weatherdeck layout of a warship is an important example. Figure 15.6 shows the minimum length required for a typical frigate upon which may be imposed working from the stern: the need to land a helicopter, hangar it, fit a missile system with adequate arcs of fire controlled by a radar guidance system with similar arcs, machinery uptakes and downtakes, a second missile system, mast acting as an aerial, bridge, gun and forecastle. It is not infrequently the case that the length of the ship is governed by such considerations resulting in both mass and volume to spare—although not often very much. If there is excessive mass and volume resulting then ingenious ways of reducing length are considered; for example, retractable hangars, hangars between split funnels, guided weapons efflux over the stern. There is however not usually much pressure to reduce length because length is generally beneficial to the ship's hydrodynamic performance.

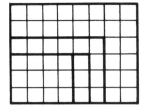

Fig. 15.7 Container stacks

Container ships are similarly constrained by the linear dimensions of their payload. The additions of standard container dimensions do give designers of such ships choices in beam and depth close to the steps shown in Fig. 15.7 imposed by their internal stowages.

PARAMETRIC STUDIES

The stage at which the dimensions and their ratios will be considered depends much upon the nature of the investigation and the requirements of the owners. Whether or not they are considered during concept and pre-feasibility study, it is certain that they must be thought about before the ship is declared to be a viable proposition. There are large numbers of parameters of assistance to naval architects upon which we have already dwelt throughout these books. Faced by them all, a sense of proportion is not easy to acquire. However, there are a few which are very dominant and rather more which are especially helpful in directing the choices before the detailed calculations are embarked upon.

So far, we have derived from economic or military arguments the payload, the appropriate volume and mass of the ship and constraints upon specific linear dimensions. Now, the dimensions must be derived to achieve the best compromise among many conflicting features—speed, endurance, strength, stability, seakeeping, manoeuvrability, cost, production, architecture, protection, habitability, signature suppression, survivability, logistics, reliability, etc. In making such choices, it is necessary to have constantly in mind the effect of changes to the parameters, qualitatively and quantitatively and it is necessary to remind ourselves of some work discussed throughout these chapters.

Length The longer the ship the better, in general, is the longitudinal seakeeping and the smaller the power required for a given speed. Against this, there is an increase in longitudinal structure and there may be more difficulty in achieving high manoeuvrability. Usually, with fuel prices relatively dominant, the most important consideration is the minimization of the propulsive power needed and long narrow ships find favour. There is a limit brought about by the awkwardness of building narrow compartments at the ends and values of (M) much in excess of 8.0 are not very common for that reason. Nor should greater

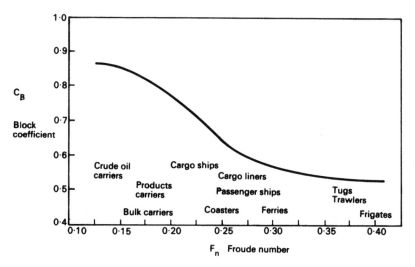

Fig. 15.8 Optimum block coefficient

length be allowed to cause a reduction in beam that would create stability difficulties or a reduction in draught that might cause slamming.

Prismatic and block coefficients These coefficients give an idea of the fineness of the ends relative to the middle part of the ship or to the product BT; for fine forms as with warships, prismatic coefficient is preferred as an aid while bluff merchant ship forms are judged usually on the basis of block coefficient. Prismatic coefficient has an important effect upon residuary resistance and optimum values have already been shown in Chapter 11, Fig. 11.18. Low prismatic coefficients mean fine ends which can give problems in confining within the ship's lines bulky fittings, like diesel generators, sonar arrays or weapon systems. Optimum block coefficients as a function of F_n have been proposed for various types of merchant ships and these are shown in Fig. 15.8.

Dimensional ratios B/T, T/D Beam-to-draught ratio is of major importance to initial transverse stability and natural period of roll. Figures of around $2-2\frac{1}{2}$ are common in weight dominated designs and about $3\frac{1}{2}-4$ are usual for warships and passenger ships. There is a slight increase in resistance as B/T increases. Draught/depth ratio T/D is extremely important to large angle stability since it determines the point of deck edge immersion. It also determines freeboard and is therefore a measure of deck wetness and it indicates the reserve of buoyancy for survivability. Frigates tend to values of T/D around 0.5. Common values of various ratios of large numbers of ships are given in Fig. 15.9 and Table 15.4.

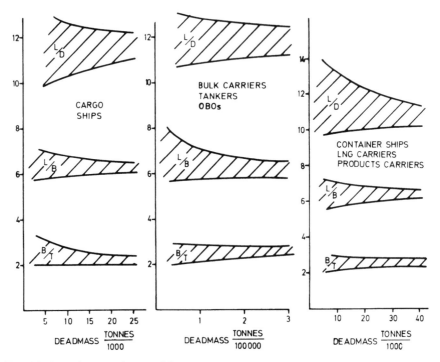

Fig. 15.9 General ranges of principal dimensions

Table 15.4
Typical warship type dimension ratios

Warship type	∇ $10^3\mathrm{m}^3$	$L/\nabla^{1/3}$	L/B	L/D	B/D	B/T	F_∇ $[U_S/\sqrt{(g\cdot\nabla^{1/3})}]$
WW2 battleship	40–60	7	7	14	1.8	2.5–3	0.8
WW2 destroyer	2–3	8–9	10	16	1.8	3–3.5	1.5
Minehunter	0.5	5–6.5	5–6	8	1.4	3.2–4	0.8
Corvette	1–2	7–8	7–8	11	1.5	3.5	1.3
Frigate	3–5	7–8.5	8–9.5	13	1.5	2.8–3.5	1.2
Cruiser	7–10	7–8.5	8–10	12	1.4	2.5–3.2	1.1
A/C Carrier	13–90	6–7.5	6–8	9	1.3	3.3–4.1	0.8

U_S = Ship speed
F_∇ = Froude Displacement Number

Ship form The effect of form on resistance has been discussed at length in Chapters 10 and 11. There are some additional features that have to be borne in mind. Flare of the ship's side at the waterline of some 15 degrees can be very beneficial in keeping $\overline{KM}$ constant with increasing displacement so that adequate stability can be maintained with variable payload. Knuckles higher up the ship's side avoid excessive flare and are often useful in throwing water clear as the ship pitches, so that spray is not whipped up by the wind. Large flat bottom areas, especially in the first 20 per cent of the length could well invite excessive slamming. Vee sections forward are beneficial to seakeeping. High sheer at the forecastle often helps to keep a dry fore-deck.

Waterplane coefficients A high waterplane coefficient C_w has a beneficial effect on seakeeping, although secondary to the effects of length. A fine angle of entrance at the bow gives a good start to the hydrodynamic flow over the ship and can be beneficial to resistance and to noise reduction.

A good relationship for stability purposes between C_w and C_p for small warships is

$$C_w = 0.44 + 0.52C_p.$$

Longitudinal centres and bulbs The effects of LCB position are relatively small on resistance but the addition of a bulb forward can be important to resistance especially at high F_n for cargo ships with a low block coefficient; however bulbs must be considered carefully during tank tests and generalizations are misleading. The centre of lateral resistance below water is important to manoeuvrability and it is useful in the early stages to have some flexibility available in the provision of deadwood aft or forefoot cutaway so that the stability coefficients can be massaged after tank tests.

Propeller While optimum design points for propellers have been discussed in Chapter 11, some more general observations are worth keeping in mind. Slow revving large diameter propellers tend to be favoured wherever possible because they enable higher QPC to be achieved, they are less noisy and they should cause less excitation to the hull. There are limits however. They must be adequately immersed and enjoy satisfactory clearance from the ship's hull and from a dock bottom. More important, there will be a speed of propeller rotation below which an additional reduction gear train will be required, causing a step change in gearbox costs. While there may be some special reason for adopting controllable pitch propellers, ducts or shrouds; vertical axis propellers and other devices, it must be remembered that a well-designed fixed-pitch helical propeller is difficult to better for efficiency.

FEASIBILITY STUDIES

We now come to the very nub of the ship design process. How can it best be represented in the mind of the designer? What sort of order can be imposed upon a process which is so highly interactive? A change to any one parameter will affect many factors and, in turn, require changes to other parameters; it is virtually impossible to change one dimension or parameter without significant effects upon many dependent variables. As we have already discussed, the process is iterative and convergence upon the final solution can only be achieved by going over the elements repetitively until they all match.

There are many ways of representing this convergent process but one of the most evocative is the design spiral (Fig. 15.10). This conveys both the interactive and the iterative nature of the whole ship design. It is presented here because it is at the feasibility stage that most of the factors come into play for the first time. Generation of the need for the ship is represented at the centre of the spiral from economic or military argument. From that, the spiral begins first with the identification of the characteristics of the payload or cargo and the complement. The prime parameter is chosen and the concept phase of the design process may proceed as already described. Some sectors of the diagram may be left for later turns of the spiral.

The order in which factors are considered and, indeed, which ones are omitted is a matter of judgement but broadly follows that shown. The number of circuits made until satisfactory convergence has been achieved will also vary with each ship design but they will be least when the most appropriate prime parameter has been chosen. If a less appropriate prime parameter has been chosen, convergence will still occur but it will take longer. Feasibility study will certainly take more than one circuit and the stage at which the ship may be considered viable is also a matter of judgement. It is often useful to calculate the effects of varying one specific parameter on the parameters and performances represented in other sectors of the spiral, simply to discover how influential it is, e.g. to see how small changes in each dimension affect stability, strength, seakeeping, survivability, cost and propulsion. This variational analysis gives a designer a 'feel' for the design, assisting his judgement as he proceeds.

One especial difficulty in warship design is in the assessment of the complement. Despite many rules, assessment of the necessary complement has a high opinionative element affected by the standards of good ship husbandry, cross training, command philosophy, contract hours of personnel and calls on the crew in emergencies. Moreover, taking as it does 25 per cent of the volume of the

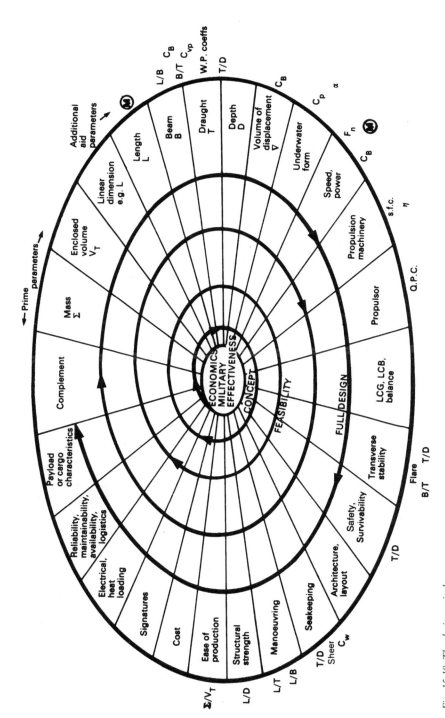

Fig. 15.10 The design spiral

ship, it has a profound effect on ship size and cost. Merchant ships also may find a small crew economical and a bigger investment in automation necessary. Wise designers will therefore set up very early in feasibility, a dialogue with those concerned to establish firmly what accommodation should be provided.

At the full design stage changes will still occur but by this stage the changes will be more in the nature of adjustments and the interactions will be much reduced. At the design production stage the nature of the activity changes and the spiral can be discarded. Design then becomes an important exercise in communication between the detailed creative activity in the shipyard design office and those who order the materials and build the ship.

Figure 15.10 also shows around the periphery some of those aid parameters which have been discussed in the previous section. It is not a complete list and students will wish to augment this *aide-mémoire* for themselves.

The object of the feasibility stage is to ensure that the elements of the ship design form a matching set. In considering such matters as survivability and strength it is necessary also to decide upon the standards to which the ship design is to conform. These will depend upon the scenario of operations prescribed and the constraints, like cost, which are imposed. In warship design, the days when nothing was too good for the fighting man have long passed and the decisions on what standards of habitability, protection and weapons can be afforded are difficult and painful. Thus, at the feasibility stage the basic compromise among all of the conflicting demands upon a ship is struck.

At this stage also items which require development or which involve any technical risk are identified, programmes are drawn up and preparations are made for the major commitment of full design development.

FULL DESIGN

It would be foolhardy to assume that nothing is going to change after the feasibility phase. As details are worked out the need for adjustments becomes apparent and the consequences of quite trivial matters can sometimes be extensive. Furthermore, some of the assumptions implicit in the feasibility study will not develop as expected, especially those with high technical risk, and the owner or the Naval Staff may well have a change of mind as market forces or military threat change. The full design phase retains its dynamic character but is, nevertheless, the stabilizing period for the design.

During this time, the compartments which have hitherto been spaces on a layout are considered individually, piping systems and other services are designed and inserted into the ship, electrical distribution evolves, data highways and the flow of information, people, material and command are all developed. The number of designers required increases by an order. As these matters are developed, conflict is inevitable and adjustments to the general arrangement layout to alleviate the problems are made continually. The more complex the ship, the more difficult and prolonged is the process. Tank tests during this time will have provided the final 10 per cent accuracy on the powering required and will have demonstrated the need for adjustments to underwater form, propeller dimensions and other hydrodynamic matters.

Although most, if not all, of the procedures used in developing the design are now, or could be, computerized, it is instructive to describe some of the manual methods used previously. This can give a clearer understanding of the principles involved. Generally, the computer has merely enabled them to be carried through in more detail, more accurately and in a shorter time scale. That is to say the computer has not changed what the designer is aiming to achieve but, rather, the means of achievement.

It has already been shown that definitions are important and must be precise and comprehensive. There must be no room for ambiguity. Consistency is vital if databases of information are to be used intelligently. Thus weight and cost data must be broken down in a disciplined and well-defined way. These divisions should be consistent with the systems and capabilities used to define the ship. Information on equipments must precisely define their form fit and function.

As the design progresses decisions will be made and, as a result, elements of the proposed ship will emerge. It is important that such decisions, and their results, are recorded in such a way that they can be readily retrieved. Put another way there must be an audit trail which can be followed later to establish what happened, why and who was responsible. Design records or logs can contain the minutes of all important meetings and conference decisions. In the case of warships a *Ship's Cover* was always produced and good ones are invaluable in following the thought processes that led to designs being configured in a particular way. Clearly their value depends very much upon the skill of the compiler.

From the beginning of the design process the naval architect is aware of how one element of the design interacts with another, often requiring a compromise between conflicting desires. This can apply at all levels of detail. Thus:

- a change in upper deck layout which involves moving the funnels will impact upon the positioning of the machinery low in the ship and the uptakes, downtakes and removal routes;
- the level of manoeuvrability demanded may necessitate fitting twin shafts or some form of transverse thruster;
- changing the helicopter to be carried will impact upon the flight deck size and layout, the hangar, support services, the strength of the flight deck and possibly the supporting transverse bulkheads.

One way of illustrating these interactions is the *design influence diagram* (see Fig. 15.13).

Many people are involved in the detailed development of the design. Each must be aware of the decisions taken by others that affect what they themselves are doing. For some of these decisions the use of what may be termed *master general arrangements* is useful. Thus one master can show all doors and hatches, indicating in which direction they open. Another can show the size of stiffeners on decks and bulkheads, and for the latter, which side of the bulkhead they are placed. These will avoid duplication, or the possibility that the designers of the compartments on both sides of a bulkhead will assume that the stiffeners and doors do not take up space within their compartment.

Other masters can be used for the routing of each pipework, cable or trunk system. Trunks can be sized to indicate the affect on the deckhead height available.

Very often a compromise must be struck between various conflicting requirements. This is one way in which the naval architect's skill, or 'art', is brought into play.

There is no absolute 'right design' or even 'optimum design'. There are a range of *optimization* techniques available but, powerful as these may be, they are limited to specific design aspects. Even those of widest scope do not embrace all design considerations. The designer will find them very useful but the final overall design balance is a personal one.

There are available to the designer many aids and processes that range in usefulness from valuable to handy upon which to draw. Let us look at some of these.

Scheduling In all, some 20,000 drawings may be required to describe a complex warship. Work instructions and machine computer tapes are derived from them. This great bulk of work must be arranged in logical order according to a schedule which accords with the production sequences. While less complex, the merchant ship design has similarly to be controlled.

There are many aids available to assist scheduling. Bar charts and bring-up systems are typical. One of the most powerful is the network which shows each event connected to those which must precede it. Figure 15.11 shows a fragment of a network schedule. There are paths through it which determine the total length of time for the whole process, called critical paths. It is these that excite the attention of the good manager. Network scheduling can quite readily be programmed for a computer and daily comparisons of the schedule intent with achievement are common as a control tool. Time, material, cost, labour, bought-in items, parts lists, tools, all normally form part of the scheduling events and processes and very comprehensive management control systems for shipyards are now available.

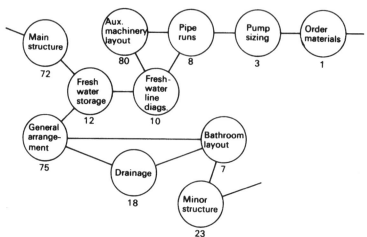

Fig. 15.11 Fragment of a drawing network schedule

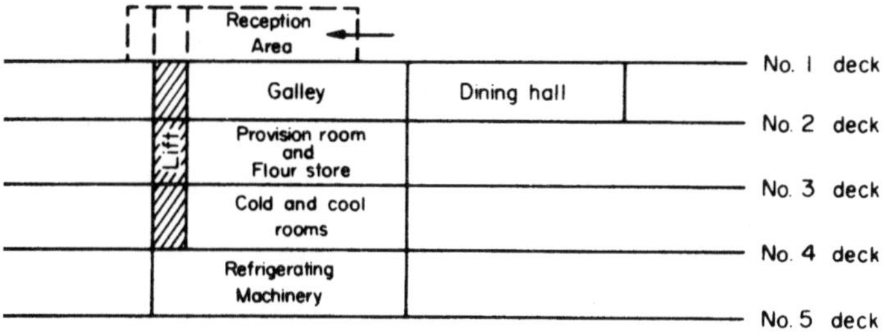

Fig. 15.12 Catering arrangements in destroyer

Flow Diagrams Simple flow diagrams to assist local arrangement in a ship are exemplified by Fig. 15.12. This ideal positioning of the commissariat arrangements in a destroyer is derived from common-sense consideration of the flow of goods in replenishment and food preparation. Similar arrangements may be derived for messdecks and bathrooms based upon considerations of comfort, noise, heat, safety, escape, convenience and the flow of people to toilets, action stations, canteen, etc. Where there are particularly difficult areas, land architects have used devices like trip frequency matrices or diagrams on which lines representing each flow expose bottleneck areas. Flow of information for weapons control is vital to a warship, even though it is made easier by the introduction of data buses. Flow of fluid and electrical power are pursued by the processes of Chapter 14.

Virtual reality, mock ups and models Very congested or complicated spaces (e.g. command centres) where there are important man–machine interfaces used to be modelled full scale, usually in wood. Nowadays *virtual reality* techniques are likely to be used (see computers in Chapter 2) using the ship definition contained within the CAD system. Whether virtual reality, mock up or model is used the aim is to bring together operators, equipment designers and human factors experts to achieve a layout of maximum efficiency.

The next most important area of the ship to be modelled virtually or physically, is that of the main propulsion machinery spaces. Physical models, if used, are likely to be at a scale of one-tenth or one-fifth full scale. At that scale it is possible to gain access to nearly everything. It is used firstly for the design of the pipe runs, ventilation trunking, placing of ancillary equipment to check access and removal routes and sometimes later for production purposes. Clear plastic and coloured pipes and fittings make comprehension easy. Smaller scale wooden models are often made for the anchor arrangements and, in copper, for assessing electromagnetic radiation efficiency.

It is becoming more common now to develop full-scale the total weapons system ashore to assess its compatibility and, even, for production of cable runs and other services. This shore test facility is a major investment of many millions of pounds and needs to be judged on the basis of the gains in operating efficiency and amortization of cost to the ships under production.

During production of a ship, the intentions expressed by the various layouts can be brought together in chalk in the bare compartments to produce 2D lineouts. Where justified, wires can be run through a space to create a 3D lineout. Sometimes, the desired layout is left until these lineouts are available. They are invaluable in making last-minute checks that all of the drawings—often produced by many separated authorities—are indeed compatible. Crossing pipes and ventilation trunking, unacceptably low deckheads, interferences and plain forgetfulness are uncovered and corrected. Again virtual reality techniques can be used in place of physical line outs.

Work Study In principle, a work study practitioner considers a certain process, perhaps design, fabrication, installation or maintenance, and asks:

What is achieved?	*Why* is it necessary?
Where is it done?	*Why* there?
When is it done?	*Why* then?
By whom is it done?	*Why* by that person?
How is it done?	*Why* in that way?

In asking 'Why?' the practitioner also seeks alternatives and then assesses the advantages and disadvantages of the various solutions to decide which is the best. This critical examination of the facts often reveals that jobs are unnecessary or are carried out in a certain way by certain people for purely historical reasons. The end product should be a better way of achieving the desired aim. Better may mean that the process is cheaper, quicker, requires less people or is in some other way superior depending upon the terms of reference under which the study was conducted. Work study does not need a highly specialized practitioner; all engineers and technologists should use the tool when it is likely to be profitable. It is often an expensive tool and it should itself be the subject of cost-effectiveness enquiry before being employed.

Value Engineering This is also a questioning process but must be quantified to be really useful. It is readily applied to making economies in the production process. An element of the design is critically examined with a view to achieving the same function more cheaply.

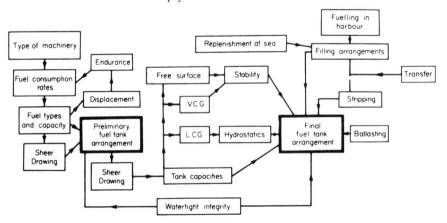

Fig. 15.13 Design influence diagram for fuel tanks

COMPUTER-AIDED DESIGN (CAD)

Initially it was the computer's ability to carry out repetitive calculations rapidly and accurately that made them attractive for much of the traditional design work such as calculation of displacement from a table of offsets and then on to stability. Since then they have had a much more fundamental impact upon the work of the naval architect. In outline terms the following comments indicate what can be achieved.

- A ship's three-dimensional form can be defined accurately, which is the starting point for design and manufacture. Books are available on techniques that can represent and manipulate fair shapes such as the hull by patches, splines, polynomials, elements, etc., within the constraints placed on it by the designer.
- From the hull form can be derived the hydrostatic and stability characteristics, which affect the static behaviour of the ship in still water.
- Computational fluid dynamics techniques can be used to study flow past the hull and its appendages, and the wave system generated.
- Defining the seaway leads to a determination of the ship responses in terms of overall motions and structural behaviour. Finite element analysis methods can be used for global and local calculations.
- Similarly dynamic responses can be determined in response to control surface movements to establish the directional stability and manoeuvrability.
- Layouts can be studied and the computer can generate automatically the areas and volumes of compartments and tanks. The layouts can provide the basis for a computer-generated 'walk through' of the design showing how spaces will look (colour, texture as well as spatial layout can be shown) to assist both the designer and prospective owner. Accessibility and lines of view can be assessed.
- As decisions are made on structure, equipments, systems and fittings, weight distributions can be kept up-dated and, with them, the stability, strength and other ship design properties.
- Power calculations and propulsor designs can be produced.

As discussed elsewhere, design is an iterative process. The computer enables those iterations to be made more rapidly and accurately, and in greater detail than was previously possible. It remains important, however, that assessments at each iteration are at levels consistent with the firmness of the design. Some calculations have only been possible as the power of computers has increased; others have become economic as costs have dropped; applications are now available on desk-top computers which once needed a large main frame.

At first individual calculations were computerized. These were progressively adapted to form suites of integrated programs where output from one provided an automatic input to others, with many programs interacting with each other. These integrated suites of programs are known as *computer-aided design (CAD) systems*. Some calculations are still carried out separately from the main suites of programs. For instance, CFD calculations of flow around a hull are usually studied separately although using inputs from programs defining the hull shape.

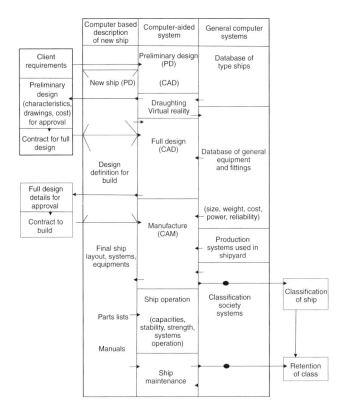

Fig. 15.14 Schematic of computer-aided processes

There are now CAD systems for a wide range of applications—large displacement vessels, small craft, surface warships, submarines, container ships, multi-hull vessels and so on. Some deal with the preliminary design phases and others relate to the full design process. And the usefulness of the systems is not restricted to the design process. The design outputs can feed directly into the shipyard and be integrated with their programs for *computer-aided manufacture* (CAM). Thus the form is defined digitally and there is no need to supply tables of offsets. The structural information and lists of equipment, fittings, cable runs and piping provide a data base for feeding directly into material ordering systems. Shape information provides the input for computer controlled cutting machines, associated with nesting programs for maximum economy of material. These CAM applications lead to greater accuracy of fit making fabrication easier and resulting in less built-in stress. They can considerably reduce the manpower and time for production.

In the same way, design and manufacturing information feeds into maintenance manuals and routines and can be used as the basis for recording the state of structure and systems during service. Such information assists surveyors to keep up with the deterioration of the structure during the life of the ship. Similarly the information provides a base for enabling ship staff to keep track of stability and strength during loading, unloading and operation of the ship. In fact there is virtually no activity in the design office, shipyard or on board ship that cannot be controlled, or at least influenced, by computers. Whether such systems are economic and appropriate must always be considered. Because an activity can be computerized it does not follow that it should.

The pace of change in the field of CAD/CAM has been, and continues to be, rapid. Thus it is not profitable, in the context of this book, to discuss current systems with their relative merits. The reader should refer to the technical press for information on developments. As an example the Royal Institution of Naval Architects produces, in its journal *The Naval Architect*, a regular review of CAD/CAM developments.

Design for the life intended

There are many aspects of the intended life of the ship which were discussed earlier in this chapter that strongly influence the style and standards of the ship. The first considerations are for efficient use and economical production but so also must the in-life support and the intentions over modernization be considered. While this book is not intended to be instructive on the important subject of production, some reference is needed in so far as it affects design. Let us then look briefly at design for use, design for production, design for availability, design for support and design for modernization.

DESIGN FOR USE

It is important that systems and equipments should be easy to use. The term 'user friendly' is often used in everyday life and is a descriptive one. In ships, easily understood and used equipments mean less specialized training for operators. Equally important they are more likely to be used effectively when operators are under the stress of an emergency or in action.

Where a human being is involved a designer must take account of the way people function physically and mentally. In the general sense naval architects have always taken account of human factors in their designs, e.g. the pull a person could exert on a rope, the sizing of furniture, the provision of minimum ventilation standards. However, the efficient blending of human and material in design can only be achieved by a more positive approach and a more formal application of human factor principles. This involves calling upon the professional expertise of the physiologist and psychologist as well as the engineer and scientist.

In some areas of design this has been widely recognized in recent years, e.g. in the design of control consoles. Even in this relatively simple example several distinct problems need to be tackled, viz.:

(*a*) Ergonomics. The size and layout of the console must be such that all dials can be read; all controls must be readily reached, levers requiring fine adjustment must be operable with minimum force; and so on.

(*b*) Design 'philosophy'. The design of layout must help the operator in understanding the results of actions. Thus for a machinery control console a control panel which represents diagrammatically actions within the machinery (e.g. flow of steam) is useful. So also are groupings of dials indicating physical state with controls which change those states; positions of controls for valves indicating whether the valve is open or shut; dangerous conditions immediately obvious by red or flashing lights, aural alarms, etc.

(*c*) Training. The need to train for routine tasks is obvious. The training for emergency conditions is also vital because of their importance when they occur and in the fact that, hopefully, they occur only infrequently. Realism is essential.

All too often, a designer of an equipment or system assumes that the user will think and act as the designer would. The different levels of experience and ability of the designer and user are too often reflected in the operating instructions and handbooks issued so making it difficult for the user. Apart from this, however, a proper analysis of what a person does in certain circumstances, and why it is done, often shows that popularly held concepts are invalid. A common finding is that people tend to see what is expected and ignore what is not expected. This may be by a biased selection from the total data available or even by imagining data that is not there. Radar operators may generate bogus tracks or ignore obvious contacts. Thus, for a control console, it is necessary to determine the information to be presented to a controller, what form it should take (e.g. digital or analogue), what should be the result of any actions (e.g. should they directly modify the machinery state) and to establish how an operator will know when the control system itself, including its feedback, is at fault rather than the system being controlled.

In most systems, once the totality of actions has been deduced, a choice exists as to which the human should do and which should be left to the machine. In deciding, the relative strengths and weaknesses of both must be considered. Thus a machine is good at repetitive calculation whereas a human may make errors, particularly when tired or under stress. The human is better at pattern recognition than a machine. Imagine programming a machine to pick out a relative on a photograph which a human can do quickly. In some cases the division between the human and the machine is fixed but increasingly the interaction between the two is varied to suit the need. Thus in controlling a ship it may be desirable to give the helmsman more of the total task when conditions are quiet (e.g. on an ocean passage) to encourage alertness and to allocate more to the machine when the operator would otherwise be overloaded (e.g. operating in a congested seaway in bad weather).

It was the advent of the computer that provided the human factor practitioner with the greatest problems and greatest challenge. It offers now scope for the development of systems with artificial intelligence, with the human and the machine effectively having a dialogue and supporting each other. This may be for training purposes using part-task simulators or in the operational role.

Even without artificial intelligence as such, the growing capability of the computer for a given size and cost opens up prospects of decision or training 'aids'. These will often appear similar to 'games' giving the human a chance to study 'what happens if' so facilitating selection of the action that leads to the desired outcome. Many of these potential applications involve probabilities. Few outcomes of high-level actions are certain, depending upon circumstances of the environment and what actions others may or may not carry out. Most individuals find probabilities hard to conceive in other than relatively simple cases. In multi-variant problems the best method of presenting data for human decisions is by no means clear, although in general terms it has been shown that graphical rather than alpha-numeric displays are preferable.

DESIGN FOR PRODUCTION

A good designer will have a feel for features which are likely to prove difficult to build, and which will, therefore, be expensive of time and money. Without that feel it will not be possible to arrive at sound compromises so necessary in ship design. Ideally designers will have had some experience of production. As an example, plating curved in only one dimension is much cheaper to produce than plating curved in two dimensions, but for the hull there may be a resistance or radar signature penalty. For many years now warships have dispensed with sheer and camber on decks other than weather decks. In steel structures the more welding that can be done by machine, the cheaper and more reliable it will be. The longer the straight runs of welding the less time will be absorbed in setting up the machines. Using stiffeners in one direction only, avoiding inter-section of welds and intercostal members will help, as will the use of steels which do not need special pre-heating procedures and the use of standard sections. Using swedged or corrugated bulkheads can avoid the need to weld stiffeners to plating. Confined areas must be large enough to provide access for fabrication and application of protective coatings. Grouping system compon-ents to reduce the length of cable and piping runs will reduce cost and weight. Some design features will be universally advantageous. Others will depend upon the production methods used, and equipment available, in the building yard. In general, the more pre-outfitting that can be done before the main sections of the ship are joined together, the better. Designers need to be in continual dialogue with the production team to provide break points which are compatible with the most economic production process.

DESIGN FOR AVAILABILITY

Availability has been seen earlier to be one of the main constituents of effect-iveness or value. In turn, it is dependent on reliability and maintainability. These terms possess a range of meanings, the vernacular no less important than the mathematical, presently described. Reliability engineering developed slowly after it was first proposed by Sir Alfred Pugsley in 1930 but it received great impetus from the space programme 30 years later.

Let us consider the time taken for a large population of similar devices tested in identical conditions to fail. The number of failures in each interval of time

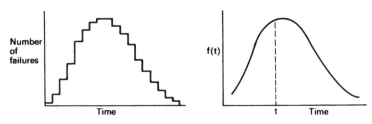

Fig. 15.15

may be plotted as a histogram. With sufficiently small intervals of time the number of failures divided by the total population may be represented by a smooth curve $f(t)$ called the probability density function.

If we now regard the behaviour of any one of the items as represented by the behaviour of the total sample, it is clear that the probability of failure up to time t is

$$F(t) = \int_0^t f(t)\, dt$$

Because the total area of the curve must be unity, the probability of not failing is

$$R(t) = 1 - F(t) = 1 - \int_0^t f(t)\, dt$$

This is the reliability. It is defined formally as the probability of a device performing adequately for the period of time intended under the operating conditions encountered. Note the dependence on specifying the environment. The integral curve $F(t)$ is the cumulative failure distribution and $R(t)$ is the reliability distribution function. There is one other important parameter $z(t)$, the instantaneous hazard rate or failure rate which can be shown to be

$$z(t) = \frac{f(t)}{R(t)}$$

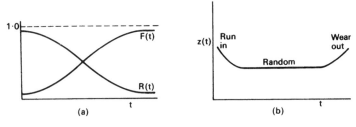

Fig. 15.16 *(a) Reliability functions, (b) Bathtub curve*

Testing or experience shows that the reliability functions may follow a specific statistical distribution. Wearout, for example, is often found to follow a Gaussian law, the probability density function being

$$f(t) = \frac{1}{\sigma\sqrt{2\pi}} \exp -\frac{1}{2}\left(\frac{t-\mu}{\sigma}\right)^2$$

where μ and σ are the mean and standard deviations.

A most important distribution related often to complex equipments subject to random failure is the exponential for which

$$f(t) = \frac{1}{\theta}\exp\left(-\frac{t}{\theta}\right) = \lambda\exp(-\lambda t)$$

$$R(t) = \exp\left(-\frac{t}{\theta}\right) = \exp(-\lambda t)$$

$z(t) = \frac{1}{\theta} = \lambda$, a constant, the failure rate. θ is its inverse known as the mean time between failures (MTBF). It is this which accounts for the main portion of the bathtub curve of Fig. 15.16(*b*). This type of curve typifies many systems in which, after an early period of running in, it enjoys a very long period of roughly constant failure rate until, as it ages, wearout makes the hazard rate rise again.

The exponential distribution is actually a special case of the Poisson distribution which may, in order to assess the number of spares to be carried, be used to predict the number of failures k in a stated time

$$p(k) = \sum_{i=0}^{i=k} \frac{(\lambda t)^i}{i!}\exp(-\lambda t)$$

The mathematical convenience of the exponential form makes it a temptation to apply to every case. It is important therefore to be assured of its relevance. Where it is applicable, it is salutary to examine the probability of failure; for an MTBF of 3000 hours for example the reliability is

Mission time (hours)	3000	750	120	24	4
Reliability	0.368	0.779	0.961	0.994	0.999

while the probability of there being failures in 3000 hours from the Poisson function is

Number of failures k	0	1	2	3	4
Probability	0.37	0.26	0.08	0.019	0.004

The overall reliability of a group of components which go to make up an equipment can be computed by combining reliabilities by laws very similar to those governing the combination of resistors in parallel or series. The components may not be physically in series but if their dependence on each other is sequential they will be functionally so and the overall reliability is the product of each, $\Pi R(t)$

$$R_s(t) = \prod_{i=1}^{n} R_i(t)$$

If, for example, there were 10 components each with a reliability of 0.99 placed in series, the overall reliability would be $0.99^{10} = 0.905$. If in parallel, equally loaded, the components' unreliability is similarly combined, so that

$$R_s(t) = 1 - \prod_{i=1}^{n} F_i(t) = 1 - \prod_{i=1}^{n}(1 - R_i(t))$$

In the above case, units in parallel would then have a reliability of

$1 - (0.01)^{10}$ which is almost unity.

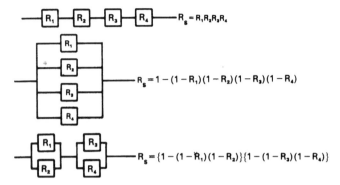

Fig. 15.17 Some system reliability models

This gives the clue to reliability modelling. Systems of components whose individual reliabilities operating in similar environments are known from data bases can be built up from knowledge of their functional arrangements. Functional dependency diagrams like Fig. 15.18 assist. The reliability modelling then enables the overall reliability to be calculated and improved, if necessary by the incorporation of redundant equipments in parallel where there are weak points.

  Reliability modelling is important not only to the assessment of risk but to adequate quality assurance and the rational scaling of spare parts. Maintainability in addition to its pragmatic meaning of ease of maintenance, is suscep-

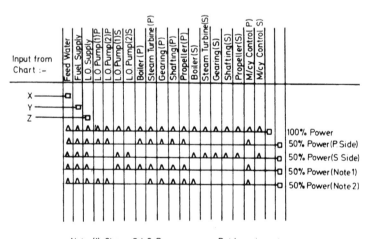

Fig. 15.18 Part of speed characteristic dependency diagram

tible to mathematical treatment in which the variate is the time to repair for a population of faults.

Availability at its simplest is the ratio of the time when a system or equipment is 'up', i.e. able to work to the total time. Redundancy in a system obviously raises availability and hence overall effectiveness.

DESIGN FOR SUPPORT

This is not a major influence on merchant ship design but for the dense warships it is extremely important. Reliability and maintainability studies, already discussed, identify those equipments which may require removal during the operational life of the ship. Gas turbine change units, for example, are planned to be replaced at a specified life and occasionally need replacement at unexpected moments. The requirement is to change them within 48 hours and removal routes have to be designed into the ship (often up the air downtakes). Diesels usually need portable plates through which they may be unshipped. Access around equipment for maintenance has to be planned carefully, including room to withdraw rotors, and maintenance envelopes are usually shown dotted on detailed layout drawings. Propeller shaft withdrawal and rudder removal in dry dock have to be thought about. Room has to be left in the layouts for spares close to where they may be needed, especially if they are heavy or bulky.

Electronic equipment is often subject to a repair-by-replacement policy and standardization here plays an especially important role. Standard boards or cards of electronic elements assist in easing a vast logistic problem. Some weapons elements must also be changed during weapons update periods, often by ship's staff, so that some parts of the weapons system keep pace with advances by a potential enemy. Such equipment usually needs to be kept in a controlled, dry environment so that ancillary equipment has to be reliable and easily maintainable.

Impending failure can be predicted by various health monitoring devices based, for example, on spectroscopic examination of oil or vibration measurement. This monitoring equipment is often designed into the control consoles for the machines.

DESIGN FOR MODERNIZATION

A warship may last up to 30 years, during which time its weapons become obsolete. Taking a warship to pieces to replace whole weapons systems is time-consuming and expensive, especially when the system depends on elements scattered throughout the ship. Although this has long been recognized, only recently have advances been made which can substantially isolate a weapon system from the ship platform itself. This type of isolation has appeared under the general heading of modularity. One specific example is known as the MEKO system (Germany). While the principle is simple, namely to provide a self-contained module housing a weapon system, practice has been bedevilled by the need for that system to draw upon information from the gyro compass, the surveillance radar, the command system (AI0) and many other parts of the ship. This problem has now been solved by the introduction of data highways or data buses which are able to carry packaged information electronically around the ship to be drawn off where required. It is likely therefore that, although modularized ships do tend to be somewhat larger and, initially, more expensive, they will become the normal way of designing large warships. Other services for the modules like chilled water and electrical supplies may still be drawn from a central source to avoid the expense of duplication.

Zoning of warships also aims, as far as is practicable, to keep functions of a ship within zones that can be sealed off. They form not only fire, smoke and watertight subdivision but sections of the ship which are independent of others to reduce vulnerability to action damage and to ease modernization. It is not, of course, possible to make them totally independent because some systems, like electrical power, are almost certainly to be distributed ship-wide. Nevertheless, it is worthwhile where it can be done economically.

THE SAFETY CASE

Design merely to meet regulations is no longer acceptable. Nor can it be assumed that a ship and its individual parts are either safe or unsafe. There are grades of acceptability which must be individually considered as compromises among the conflicting demands of profitability, facility of operation, social acceptability, personal danger and potential litigation. Such compromises are arrived at through formal procedures which are grouped together under the description of the Safety Case and may be implemented within a ship or equipment safety management system. While some aspects are conducive to a numerical approach, decisions on the standards to be adopted are often matters of judgement.

There are two elements to perceived risk; the likelihood of occurrence and the consequences of that occurrence. Judgement is based on the compound of these two elements. Thus, a low likelihood of occurrence but a catastophic outcome would often be regarded as unacceptable, while a likely event resulting in a trivial outcome would be judged on the basis of its nuisance or economic impact. Of course, behaviour of operators and the environment within which they must work are inextricably mixed with the event.

Formal steps in the development of a safety case are clear:

(i)   identification of all potential hazards and the likelihood of their occurrence in defined circumstances and environment;
(ii) identification of the consequences of each event;
(iii) establishment of a system for controlling the occurrence and its consequence, including escape and rescue, if relevant;
(iv) issue of codes of practice and communications which constitute a safety management system;
(v)   institution of auditing procedures for the safety management system.

There are many techniqes available which help in the assessment of safety, for example reliability engineering, failure mode and effect analysis, fault tree diagrams, dependency diagrams. These may often expose the need for clear

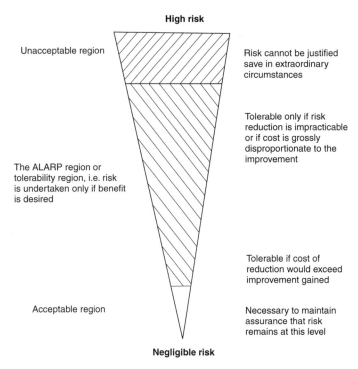

High risk

Unacceptable region

Risk cannot be justified save in extraordinary circumstances

Tolerable only if risk reduction is impracticable or if cost is grossly disproportionate to the improvement

The ALARP region or tolerability region, i.e. risk is undertaken only if benefit is desired

Tolerable if cost of reduction would exceed improvement gained

Acceptable region

Necessary to maintain assurance that risk remains at this level

Negligible risk

Fig. 15.19 ALARP guidance

information for operators or formal training, as well as equipment duplication or interlocking which physically makes dangerous operating procedures difficult.

Among the aids to reaching a decision on the acceptability of a system is the ALARP principle. This is an acronym for 'as low as is reasonably practicable'. Words which may help in the decision process are shown in Fig. 15.19. It is of little comfort to a designer to know that, despite their earnest endeavours, events subsequent to a catastrophic event will question their judgements as to what was an acceptable compromise among all the conflicting demands. That, however, is no reason for avoiding the problem.

The whole panoply of safety assessment has to be reserved for behaviour of the ship subject to major hazards such as flooding, fire and explosion, loss of steering, fracture of principal structure and capsize in a seaway. Nevertheless, the general approach to safety assessment is applicable to all equipments and systems in a ship, while inspiring the ship's company in contributing to the safety consciousness of all the crew.

Conclusion

This chapter has outlined the process of ship design. There is no substitute for the experience which application to a real ship design project can bring. Illumination and understanding derive from attempting the design of a ship in accordance with the philosophy and processes with which this book has been so sketchily dealing. It is not an arid application of engineering rote; it contains a great deal of art and artfulness, judgement and management. Beware of the facile claims of breakthroughs in ship design! Quoting from a similar engineering endeavour:

'The design required a series of difficult decisions, interrelated and far from the deductive, linear progression often imagined by people without experience in engineering.

The sequence of decisions here is typical of the trade-offs between simplicity of design, availability of material and practicability of construction that make engineering an art rather than simply—or merely—applied science.'

16 Particular ship types

Passenger ships

At one time, passenger ships were typified by passage across the Atlantic. Deposed by air traffic, the long sea voyages now belong to the cruise ships of the tourist trade and little else. Almost all other passenger traffic by sea is confined now to ferries on relatively short journeys. Even the definition of a passenger ship is no longer related exclusively to those carrying more than twelve people who are not members of the crew. A maze of international regulation has grown up through IMO to present exemptions and exclusions reminiscent of the old tonnage rules. Passenger ships are certificated by the United Kingdom in accordance with the following classes:

Class I. Ships engaged on voyages (not being short international voyages) any of which are long international voyages

Class II. Ships engaged on voyages (not being long international voyages) any of which are short international voyages

Class II(A). Ships engaged on voyages of any kind other than international voyages

Class III. Ships engaged only on voyages in the course of which they are at no time more than 70 miles by sea from their point of departure and not more than 18 miles from the coast of the United Kingdom, and which are at sea only in fine weather and during restricted periods

Class IV. Ships engaged only on voyages in partially smooth waters, or voyages in smooth and partially smooth waters

Class V. Ships engaged only on voyages in smooth waters

Class VI. Ships engaged on voyages with not more than 250 passengers on board, to sea, or in smooth or partially smooth waters, in all cases in fine weather and during restricted periods, in the course of which the ships are at no time more than 15 miles, exclusive of any smooth waters, from their point of departure nor more than 3 miles from land

Class VI(A). Ships carrying not more than 50 passengers for a distance of not more than 6 miles on voyages to or from isolated communities, islands or coast of Scotland and which do not proceed for a distance of more than 3 miles from land.

The Merchant Shipping Regulations give the legal definition of many of the terms. Design of passenger ships is then dominated for each class by strict regulations concerning:

watertight subdivision
fire boundaries

freeboard
life-saving appliances
transport of dangerous goods

While basic regulations exist for each of these features, exemptions are aimed at trading off one against the other in order to accommodate different architecture. A lesser standard of watertight subdivision for example might be permitted in some classes provided more life-saving appliances were carried. These trade-offs are nowhere more complicated than in the design of RoRo ships.

In the early part of the 20th century, passengers needed to be conveyed as rapidly and as comfortably as possible between continents. Great trans-atlantic liners served this need in varying degrees of luxury and, it must be said with varying degrees of safety, at least until the 1930 Merchant Shipping Act recognized the frailties apparent from such disasters as the *Titanic*. That legislation led to ships able to survive flooding of two or more main compartments and to the provision of more acceptable firefighting facilities and life-saving equipment.

After the Second World War, there emerged slowly a market for ships which did not follow one specific line. A public demand for cruising to many different places in pursuit of holidays and cultural interests led to fleets of cruise ships. For a while, economies of scale drove the size of these ships up and up. At present the largest ships on order have a gross registered tonnage of 136,000. These ships provide a holiday experience in their own right and for some people the ports visited are of secondary importance. However, such large ships are restricted as to the ports they can visit and often embarkation and disembarkation times are long. For these reasons the medium sized ship remains popular. Nevertheless ships carrying several thousand passengers with a passenger to crew ratio of 3:1 are common. Some ships are now being designed to carry the number of people that can be carried by a large aircraft. Thus the aircraft that deposed the ocean liner are now an integral part of the cruise business. More recent ships tend to have higher speeds to enable greater distances to be covered in a given time. Some are relatively small to enable them to visit small islands away from the main tourist spots.

Safety arrangements and evacuation procedures for huge numbers need special attention. Architecture to appeal to clients has led to the introduction of massive public rooms and to atriums that create spectacular compartments through a dozen decks. These demand some close attention to firefighting and movement of large crowds. Rapid evacuation of passengers and crew from a considerable height above the sea has led to escape chutes and self-inflatable life-rafts. Disabled people require special attention, not only in an emergency but for normal movement around the ship. Swimming pools, gymnasiums, open areas for use in competition and sport are all normal requirements. These are just some of the problems that are added to the proper attention to regulatory demands in the design of such fascinating ships. While safety provision to required standards remains paramount, there arises conflict among safety and comfort; zoning of ventilation, sills to watertight doors, fire and

Table 16.1
Comparison of large RoRo and Cruise Ship

	RoRo	Cruise Ship
Length BP, m	146	224
Beam, m	26	31.5
Draught, m	6	7.75
Gross Tonnage	27000	70370
Deadmass, Tonnes	5350	7000
Displacement, Tonnes	14000	35800
Propulsion power, MW	18	28
Side Thruster, MW	1.8	9
Speed, knots	20.5	22.3
Passengers	2120	2634
Cabins	217	1024
Crew	141	920
GRT/Passenger	12.7	26.7

smoke barriers, freedom of movement for passengers and crew all lead to compromises that will have to be debated at the design stage. Nevertheless, as Table 16.1 shows, the cruise ship is relatively capacious compared with the RoRo ferry.

At the smaller end of the market, the regulations for vessels in Classes III to VI have become more stringent in the UK following the *Marchioness* disaster on the Thames. Most new designs in the future will achieve a one-compartment subdivision. Reliance on rescue from shore up to 60 miles away does demand an abundance of personal life saving appliances and a time to sink after an accident which is prolonged. Even for operation close inshore, designers should never forget that the sea is hungry, and sometimes very cold.

Ferries and RoRo ships

The concept of an integrated transport system that gave rise to container ships is applicable equally to Rollon-Rolloff ships and to ferries. It is often economical and quicker to transport people and goods from shore to shore in their own vehicles. Trains, lorries, cars, caravans and coaches may have to be embarked, conveyed and disembarked in a safe and efficient manner, while their passengers and travellers on foot require to be looked after in a manner compatible with the distance and fare.

A ship, of course rises and falls with the tide so that the facilities which permit embarkation need vertical adjustment and, indeed, restraint. Trains are long and inflexible so that long hinged bridges on to the ship will allow slow movement to matching rails in the ship. As it is embarked to one side, the ship will heel and a rapid heel compensating system must transfer fluid to the opposite side. Wheeled vehicles can be dealt with more slowly but heel is quite noticeable as a heavy lorry is embarked. Fore-and-aft transfer of weight to ensure that there remains adequate freeboard at the bow must also be available.

Stability of the ferry must, of course, be under continuous review during loading, often with the help of a portable computer.

Matching of ship and shore depends on local trading conditions, varying from a simple ramp let down on to a beach to a road system that mates with the ship, sometimes even at two levels to speed up operations. Indeed, capital investment in such facilities is a significant contributor to that economic evaluation which must precede the definition of ship and shore as an integrated system before design begins. Aspects of the ferry will be dominated by the specific requirements for its class of operations; thus, Class III or IV ferries may well be excused from supplying extensive lifesaving or evacuation facilities. Domestic facilities too will depend on length of voyage and local sea conditions.

The most common RoRo ferry has evolved during the last fifty years. It is a sad fact that regulation of safety lags much behind innovation and must rely so often on the lessons of failure. A string of disasters starting in the 1950s with the loss of the *Princess Victoria* and working their way through *Herald of Free Enterprise* and *Estonia* in the 1980s among many others, finally persuaded the authorities that regulation to improve safety must be introduced. IMO brought out STAB 90 in 1990, beginning the process of defining minimum standards for all RoRo passenger ships, for adoption early in the 21st century. Such standards are now regarded as adequate although all nations, even those who signed the Convention, have not been able to enforce them and tragedies are still occurring.

The basic problem is very simple. By their very nature, RoRo ferries need large open spaces. If, for any reason, these spaces become flooded, the integrity of the ship is threatened and foundering can be very rapid indeed. The car decks can become flooded if the doors are breached or if a side collision allows the sea to enter. The regulations now require the ship to sustain collision damage which penetrates to a depth of one fifth of the beam of the ship. Two watertight doors never to be opened unless the ferry is secured alongside must also be fitted forward and aft. Often, the inner door doubles as the ramp to the shore.

Meeting the survival requirement should the car deck become flooded is also relatively simple, although it may reduce capacity slightly. A belt of buoyancy is provided by watertight compartments each side of the cardecks above and below the normal waterline. The inner barrier provides additional protection against a collision and the contained compartments give buoyancy as the ship heels reducing the tendency to capsize. This tendency was very apparent in the *Herald of Free Enterprise* disaster. The estimation of time to founder is an important feature of the safety case and regulations require the ship to be evacuated within 30 minutes of the order to abandon ship. This requirement has a significant effect upon lifeboat arrangements, leading to a lower lifeboat embarkation deck and nested lifeboats. Figure 16.1 shows a typical RoRo cross-section. Machinery uptakes may sometimes be split to use the watertight side spaces and two funnels are not uncommon.

Many ferries are now fitted with side thrusters to assist them in coming alongside while others have adopted vertical axis propellers which serve a similar purpose. Diesel-electric propulsion using pods at the propeller have

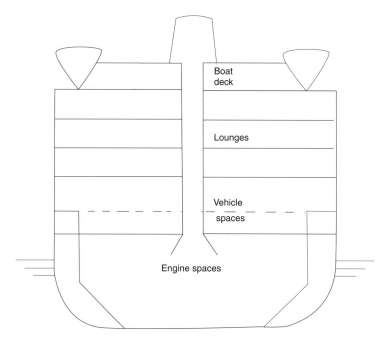

Fig. 16.1 Typical RoRo cross-section

also been fitted. Economic analysis has introduced a trend towards higher speeds that require gas turbine propulsion while other designs have used surface effect vehicles as RoRo ships.

Freight RoRo vehicles are basically similar to passenger ferries but are also trending towards higher speeds for economic reasons. Their cargo is somewhat more predictable and less variable in dimensions and can lead to their exclusive use on lines. Transportation of cars for trade is typical.

Aircraft carriers

The characteristics of an aircraft carrier are profoundly affected by the type of aircraft that it is required to operate, which may be fixed wing, deflected jet, vertical take off or helicopter. Unless the types and numbers of aircraft are known with some precision, the aircraft carrier will be larger and more expensive than it need be; there is a high price to pay for flexibility.

Fixed wing carriers are complicated ships, often of 2000 compartments and carrying 4000 crew. As well as all the domestic, navigational and machinery requirements associated with all surface ships, the aircraft carrier must operate, direct and maintain perhaps fifty complex aircraft. Fixed wing aircraft are catapulted by one of several catapults up to 100 m long at the fore end of the flight deck while the ship is steaming head to wind. Because they normally require a length for landing not available to them on a ship, the aircraft are retarded on landing by an arresting gear; a hook on the aircraft is directed on to

a wire stretched transversely across the flight deck which is connected to a
damping mechanism below. For both physiological and practical reasons,
accelerations and decelerations higher than 5 or 6 *g* cannot be achieved and
this gives minimum possible lengths for catapults and for arrester wire pull-out.
To give flying speed of 120 knots to a 30 tonnef aircraft, for example, a catapult
would need about 30 m of constant 6 g acceleration and 30 MW power, some of
which is contributed by the aircraft. Angle of descent of the aircraft and
clearance over the stern, spacing and pull-out of arrester wires, centring gear
and length of catapult and bridle catching gear thus all contribute in dictating
minimum flight deck length.

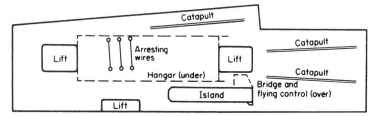

Fig. 16.2 Flight deck

Also accommodated by the flight deck are aircraft deck parks, fuelling
positions, weapons areas, servicing positions, helicopter operating areas, landing aids and six or more aircraft and weapons lifts. The island itself, tradition-
ally on the starboard side, houses the bridge, flying control, action information
centre, flight deck personnel and the long range radars and communications
equipment for detecting and controlling aircraft many hundreds of miles away.
Layout of the flight deck, in fact, determines the length of the ship (and thence
displacement) fairly closely and the naval architect usually struggles to keep
island size and flight deck length to a minimum. With an aircraft specified,
however, scope is limited.

The hangar below decks needs to be as wide and as long as possible and at
least two decks high, even to accommodate aircraft with folded wings. Not only
does this cause difficult access and layout problems for the rest of the ship, but
it gives rise to some formidable structural problems, particularly if the hangar is
immediately below the flight deck when the wide span grillages must support 30
tonnef aircraft landing on at high vertical deceleration.

At least thirty different piping systems are required, including flight deck fire
main, fuelling, defuelling, air, hydraulics and liquid oxygen. Widespread maintenance facilities are required near the hangar for both aircraft and weapons.
Underwater, a side protection system is fitted against mine and torpedo attack
and armour is disposed around the vitals. The flight deck itself normally
constitutes armour, although in some older carriers, this was relatively thin
and the hangar was left open below it, armour being provided at hangar deck
level.

Boats are stowed in pockets in the ship's side in order to keep the flight deck
clear, together with items of ship's equipment, fairleads, capstans, etc. Accom-

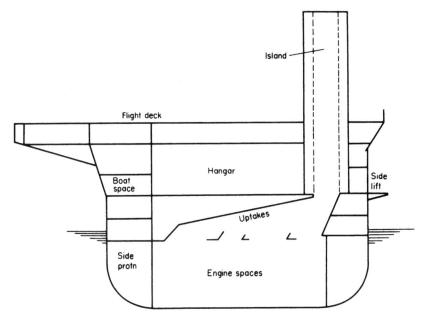

Fig. 16.3 Aircraft carrier section

modation arrangements in the ship are conventional except that additional spaces for aircrew such as briefing rooms, flying clothing cloakrooms and air intelligence spaces are needed.

Such is the power needed—over 150 MW for large carriers—that steam machinery plant is usual; a nuclear reactor avoids the need for frequent refuelling. A fleet train is needed for regular provisioning, of course, but an oiler is vulnerable.

These complicated ships are expensive to build and run and are now confined to quite few countries. However, ships designed to carry helicopters and 'vertical short take-off and landing, VSTOL' aircraft are now common. Deflected jets, pioneered in the Harrier, enable such aircraft to take off and land vertically, to hover and even to fly backwards. The vertical evolutions do use a lot of fuel which could be put to better use in extending the range. As a result, ramps have been devised of a shape which permit such aircraft to take off under their own power with a very short run. The profile of the ramp is critical in imposing a nose-up attitude without allowing the forces which the ramp imparts to the undercarriage to cause it damage. With the need for long catapults and arresting gear obviated such ships can be smaller and cheaper. The structural design is a particularly important aspect of aircraft carrier design. Because they are intrinsically asymmetric and require major discontinuities to accommodate lift wells, there is a tendency to torsional vibration with nodes at the structural weaknesses. A dynamic analysis to determine the modes is now possible.

Detection of submarine contacts by sonobuoys and the prosecution of attacks by homing torpedo can be done effectively by helicopters many miles from the parent ship. Escort destroyers and frigates are able to carry one or perhaps two helicopters of medium capability but there is a need for some ships to accommodate larger numbers. Helicopter carriers with a dozen or so aircraft can be adapted into relatively simple ships capable of arming, operating and maintaining the aircraft as well as accommodating the necessary personnel. This last task is not negligible; it is not unusual for as many as 25 people to be required for each helicopter and such ships are space demanding. The ships may also act as a garage for the maintenance of the escorts' aircraft.

One interesting way of getting helicopters or VSTOL aircraft to sea quickly without a purpose-built ship is by using standard containers. Using a normal container ship the containers may be stacked like children's building bricks to make a hangar while the containers themselves can accommodate spares, personnel, servicing facilities, communications, weapons and operations spaces. Purpose-built parts are required for the hangar roof and for bridging the gaps between containers. The total kit to achieve such a fitment is quite large and a 250 m container ship may be able to house only eight or ten aircraft, with two spots for take-off and landing. Nevertheless if there has been proper preparation in advance it can represent a very useful way of augmenting aviation capability in a national emergency by taking ships up from trade. The ships themselves will of course need a certain amount of modification by removing vertical obstructions to safe flying. Figure 16.4 shows a possible arrangement.

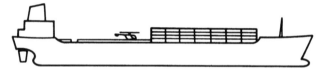

Fig. 16.4

Bulk cargo carriers

Cargo which may be carried in bulk includes oil, ore, chemicals, vegetable oils, molasses, liquefied gas, coal, grain and forest products. Even cars and containers present a homogeneous cargo whose effects upon the ship are reasonably predictable.

Economic arguments indicate that most oil should be carried worldwide in very large crude carriers, VLCCs and some 500 such vessels ply the oceans. They may carry as much as 300,000 tonnes of crude oil in holds arranged perhaps six longitudinally and three or four abreast. The block coefficient is often around 0.8. Machinery aft is usually diesel driving a single shaft and a boiler produces steam for domestic use, for heating the bunker oil and for steam cleaning cargo tanks. There is also an inert gas system for cargo tanks to prevent the build up of an explosive mixture above the cargo. There is a superstructure aft with bridge and accommodation and a central walkway along the upper deck to protect personnel from the effects of a very small freeboard. Loss by structural failure is now fairly rare even though cracking is not and should be dealt with in good time. Most loss is attributable to collision or grounding giving rise to severe environmental pollution. In 1990, the USA introduced an Oil Pollution Act which required all oil tankers using their waters to have double skins and IMO followed in 1993. Unhappily, evidence suggests that this would not have helped to reduce pollution in many of the major environmental disasters that the world has suffered and designers have sought more effective measures.

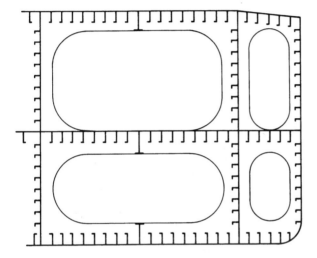

Fig. 16.5 Modern mid-deck tanker midship section

One way of containing at least some of the cargo after a grounding is to subdivide the ship by additional oiltight decks. Spaces below would be maintained by regulated air vents at atmospheric pressure so that, should the bottom be pierced, there is a pressure differential forcing the cargo inwards. It is then allowed to weir into empty tanks. There are several variations on this theme, which could work, at least partially until there is structural disintegration, even in the dynamic fluid conditions that prevail. Of course, an increase in cost and a reduction in payload is inevitable.

Losses of ore carriers caused great anxieties during the 1980s. Many of them disappeared without trace, presumably by structural failure. There is evidence to suggest that the scarphing of the midships structure towards the ends into a reduced section modulus was not, in the 1970 designs, carried out with sufficient care and that local stress concentrations caused cracking which propagated fast. Some 10 to 15 per cent of the length from the stern has been shown to suffer high stresses in a seaway. Damage to structure by the huge grabs used to unload cargoes exacerbates the problem while some cargoes generate highly corrosive fluids that further damage the structure with time. These matters have led to strong pressures for more high quality steels in critical parts of the ship. More intensive survey is now adopted and cooperation within IACS has led to

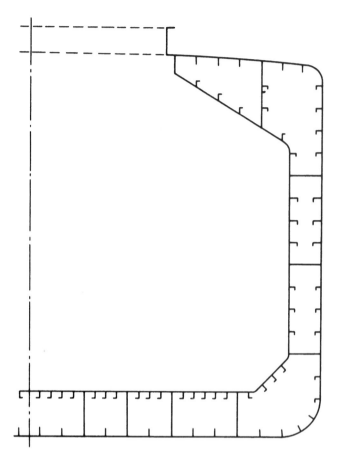

Fig. 16.6 Traditional bulk carrier midship section

new regulations which provide good structural management, including hull condition monitoring.

Slamming can be troublesome in bulk carriers and masters need to be advised on how to minimize these effects; it tends to be a highly tuned condition so that small changes in course or speed are usually effective in reducing its incidence. Of particular importance to the bulk carrier is the sequence of loading and unloading. It is relatively easy to cause structural damage by large shear forces between full and empty tanks as well as unacceptable still water hull flexure. Masters must be supplied with examples of these effects, often programmable readily on a microcomputer, to guide decision making. Designers may also wish to present their classification societies with a direct structural analysis rather than simply following their rules. This is entirely acceptable and can lead to improved structure.

Grain carriers, often affected by the need to sail the Great Lakes of North America are common. Grain carriers introduce special problems of stability due to the free surface effects of the grain. While shifting boards reduce the effects, large holds make these difficult to fit; covering the surface with bagged grain damps the free surface but does not remove the problem. Masters are required to take the ullages (i.e. height above the free surface) of the grain and to apply these to diagrams supplied by the Authorities (see Chapter 14) for determining revised stability data. Expected angles of heel of the ship should not exceed 12 degrees due to possible shifting of the grain (see also Chapter 5).

A type of bulk carrier of increasing importance is the liquefied natural gas carrier. Natural gas, predominantly methane, is given off in vast quantities in a few areas of the world, particularly at oilfields. For use in other countries, it needs to be transported economically. The gas is first liquefied by compressing it and cooling it to temperatures around minus 100°C in which condition it is pumped into transit tanks. Such tanks are isolated from the ship's structure by very thick insulation and the ship is fitted with double bottom and side protection. During its passage across the sea, certain of the gas boils off naturally either to waste or it may, economically, be used to help drive the ship, perhaps by gas turbine. To minimize this loss and avoid carrying refrigeration machinery, the ship needs to be fast and reliable and the characteristics of the ship are, again, determined by the economics of operation.

Material for the liquefield gas tanks must not be brittle at these very low temperatures—which render even rubber brittle. Suitable materials are certain aluminium alloys and some nickel steel alloys. It is very important, of course, to prevent the mild steel of the ship from being reduced in temperature by leakage of the liquid, whereby it may become brittle (see Chapter 5 on brittle fracture).

Every cargo brings its particular problems but none more than special liquid cargoes such as molasses, sulphuric acid and sulphur, including thermal stresses, explosive vapours and corrosive materials.

Submarines

Submarines are vehicles designed to operate principally at considerable depth. Most applications to date have been to warships. Commercial applications have been oceanographic research vessels and small vehicles for laying pipes and servicing well-heads on the sea bed. Submarines are uneconomic for general commercial work.

Naval submarines were originally of limited capability. They were very good at attacking shipping with torpedoes, their invisibility enabling them to approach merchant ship targets unobserved. Nowadays, modern cruise missiles enable them to attack accurately land targets well inland. Other missiles mean they can engage surface ships or aircraft. The submarine is an ideal vehicle for landing small groups of special forces personnel on defended beaches. Thus the submarine has become more of a general purpose vessel than one with a single, albeit vital, mission.

Because the vessel has to operate on the surface and submerged all the usual naval architecture problems have to be studied for both conditions. Some have to be studied during the transition phase.

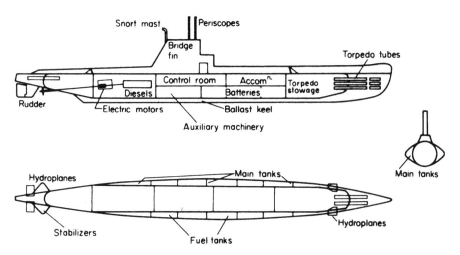

Fig. 16.7 Diagrammatic arrangement of conventional submarine

The general design of a conventional submarine is illustrated in Fig. 16.7. Major differences compared with a surface ship are:

(a) the shape, which is conditioned by the need to have efficient propulsion submerged;

(b) the enclosure of the main portion of the vessel in a pressure hull which is usually circular in cross-section to enable it to withstand high hydrostatic pressure at deep diving depths. The circular section means greater draught generally than a surface ship of the same displacement. It also requires that a docking keel be provided, unless special cradles are available, and a top casing for men to move around on in harbour;

(c) the hydroplanes, for controlling depth and trim angle; usually two sets are provided, one aft and forward or on the bridge fin;

(d) tanks, usually external to the pressure hull, which can be flooded to cause the vessel to submerge;

(e) a dual propulsion system. The submerged propulsion system is usually electric drive supplied by batteries and surface propulsion is usually by diesels. The batteries need frequent recharging, which means that a conventional submarine has to operate on the surface or at periscope depth for considerable periods. These disadvantages are overcome in nuclear submarines or in vessels with air independent propulsion;

(f) periscopes and sensor masts to enable the vessel to operate close to the surface;

(g) a special air intake, the snort mast, to enable air to be taken in when operating at periscope depth;

(h) special means of controlling the atmosphere inside the submarine. Apart from the normal conditioning equipment, carbon dioxide absorbers and oxygen generators are provided.

Comments on some aspects of submarine design are:

(a) *Hydrostatics*. Although the hydroplanes can take care of small out-of-balance forces and moments, the vessel when submerged must have a buoyancy almost exactly equal to its weight and B must be vertically above G. For reasons of safety, in practice the buoyancy is usually maintained slightly in excess of the weight. The capacity of tanks within the pressure hull for adjusting weight and longitudinal moment is limited, so that initial design calculations must be accurate and weight and moment control are more critical than for a surface ship. In the latter case, errors involve change of draught and trim from the design condition. If a submarine is too heavy it will sink and if too light it will not submerge. If B and G are not in the same vertical line, very large trim angles will result as B cannot move due to movements of 'wedges' of buoyancy. The term trim here is used in the conventional sense. In submarines, fore and aft angles are usually termed pitch and the term trim is used to denote correct balance between buoyancy and weight. This critical balance between weight and displacement means that if weights are ejected, e.g. a torpedo, then a carefully metered quantity of water must be taken on board immediately to compensate.

(b) *Stability*. The stability of the submarine for heel and depth when submerged were discussed in Chapter 4. In the submerged state, longitudinal and transverse $\overline{BG}$ are the same. On the surface, the usual calculations can be applied but as the submarine dives the waterplane reduces considerably as the bridge fin passes through the surface. In this condition, B may still be relatively low and a critical stability condition can result.

Submarines are subject to special tests, the *trimming and inclining experiments*, to prove that the hydrostatic and stability characteristics are satisfactory in all conditions. The correct standard condition is then achieved by adjustments to the ballast keel.

(c) *Strength*. The pressure hull must be able to withstand the crushing pressures at deep diving depth. The hull is substantially axisymmetric to minimize bending stresses in the highly loaded circumferential hoop stress direction. In the absence of significant longitudinal bending longitudinal stiffening would be inefficient and the structure is treated as a ring stiffened cylinder.

An approximation to the effect of depth of operation on the weight of hull can be obtained by considering the simple case of a circular cylinder which is a good shape for withstanding external pressure. A sphere is better, but this form is used only for certain research vehicles. If stiffening is ignored and hoop stress is used as the design criterion then, for a given material, the permissible stress will be constant. Ignoring the ends of the cylinder, the thickness of hull plating, and hence hull weight, required will be proportional to pressure multiplied by the diameter. For a given diameter, the buoyancy is constant so the ratio of structural weight to buoyancy increases linearly with increasing pressure, i.e. depth. There will be a depth when there will be nothing available for payload. For a given depth of submergence, the ratio remains constant. If the diameter is allowed to increase, then at a given depth the hull weight increases as the diameter whereas the buoyancy increases as the square of the diameter. The proportion of the buoyancy devoted to structure is inversely proportional to the diameter. Other weights, e.g. machinery, crew, etc., are usually relatively less for a larger ship which should have a higher deadweight/displacement ratio.

Design is usually carried out assuming axial symmetry of structure and loads. This idealization enables approximate and analytical solutions to be applied with some accuracy. Subsequently detailed analyses can be made of non axisymmetric features such as openings and internal structure. The dome ends at either end of the pressure hull are important features subject usually to finite element analysis and model testing.

Initial analysis of a pressure hull with heavy transverse bulkheads is as a uniformly loaded ring stiffened cylinder, the longest compartment being taken as the most critical. The maximum mean plating hoop stress occurs circumferentially mid way between frames. The maximum longitudinal stress occurs on the inside of the plating in way of the frame, important because it is an area of heavy welding.

Buckling of the hull is possible and the following are assessed:

(*a*) Inter-frame collapse, i.e. collapse of the short cylinder of plating between frames under radial compression. Such a failure is likely to occur in a large number of nodes.
(*b*) Inter-bulkhead collapse, i.e. collapse of the pressure hull plating with the frames between bulkheads. This is sensitive particularly to the degree of out-of-circularity in construction.
(*c*) Frame tripping.

The design is developed so that any buckling is likely to be in the inter-frame mode and keeping risk of collapse at 1.5 times the maximum working pressure acceptably small. The effects of frames, shape imperfections and cold working residual stresses are allowed for empirically. Small departures from circularity can lead to a marked loss of strength. In one case, the pressure causing yield at 0.25 per cent shape imperfection on radius was only half that required for perfect circularity.

(*d*) *Dynamic stability*. This has already been discussed in Chapter 13 in some detail. The limited diving depth available for reasons of strength reduces the time available for corrective action should the vessel suddenly take on a bow down attitude. For example, at 20 knots and assuming that the vessel is already at 50 m depth with a collapse depth of 200 m, a 30 degree angle means that the vessel reaches her collapse depth in about 30 sec. If the depth of water available is less than collapse depth, as it would be in many coastal areas, then the time available is even less.

(*e*) *Powering*. For a given displacement, a submarine has a greater wetted surface area than a surface ship. This means a greater frictional resistance, which, for comparable conditions, means that the submarine must operate at depths where the wavemaking resistance is substantially reduced. In practice, this means operating at depths of the order of half the ship length or more.

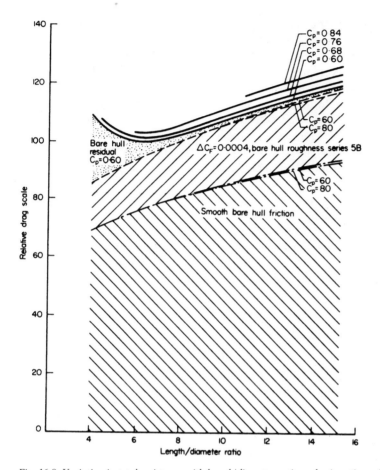

Fig. 16.8 Variation in total resistance with length/diameter ratio and prismatic coefficient

The importance of length/diameter ratio and prismatic coefficient on submerged resistance is shown in Fig. 16.8. The USS *Albacore* provided much useful data on submerged performance at high speed. It had a length/beam ratio of about 7.4 compared with the ratio of about 10 in Second World War submarines. It also had a single screw which gave high propulsive efficiency.

Sources of power in a conventional submarine have for many years been the lead–acid battery driving an electric motor for underwater propulsion and the diesel engine for charging the batteries or propelling the boat on the surface. Air is drawn down the conning tower or the snort mast. Other methods have been tried, notably the steam submarines of the 1930s and the Walter turbines using oil and high-test peroxide in the 1950s. Various alternatives to lead–acid are being introduced successfully to make much larger powers available. Even larger powers might derive from the research into hot batteries which naturally bring other problems or from fuel cells which benefit from the exothermic reaction of oxygen and hydrogen brought together to make water.

Over the years batteries have been improved to provide greater endurance underwater. However, the combat efficiency of a submarine is dependent on its ability to remain submerged and undetected, so much effort has recently been devoted to developing air independent propulsion (AIP) systems to provide some of the benefits of nuclear propulsion without the great expense (see Chapter 14). Solutions such as closed-cycle diesel engines, fuel cells and Stirling engines are being considered. The systems still require a source of oxygen such as high-test peroxide or liquid oxygen. Some have considered an artificial gill to extract dissolved oxygen from seawater. Fuel sources for fuel cell application include sulphur free diesel fuel, methanol and hydrogen.

Providing a submarine with a propulsion system to enable it to remain submerged for long periods necessitates making provision for better control of the atmosphere for the crew. The internal atmosphere can contain many pollutants some deriving from the new materials carried and others becoming important because they build up to dangerous levels over an extended period. A much more comprehensive system of atmosphere monitoring and control is needed than that in earlier conventional submarines.

COMMERCIAL SUBMARINES

So far commercial applications of the submarine concept have been limited to relatively small vessels although some have dived very deep. Many have been unmanned, remotely operated vehicles. Most of these applications have been associated with deep ocean research or exploitation of the mineral wealth of the oceans. Small submersibles are also used to rescue the crews of disabled sub-

Table 16.2
A range of submersibles

Name	Diving Depth (m)	Length (m)	Width (m)	Height (m)	Displacement (Tonnef)	Crew
Deep diving manned submersibles						
Sea Cliff	6000	7.9	3.6	3.6	24.0	3
Aluminaut	4600	15.4	3.0	5.0	67.6	6
Alvin	4000	7.6	2.4	3.9	16.7	3
Sea Turtle	3000	7.9	3.6	3.6	24.0	3
Cyana	3000	5.7	3.0	2.1	8.5	3
Deep Quest	2440	12.2	5.8	4.0	52.0	4
Shinkai 2000	2000	9.3	3.0	2.9	25.0	3
Unmanned submersibles						
CURV III	6700	3.05	2.13	2.13	4.90	Tethered
AUSS	6000	5.2	1.27	1.27	1.27	Free
Argo	6000	4.8	1.0	1.18	1.59	Tethered
Angus	4000	4.27	1.83	1.52	2.45	Tethered
SAR	4000	4.57	1.22	1.22	3.63	Tethered
SBT	1400	4.0	2.8	3.0	8.00	Tethered

marines or for investigations of shipwrecks (see Table 16.2). Another application has been for the leisure industry where submersibles take people down to view the colourful world below the sea. These naturally tend to operate in areas where water clarity is high and the fish life is abundant. Submersibles carrying 40 passengers are in service in Florida and the Caribbean.

In the above type of operations the submersible may be the only way of tackling a problem, e.g. the servicing of an oil wellhead in situ. In the leisure application, very special economic considerations apply. The carriage of bulk cargoes by submarine is unlikely to become commonplace because of the extra costs of building and operating submarines. Because the pressure hull must be cylindrical for strength efficiency, draughts for a given internal capacity are likely to be much greater than the corresponding surface ship. This complicates docking and restricts the harbours and routes such vessels can use. Special trans-shipment arrangements might be necessary. Submarine building costs are likely to be several times that of the corresponding surface ship, reducing as overall size increases. Safety would present special problems as the vessel would have very little time to respond to an emergency before exceeding its collapse depth or hitting the seabed.

To operate submerged in the ballast condition, it must be possible to introduce ballast water equal in weight to the cargo carried. This leads to a desire for a high density cargo. It would not be economic to cut large openings in the pressure hull so the cargo would ideally be capable of being loaded and discharged rapidly through relatively small openings.

Some have argued that surface units will be so vulnerable in a future hightech war that only submarines could be used with any reasonable chance of reaching the desired destination. So far such pessimism has not been borne out in the major conflicts of recent years. Also the cessation of the Cold War makes the type of scenario envisaged in such thinking less likely. Certainly the peacetime penalties associated with the construction and operation of these vessels are too great to make it likely that any country would embark upon any significant build programme.

A more likely scenario, although one not yet accepted, is the use of commercial submarines to obtain the capability to operate under ice. This might be to exploit minerals on the ocean floor or to obtain access to areas normally cut off by extensive ice fields. Such vessels would need to be nuclear powered or use some other form of air independent powering.

Container ships

There has been a revolution in the transportation of goods throughout the world. Goods may now be collected at their point of origin by lorry or train and taken to the port for sea transit. There they are deposited within a system of gantries or cranes which take them to the ship or to temporary storage until they can be embarked. The containers are standard worldwide and are called TEUs. Millions of the TEUs are available for hire in the knowledge that they will fit into ships specially built to receive them and take them to their destinations, where the reverse process takes place.

The ships themselves adopt various systems to hold the containers. The ships are like hollow shoeboxes, stowing containers below a single deck which has very large hatches. More of them may be carried above the deck in stacks. There has evolved also an open container ship constructed like a double-skinned U without a deck. The containers may be locked together by fittings, such as twist locks and are lashed so that they are unable to move even in severe weather. Alternatively, ships may be fitted with long vertical stanchions throughout so that a container can be housed between four of them. Fig. 16.9 shows a typical cross-section of an open container ship.

A huge range of ships now exists. Small coasters may carry a few containers on deck while at the other end of the scale, ships 350 metres long carrying 10,000 containers are not uncommon.

In large ships there may be 20 containers athwartships in stacks 20 high, of which a quarter at least may be above deck. Closed ships may also have stacks five or six high above the hatches and lashed in various ways to avoid movement in bad weather.

Regulations have, of course, evolved to make these ships relatively safe and losses of complete ships are rare. Containers, however, are regularly lost overboard and present a significant hazard to small ships, especially yachts, and a drain on the insurance market. Forces on the lashings in rough weather are high and the efficiency of the devices depend much on the vigilance of the crew (twist locks alone are insufficient).

Anxieties over safety have been expressed by Vossnack who points out that the statutory freeboard is extremely low, thereby keeping the underdeck tonnage (and harbour dues) low. As shown in Fig. 16.9 this gives an angle of deck edge immersion and uncontrollable flooding often around thirty degrees. Moreover, minimum crew numbers based on gross tonnage raise doubts about their adequacy in foul weather.

Further development of the carriage of standard packaged cargo such as motor vehicles may be expected. There have been studies to provide a faster service across the oceans. Demands on the design for high speed ships could be considerable in terms of machinery and air intakes for gas turbines which would reduce the number of packages that could be carried. Structure to deal with the enhanced impact loads at the bow and on the bottom would also have to be more substantial. Of course, the final decision would rest upon the predicted economy of trading off an amount of cargo against its earlier delivery.

Frigates and destroyers

These vessels cover a range of displacement from about 2500 tonnef to 6000 tonnef while having the same general roles. Lengths will be in the range 100 to 150 m, with length to beam ratio of about 8:1. The larger ships can fulfil more of these roles and operate effectively in more severe sea conditions. The two titles are imprecise, the word 'destroyer' originating as 'torpedo boat destroyer' a hundred years ago.

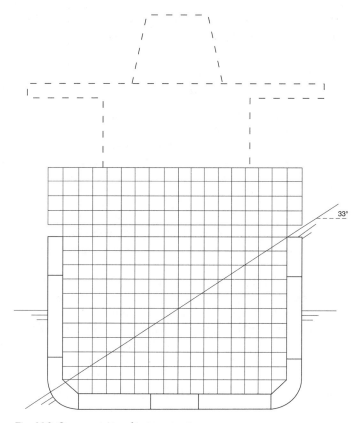

Fig. 16.9 Open container ship cross-section

The roles covered by these ship types include:

- Anti-air Warfare (AAW)
- Anti-submarine Warfare (ASW)
- Anti-surface Warfare (ASuW)
- Task force protection
- Self defence
- Shore bombardment

Because they cover a wide range of functions these vessels are often described as 'maids of all work'. All carry a gun for shore bombardment and for use against pirates or drug runners. Some are intended to provide one primary capability with limited secondary abilities in other roles. Compared to general purpose ships these 'specialist' vessels have the advantages that:

- individual ships can be smaller;
- each ship is less expensive and more hulls can be afforded on a given budget enabling a military presence to be exerted in more areas at any one time;

- the smaller ships tend to have lower signatures;
- the loss of one ship is less damaging to the total military capability of the task force although the loss may be critical to the particular mission being undertaken.

The advantages of the general purpose ship are:

- if both AAW and ASW functions, say, are needed then one ship suffices which is less expensive than two specialist hulls;
- the larger ship provides a more stable platform for the weapon systems and those systems are less likely to be degraded by adverse sea conditions;
- there is more scope for providing some duplication of, or protection to, vital services;
- the manpower requirement for a given total military capability is less.

The following remarks relate mainly to the larger ships in the range and some features will not be provided in smaller hulls. A typical profile is shown in Figure 16.10.

For their size, these vessels have a high level of military capability. Indeed, some 60 per cent of the cost of a frigate is devoted to its fighting capability, compared with 25 for its mobility function and 15 for its float function. Because of this, careful attention must be paid to their vulnerability. Susceptibility to attack is reduced by giving them low signatures—radar cross-section, acoustic, infra-red, and magnetic. They cannot be made invisible but low signatures make their initial detection more difficult, makes it harder for incoming missiles to lock on to the target and makes it more likely that countermeasures will be effective.

The design must cater for some enemy weapons striking the ship. The most likely attack scenarios must be analysed and the layout and structure arranged

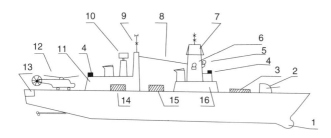

Key: 1. Bow sonars
      2. Gun
      3. Long range AA
         or ASW system
      4. Close-in weapons
      5. Navigation radar

      6. Satellite communications
      7. Phased array radar
      8. Roof aerial
      9. Communications
      10. Search radar

      11. Hangar
      12. Helicopter
      13. Towed array sonar
      14. Torpedo system
      15. Ship/surface
          weapon system
      16. Command and
          control

Fig. 16.10 Typical frigate profile

so as to limit the extent of damage and increase the chances of the ship being able to maintain at least partial weapon capability for those weapons needed for the mission. Power supplies and other vital services must remain available. Typical analysis methods were touched upon in Chapter 5. Zoning can be used to reduce the spread of flooding and fire and to make vertical sections more autonomous as regards most essential services. Double skin bulkheads can help contain blast and reduce splinter penetration. Box girders under the upper deck can increase the residual longitudinal strength after damage and provide protected runs for vital services. Siting important compartments low down, or well away from the ship side, makes them less vulnerable. Providing separate machinery spaces, separated if possible, for each shaft reduces the risk of the ship being left helpless in the water.

One lesson that has been painfully learnt many times is that last minute changes to the design to reduce procurement cost or size is very likely to increase vulnerability to attack.

Vessels intended to act as escorts must be capable of relatively high speed, perhaps 28 to 30 knots. They may need to change their position in the task force screen or regain their position in the screen after prosecuting an anti-submarine attack. However, they are likely to spend a lot of time cruising at economical speeds, perhaps 14 knots. To provide reasonable economy at cruising speed combined with ability to go fast, these ships usually have a combined machinery plant. As described in Chapter 14, these typically involve gas turbines for the high speed in addition to, or in place of, diesels provided for cruising. A 6000 tonnef ship will have machinery giving some 40 MW. Block and waterplane coefficients will be about 0.50 and 0.75 respectively. To provide good manoeuvrability twin shafts and rudders are often fitted.

The weapon systems must be chosen so that the ship can act independently or as part of an integrated task force, providing defence in depth. In a task force, for instance, an aircraft carrier would provide the first line of defence from air attack by its fighters. Helicopters, from the carrier or the escorts, would provide long range anti-submarine defence. Progressively the long range, medium range and finally the close-in defensive weapon systems come into play. The escort's duty is to provide cover for the ships it is escorting and then for itself. At some point the command must decide whether, or not, to deploy decoys to seduce the incoming attack. Computer-aided command systems help to ensure good, timely, decisions are made.

A wide range of communication frequencies are required and the ship will usually have a roof aerial and excite various structures such as masts. To meet the separation and height requirements for communications, the superstructure is often in two main sections. The space in between can be utilized for replenishment at sea and weapons.

High speed small craft

There is considerable scope for debate as to what is meant by both 'high speed' and 'small'. In this section the boundaries are drawn quite widely so as to embrace a number of interesting, and often technically challenging, hull configurations and propulsion systems. Each was introduced to overcome problems with earlier forms or to confer some new advantage. Thus catamarans, and other multi-hull configurations, avoid the problem of loss of stability at high speed suffered by round bilge monohulls. They also provide large upper deck spaces. Hydrofoil craft reduce resistance by lifting the main hull clear of the water. Air cushion vehicles give the possibility of having all the craft clear. Apart from reduced resistance this provides a degree of amphibiosity. The effect of waves on performance is minimized by the Small Waterplane Area Twin Hull (SWATH) concept. Some designs are tailored specifically to reduce wash so that they can operate at higher speeds in harbours or on waterways.

The choice of design must depend upon the particular requirements of the service for which it is intended. In some cases the result is a hybrid and the number of possible permutations is very large. Also, although most applications of these concepts have been initially to small craft some are now appearing in what may be termed medium size, especially for high speed ferry service. For simplicity, in the following sections the concepts are dealt with individually.

MONOHULLS

Most high speed small monohulls have until recently been hard chine forms. A notable exception were the German E boats of the Second World War. With more powerful small engines, round bilge forms have been pushed to higher speeds and have experienced high speed stability problems. For the hard chine forms, greater beam and reduced length give improved performance in calm water but lead to high vertical accelerations in a seaway. Their ride has been improved by using higher deadrise angles leading to a 'deep vee' form. This form was used, for instance, in the Atlantic Challenger *Gentry Eagle*.

Current practice is generally to favour round bilge for its lower power demands at cruising speed and for its seakindliness, but to move to hard chine at Froude numbers a little above 1.0 because of the stability problem. One advantage of the round bilge form in seakeeping is that it can be fitted with bilge keels much more readily than can chine forms.

MULTI-HULLED VESSELS

There are many applications: sailing catamarans, off-shore rigs, diving support vessels and ferries. The concept is not new. Two twin hulled paddle steamers of about 90 m length were built in the 1870s for cross channel service. One had two half hulls connected by cross girders and driven by paddle wheels placed in the parallel sided tunnel between the hulls. The other had two complete hulls. Both ships had a good reputation for reduced rolling, rolling only 5 degrees when other ships rolled 15 degrees.

The upper decks, spanning the two hulls, provide large areas for passenger facilities in ferries or for helicopter operations. In research vessels or mine countermeasure vessels they provide space for deployment of towed bodies of various kinds. General comparisons with monohulls are difficult because it depends whether such comparisons are made on the basis of equal length,

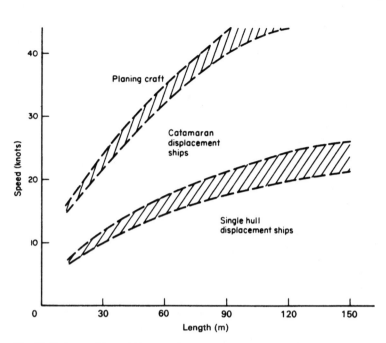

Fig. 16.11 Areas of favourable operation

displacement, or carrying capacity. What is needed are two design solutions each meeting the owner's requirements. The comparison in Fig. 16.11 is based on results for hulls of the same length and draught, the beam of the monohull being twice that of each catamaran hull. This is unfair to the monohull. The greater wetted area of the catamaran leads to increased frictional resistance but their relatively fine hulls lead to reduced wave resistance at higher speeds. There will be interference effects between the two hulls. These will be less at high separation but this may make docking difficult and lead to excessive transverse stability. A reasonable separation of the hulls is about 1.25 times the beam of each. Generally the manoeuvrability of multi-hulls is good.

The increased transverse stability and relatively short length mean that good seakeeping is not their strongest point. Improvements in this respect have been obtained in the wave piercing catamarans developed to reduce pitching, and in the SWATH designs where the waterplane area is very much reduced and longitudinal motions can be reduced by the use of fins or stabilizers if necessary.

Multi-hull designs suffer from a relatively high structural weight and to preserve payload some designs use aluminium to reduce structural weight. Wave impact on the cross structure must be minimized and high freeboard is needed together with careful shaping of the undersides. SWATH ships, because of their very small waterplane area are very sensitive to changes in load and its distribution. A system of water compensation is needed and this ballasting system can help mitigate heel due to damage leading to partial flooding of one hull.

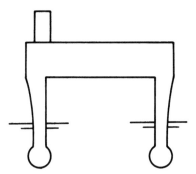

Fig. 16.12 SWATH concept

Propulsion of SWATH ships clearly invites a prime mover in each pod, or at least a propeller on each. For ships below about two thousand tonnef, the walls are insufficiently wide to permit the passage of large prime movers and designers have to conceive means of developing the necessary power in the 'tween decks and delivering it either as jet propulsion above water or to propellers on the ends of the submerged pods. Bevel gearing, conventional or superconducting electrical devices and hydraulic motors are all possibilities, although the driving motors themselves may not be readily removable for refitting.

In recent years, even for ships of significant size, such as frigates, considerable interest has developed in trimarans which have a long slender central hull with two narrow side hulls. The advantages claimed for this form are:

- reduced resistance and hence power for a given speed. (Said to be about 18 per cent less power for 28 knots in an escort sized vessel.) Greater fuel economy;
- improved seakeeping performance at high speed. Operational in higher sea states;
- large deck area, improved stability and reduced motions for helicopter operations;
- increased directional stability;
- better top weight growth margins.

Several studies have shown this configuration to have advantages for a wide range of applications from quite small ships up to aircraft carriers and cruise ships. To prove the validity of the concept the UK MOD decided to invest in a 97 m, 1100 tonnef displacement, demonstrator vessel, which was launched in 2000. This is RV *Triton* with an overall beam of 22.5 m, main hull beam 6 m, side hull beam 1 m, and maximum draught 3.2 m. Powering is diesel electric and maximum speed 20 knots. *Triton* was built to DNV High Speed and Light Craft Rules.

SURFACE EFFECT VEHICLES

Vessels which benefit from an aerostatic force are called variously cushion craft, ground-effect machines, hovercraft, surface-effect ships and sidewalls. The

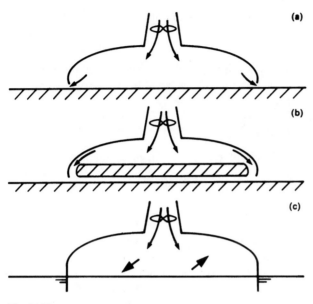

Fig. 16.13

aerostatic force is generated by a downward current of air creating an air cushion beneath the craft of which there are three general types:

(*a*) plenum chamber craft;
(*b*) peripheral jet craft;
(*c*) sidewall craft.

The plenum chamber craft is typified by the lawnmower of that design. Air is maintained in a plenum chamber and escapes around the periphery. Some rudimentary theory can be deduced to give an idea of the importance of the various parameters.

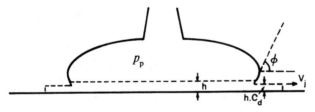

Fig. 16.14

The potential represented by the gauge pressure in the plenum p_p is converted, according to Bernouilli's law into the kinetic energy of discharge. Assuming that the discharge velocity V_j is large relative to the plenum air velocity,

$$p_p = \tfrac{1}{2}\rho V_j^2$$

The rate of mass flow of air escaping from the periphery of length l is

$$\dot{m} = lhC_d\rho V_j$$

C_d is a coefficient of discharge which is in practice dependent upon the angle ϕ. In a steady state the weight of the vehicle W is equal to the aerostatic force F,

$$W = F = p_p A = \frac{1}{2}\frac{\dot{m}^2 A}{\rho l^2 h^2 C_d^2}$$

A is the planform area. For a circular body which gives the largest ratio of planform area to periphery.

$$W = \frac{\dot{m}^2}{8\pi C_d^2 \rho h^2}$$

Thus the vehicle weight W can be supported by a fan whose necessary capacity $\dot{m}$ diminishes with the clearance h over the surface. Moreover if h is decreased during operation the aerostatic force exceeds the weight so that the body is restored to the equilibrium position, i.e. there is vertical equilibrium. By a similar argument it is clear that there is also stable equilibrium if tilt about a horizontal axis occurs.

The peripheral or annular jet craft is more common because the air flow is more controllable. Rudimentary theory is rather less accurate. However, the value of a small value for h remains and the designer is faced with the problem of achieving a good lift using a small ground clearance yet needing a large ground clearance for the avoidance of obstacles and at sea, waves. This is overcome by making the lower part of the craft elastic using a heavy rubber skirt. Much research has been needed to produce skirts which are adequately robust. Truly amphibious craft result.

Because the hovercraft is above the water it has a low lateral resistance to disturbance by wind. If it is driven by air propellers they may have to be vectorable to control the positioning in wind and the stability in manoeuvre has to be the subject of study much like that of an aircraft. Large air rudders are consequently not unusual. Where a high degree of lateral stability is needed the two side walls of a rectangular hovercraft are extended into the water. The two ends of the craft remain sealed by rubber skirts to contain the air cushion. Such sidewall hovercraft, while no longer amphibious, nevertheless retain many of the advantages of the true hovercraft. Moreover, if the walls are now thickened, they provide a vertical buoyancy force so that the aerostatic force need not be so large. The designer must effect the compromise among these features which suits the particular needs.

As a craft hovers over the water, there is an indentation in the water which obeys Archimedes' Principle, i.e. its volume multiplied by water density is equal to the weight of the hovercraft. When the craft moves, the indentation moves with it causing transverse and divergent wave systems and a wave resistance just like a displacement ship. Sea friction is of course much reduced although there is some increase in the air resistance and an addition due to the dipping skirt.

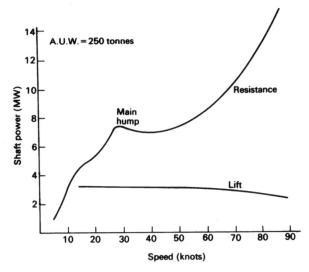

Fig. 16.15

As the craft increases speed, there comes a time when the indentation cannot properly keep up and the wave resistance enjoys a sudden reduction. The total resistance of the craft is characterized by Fig. 16.15.

High speed becomes readily possible with the sorts of power units that can be accommodated, representing one of the craft's major advantages. The resistance to motion of a hovercraft has in fact three components, each requiring study:

(*a*) aerodynamic resistance which varies as (velocity)^2 and includes components for both the vehicle and the cushion itself;
(*b*) wave-making resistance which has a peak at low speeds and then falls away to a negligible value;
(*c*) momentum resistance which varies linearly with speed. This resistance arises from the fact that the air drawn into the craft leaves it at zero velocity relative to the craft and has therefore experienced an overall change of momentum which is proportional to the craft's velocity.

Another of the important advantages of the hovercraft over displacement craft is its relative invulnerability to underwater explosion, making it a good candidate for minehunting duties. Like the hydrofoil, its payload is relatively small and aluminium alloy aircraft standard construction is often advisable, especially in small craft. With their high power-to-weight ratio, gas turbines are often preferred for the propulsion units, both for the lift fans and the driving engines, although high speed diesels are not uncommon for craft operating at around 40 knots.

Seakeeping is generally poorer at the same sea state than for many other types of craft. Limiting sea states for various types of craft are shown roughly in Fig. 16.16.

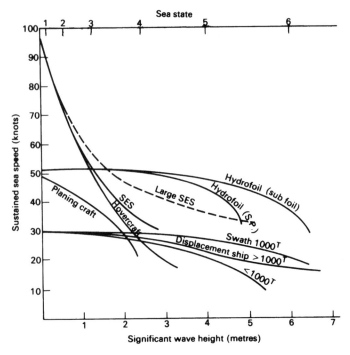

Fig. 16.16

HYDROFOIL CRAFT

A hydrofoil moving at speed through water can generate considerable lift, and if an efficient cross-section is chosen the associated drag will be relatively low. If hydrofoils are fitted below a conventional high speed craft, they generate

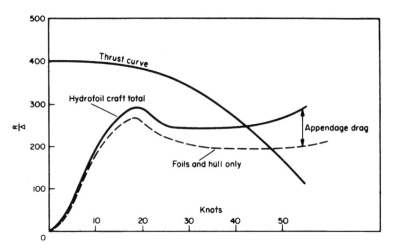

Fig. 16.17 Resistance curve for hydrofoil craft

increasing lift as the speed increases, the lift being proportional to the square of the velocity. If the craft has sufficient power available, there will come a time when the lift on the foils is sufficient to lift the hull completely clear of the water. Having lost the resistance of the main hull, the craft can accelerate until the resistance of the foils and air resistance absorb the power available. A typical curve of R/Δ against $V/\sqrt{L}$ is shown in Fig. 16.17. The hump in the curve is associated with the very high wave resistance experienced just before the hull lifts clear of the water.

After the hull has lifted clear of the water, the lift required from the foils is constant. Thus, as the speed increases further, either the angle of incidence of the foil must reduce or the immersed area of the foil must decrease. This leads to two basic types of foil system, viz.:

(a) *Surface piercing foils* in which, as the craft rises higher, the area of foil immersed reduces as it passes through the water surface;
(b) *Completely submerged, incidence controlled foils* in which the foils remain always submerged and the lift generated is varied by controlling the angle of attack of the foils.

These two systems are illustrated in Fig. 16.18.

Longitudinal balance must also be maintained, and it is usual to have a large foil area just forward or just aft of the longitudinal centre of gravity with a small foil at the stern or bow respectively. Any ratio of areas is feasible provided the resultant hydrodynamic force acts in a line through the c.g. The planform geometries are also illustrated in Fig. 16.18.

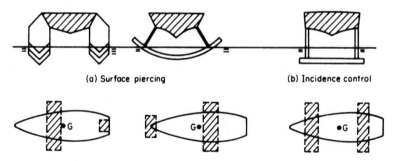

(a) Surface piercing   (b) Incidence control

Fig. 16.18 Basic foil geometries

So far a calm water surface has been assumed. To understand what happens in waves, consider a surface piercing system as in Fig. 16.19. As the craft runs

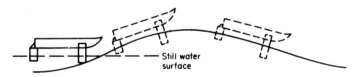

Fig. 16.19 Surface piercing system in waves

into the wave surface, the water level rises on the forward foil. The lift on the forward foil increases and this has the effect of raising the bow, keeping it clear of the wave surface. Having passed the crest of the wave, the process is reversed and the craft more or less 'contours' the waves. The more rapid the change of lift with draught on the foils the more faithfully will the craft follow the wave surface. By adjusting the rate of change, the movement can be lessened, giving a smoother ride but a greater possibility of the craft impacting the wave surface.

With the fully submerged foil system, the foil is unaware of the presence of the wave surface except through the action of the orbital motions of the wave particles. Thus, to a first order, this type of craft can pursue a level path which has attractions for small wave heights. In larger waves, the lift on the foils must be varied to cause the craft partially to contour the wave profile. In a small craft, the variation can be controlled manually but, in craft of any size, some form of automatic control is required which reacts to a signal from an altitude sensor at the bow.

It follows, that the same process which causes the craft to respond to variations in height of the water surface also provides the craft with a measure of trim stability. Roll stability will be present in a surface piercing system if the lift force which acts as the craft rolls, intersects the middle line plane of the craft above the vertical c.g. With a fully submerged foil system, roll stability is provided by means of flaps or ailerons which act differentially on the two sides of the craft in such a way as to provide a moment opposing the roll angle. This, again, is controlled by signals produced by a stable element in the craft.

Both types of hydrofoil have operated successfully for many years. Their role needs to be carefully tuned to their characteristics because, like most high performance craft they are not cheap either to buy or to run. High-speed passenger traffic in relatively calm water—up to sea state 4 or perhaps 5—has proved profitable while a 'presence' role in offshore surveillance may also be an important application. At high sea states, should the craft for any reason come off its foils, it is sometimes difficult to get up again and the craft is left wallowing in some discomfort. Impact with the water at speed should the craft come off its foils can be severe and the fore ends of these hydrofoil vessels need special strengthening and good subdivision. Aircraft standard construction is necessary in order to preserve a worthwhile payload. Aluminium alloy and fibre reinforced plastic are common.

Propulsion by water jet above water at top speeds is surprisingly efficient. The jet may be created by a diesel or gas turbine-driven high-performance pump. This avoids the need for bevel gearing for a drive down the struts or the highly angled shafting for a drive by a propeller in the water. Wind propulsion of a hydrofoil craft offers a fascinating challenge to any enthusiastic naval architect.

INFLATABLES

Inflatables have been in use for many years and, with a small payload, can achieve high speeds. The rigid inflatable is used by the Royal National Lifeboat

Institution. The rigid lower hull is shaped to make the craft more seakindly and the inflatable principle safeguards against sinking by swamping.

The RIB concept continues to develop rapidly and the craft are widely used by commercial firms, the military and other government departments. They are available in lengths up to about 16 m with speeds of up to 80 knots, although most operate in the range 30 to 40 knots. Petrol and diesel fuelled in-board or out-board propulsion units are common and some utilize waterjet propulsion. Single hull and catamaran versions are produced. The larger units come with wheel houses and in some cases are in competition with the fast planing craft.

The early RIB was a wooden hulled boat surrounded by an inflatable tube. The hull is now usually fabricated in GRP or polyethylene. A lot of research has gone into developing strong, durable materials for the collars. For further safety the collars are sub-divided. Besides being used in the leisure industry, the speed of RIBs makes them attractive to the military, the police and coastguard for life saving and for intercepting smugglers and gun runners. The offshore industry also makes wide use of them for diving work. Many are certified for SOLAS requirements as fast rescue craft.

COMPARISON OF TYPES

All the types of vessel discussed in this section have merits and demerits. A proper comparison can only be made by producing design studies of each to meet a given requirement and, as said earlier, the best solution may be a combination of more than one concept. Some requirements may point directly to one form, e.g. a landing craft capable of running up onto a hard surface may suggest an air cushion vehicle. This, however, will not be the usual situation.

Many of the craft in use today of these types are passenger carrying. The vast majority of operational SESs are used commercially for fast passenger transport and that, with speeds of over 40 knots commonplace, services can compete with air transport. Hydrofoils also enjoy considerable popularity for passenger carrying on short routes, e.g. the surface piercing Rodriguez designs and the Boeing Jetfoil with its fully submerged foil system. Catamarans are much used as high speed passenger ferries. Because of this common characteristic for carrying people, Fig. 16.20 presents an analysis of the BHP/passenger plotted against speed. This shows that hydrofoils generally need more power than catamarans and SESs and that SESs are very economical at high speed.

Offshore engineering

The spectacular recovery of gas, oil and minerals from the sea has presented naval architects with a fascinating range of problems. Drilling rigs, permanent platforms, buoyant terminals, submersible search vehicles, underwater habitats and many different servicing ships have all had to be designed to serve the industry.

In shallow waters drilling towers with their working platforms are often sunk on legs to the sea bottom, there to be secured by piles or mooring devices.

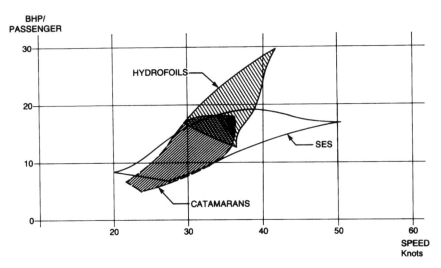

Fig. 16.20 BHP per passenger

Application of the simple hydrostatic and structural theories relevant to any floating body is straightforward.

Deeper waters and weather as fierce as the North Sea demand sterner measures. Two types of rig now predominate in such conditions, the one for exploration and the other for permanent mooring over the wellhead. The most favoured of the exploration rigs is now the semi-submersible. This comprises often two underwater cylinders each surmounted by two, three or four columns which support the drilling platform a long way above water level. Sometimes the cylinders are bent round to form a horizontal ring. Supporting a total payload of several thousand tonnes of pipes and mud they may readily exceed twenty thousand tonnes in displacement. The waterplane comprises four, six or eight separated shapes which importantly affect the hydrostatic behaviour of the rig including its static stability. The horizontal submerged cylinders are deep enough to be affected little by the surface waves and the inertia of the whole system is huge so that motion is not a problem. Structural strength, on the other hand, presents important difficulties due to both wind and wave dynamic loading and due to a wave crest above one cylinder with a trough above the other. Dynamic structural behaviour is in fact very important and the modal patterns of the structures under fluctuating excitation by the sea must be determined.

Such exploration platforms need to be mobile, often under tow by tugs but some of them are self propelled. More importantly they need to be kept still over the wellhead and in consequence, they employ vital positioning and control systems. These often comprise groups of propellers in tunnels or in retractable vectorable housings all controlled from computer systems fed by sensing devices. In addition, the rigs need to be hotels, to carry helicopters, drilling equipment, pipe racks, mud tanks, cement tanks, firefighting and life-saving equipment and diving facilities.

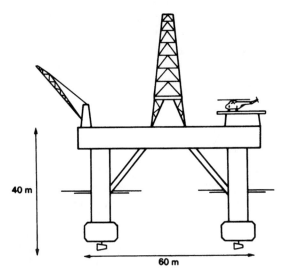

Fig. 16.21

Similar platforms with rather less elaborate arrangements can be used over the top of completed wells or, more usually groups of wells which have been capped to act as an oil terminal. Such a semi-submersible would be fixed in position by mooring cables and anchors spread out on the seabed in all directions. In preference, the oil industry today often employs the tension leg platform. This is a pontoon which would otherwise be floating freely if it were not held down by groups of hawsers fixed to the seabed. The buoyancy of the pontoon increases with the rise of tide so the pontoon has to be quite deep. The hawsers are sometimes spread to minimize lateral movement. Living quarters and the working platform are elevated high above the pontoon and the waves by lattice structures.

Vast underwater oil storage tanks are necessary where it is not possible to pipe the products ashore directly. Groups of capped wells feed the storage tank which disposes of its oil to a terminal at a sufficiently safe distance for bulk tankers to moor and take it on board. These terminals are not just simple buoys. Hostile seas cannot be allowed to interfere with the taking up of the oil which is made as automatic as is possible. The terminal might comprise, for example, a large horizontal vee into which the tanker nestles, a tall tower carrying the oil hose surmounting it.

Oil and gas pipes on the sea bottom have to be inspected regularly. From the outside this may be done by a television camera mounted on a mobile saddle over the pipes or from a submersible which may be manned. Inside the pipes inspection is performed from vehicles called slugs which may record the condition of the welds and pipe material over very many miles, logging its position with great accuracy. On the surface many different types of ship are required to service and protect the rigs, fight fires and provide search and rescue. Compression chambers on board ship or sea bottom, or habitats for housing groups of people allow divers to remain under pressure for several days, avoiding the prolonged process of decompression after a single dive.

Recovery of minerals from the seabed has hardly begun on a major scale because it is not yet an economic venture. When it does become so, there will emerge a need for special types of vehicles of all descriptions, providing yet another wealth of interest for the naval architect.

Tugs

Tugs perform a variety of tasks and their design varies accordingly. They are needed to pull or push dumb barges or pull drones in inland waterways; they are needed to pull or push large ships in confined waters and docks, and they are needed to tow large ships on long ocean voyages. Concern for the impact on the environment of an incident involving spillage of oil from a tanker (or, indeed, any other hazardous cargo or normal bunker fuels) has led to the concept of the escort tug. Tugs are broadly classified as inland, coastal or ocean, the largest of the ocean tugs approaching 1000 tonnef in displacement. They are capable often of firefighting and salvage duties and may carry large capacity pumps for these purposes.

Essentially a tug is a means of applying an external force to the vessel it is assisting or controlling. That force may be applied in the direct or the indirect mode. In the former the major component of the pull is provided by the tug's propulsion system. In the indirect mode most of the pull is provided by the hydrodynamic lift due to the flow of water around the tug's hull, the tug's own thrusters being mainly employed in maintaining the tug's attitude in the water.

Apart from the requirements arising from the above, the main characteristics of tugs are:

(a) hull form and means of propulsion designed both for a given freerunning speed and a high thrust at zero speed (or bollard pull) or economical towing speed;
(b) upper deck layout to permit close access to ships with large overhang;
(c) a towing point above the longitudinal centre of lateral pressure, usually just aft of amidships on the centre line: the towing wire is often required to have a 180 degree clear sweep;
(d) good manoeuvrability;
(e) adequate stability when the towing wire is athwartships and either veering from a self rendering winch or about to break.

Hull form is based on normal considerations of minimum resistance for the maximum free running speed which, for ocean tugs, is usually about 20 knots and for river tugs 12–16 knots. There are several restrictions to the selection of form; there is often a restriction on length, particularly for inland craft and frequently a need for minimum draught. Air drawing to propellers must be prevented, usually by adopting wide flat sections aft which give the propellers physical protection as well. A block coefficient of 0.55–0.65 is usual. The choice of propulsion unit is of fundamental importance because, like the trawler, there are two quite different conditions to meet, each at high efficiency—required free running speed and required bollard pull at zero speed or pull at towing speed.

Another way of classifying tugs is by the type and position of the propulsor units.

(a) *Conventional tugs.* These have a normal hull form and a traditional propulsion system of shafts and propellers. The propellers may be open or nozzled and of fixed or controllable pitch. These tugs may have steerable nozzles or vertical axis propellers. Some still employ paddle wheels. The main characteristics of these various propulsors are described in Chapter 10. Conventional tugs usually tow from the stern either with a tow hook or from a winch. They push with the bow.

(b) *Stern drive tugs.* These have a conventional hull form forward but the stern is cut away to provide room for twin azimuthing propellers. These propellers, which may be of fixed or controllable pitch, are in nozzles and can be turned independently through 360° providing very good manoeuvrability. Propeller drive is through two right angle drive gears and for this reason these vessels are sometimes called Z-drive tugs. They usually have their main winch forward and tow over the bow or push with the bow.

(c) *Tractor tugs.* These have an unconventional hull form. The propulsors are sited about one-third of the length from the bow under the hull, protected by a guard. A stabilizing skeg is fitted aft. Propulsion is by azimuthal units or vertical axis propellers. They usually tow over the stern or push with the stern.

In most operations involving tugs the assisted ship is moving at relatively low speed. In the escort tug concept the tug may have to secure to, run with and, in the event of an incident, control the assisted vessel at 10 knots or more. The success of such operations must depend upon the prevailing weather conditions and the proximity of land or underwater hazards, as well as the type and size of tug. Some authorities favour a free-running escort as not adding to the danger to ship and tug in the majority (event free) of operations. The tug would normally run ahead of the ship but has the problem of connecting up to it in the event of the ship experiencing difficulty. For that reason other authorities favour the tug being made fast to the escorted ship either on a slack or taut line.

The direct pull a tug can exert falls off with speed and indirect towing will be more effective at the higher speeds. One stern drive tug, displacement 614 tonnef, operational speed 14.5 knots and a static bollard pull of 53 tonnef, is capable of steering a 130,000 DWT tanker over the range 5.9 to 8.8 knots using the indirect method and below 5.9 knots using the direct method. This was on a course simulating an approach to Fawley on Southampton Water. With the tanker at 10 knots, engines stopped with rudder amidships, the tug brought her to rest in 15 minutes over an almost straightline distance of 1.25 miles.

Upper deck layout is dictated by the need to get close in to a variety of vessels and by the need to keep the towing point above the longitudinal centre of lateral pressure so that a lateral pull has a minimum effect on manoeuvrability. In a conventional tug (Fig. 16.22) the entire after half of the weather deck has only low obstructions and low bulwarks with tumble home and large freeing

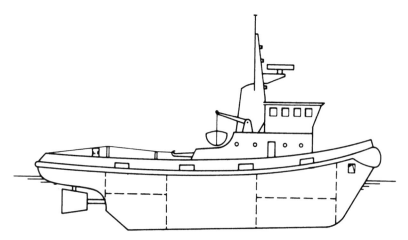

Fig. 16.22 Conventional tug profile

ports. Special towing hooks and slips are fitted. Superstructures are kept small and away from the sides where they might otherwise foul the attended ships. Hard wood fendering is fitted around the pushing areas and the structure inboard of these areas is reinforced.

A dangerous condition arises when the towing wire is horizontal and athwartships tending to capsize the tug. A self-rendering winch or a wire of known breaking strain limits the amount of the pull the tug must be capable of withstanding without undue heel. $\overline{\text{GM}}$s of 0.6 m are not unusual. Integral tug/barge systems can give good economy by creating higher utilization of the propulsion section in association with several barges. The concept has been applied to combinations up to 35,000 tonne dead mass.

Fishing vessels

Fishing vessels have evolved over thousands of years to suit local conditions. Fish which live at the bottom of the sea like sole, hake and halibut and those which live near the bottom like cod, haddock and whiting are called demersal species. Those fish which live above the bottom levels, predominantly such as herring and mackerel, are called pelagic species. There are also three fundamental ways of catching fish:

(*a*) by towing trawls or dredges;
(*b*) by surrounding the shoals by nets, purse seines;
(*c*) by static means, lines, nets or pots.

These distinctions enable fishing vessels to be classified in accordance with Table 16.3.

The commonest type of fishing vessel is the trawler which catches both demersal and pelagic species. The trawl used for the bottom is long and stocking shaped and is dragged at a few knots by cables led to the forward gantry on

Table 16.3
Fishing vessel classification

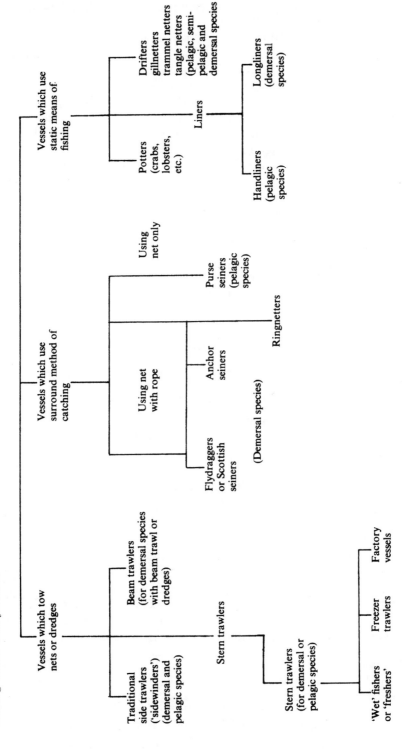

the ship. When the trawl is brought up it releases its catch in the cod end down the fish hatch in the trawl deck. Operations are similar when trawling for pelagic species but the trawl itself has a wider mouth and is altogether larger.

Trawlers suffer the worst of weather and are the subject of special provision in the freeboard regulations. They must be equipped with machinery of the utmost reliability since failure at a critical moment could endanger the ship. Both diesel and diesel-electric propulsion are now common. Ice accretion in the upperworks is a danger in certain weather, and a minimum value of $\overline{GM}$ of about 0.75 m is usually required by the owner. Good range of stability is also important and broaching to is an especial hazard.

Despite great improvements in trawler design significant numbers of vessels are lost every year and many of them disappear without any very good explanation. It is probable that such losses are due to the coincidence of two or more circumstances like broaching to, open hatches, choked freeing ports, loss of power, critical stability conditions, etc.

To give adequate directional stability when trawling, experience has shown that considerable stern trim is needed, often as much as 5 degrees. Assistance in finding shoals of fish is given by sonar or echo sounding gear installed in the keel. No modern trawler is properly equipped without adequate radar, communication equipment and navigation aids. A typical stern fishing trawler is shown in Fig. 16.23.

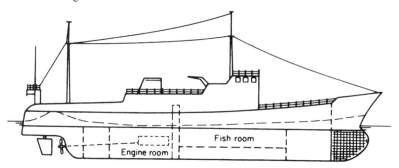

Fig. 16.23 Stern trawler

The trawler was the first type of ship for which a special analysis of resistance data was produced. A regression analysis of trawler forms for which tank tests have been made showed that a total resistance coefficient $C_R = RL/\Delta V^2$ is found to be a function of six geometrical parameters of the ship's form, L/B, B/T, C_m, C_p, longitudinal position of LCB and half angle of entrance of waterplane. From these, the power/speed curve can be produced to within an accuracy of a few per cent without the expense of tank tests.

Yachts

For many years, the design, construction and sailing of yachts has been a fascinating art about which whole books are regularly published. This is because the science is too complex for precise solution—and indeed, few yachts-men would wish it otherwise. Some tenable theories have, in fact, been evolved to help in explaining certain of the performance characteristics of sailing boats.

A yacht, of course, obeys the fundamental theory described generally in this book for all surface ships. In addition, a yacht is subject to air forces acting on the sails and to water forces due to its peculiar underwater shape—forces which

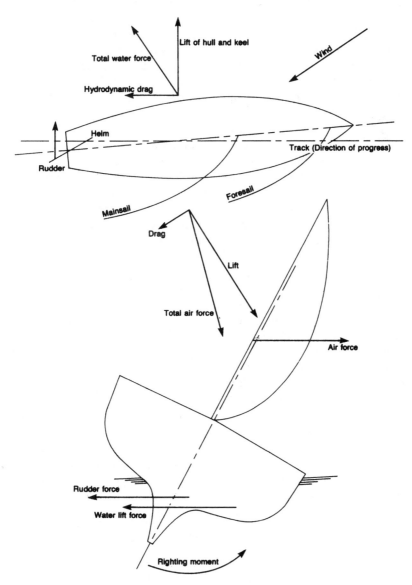

Fig. 16.24 *Principal forces acting on a boat sailing into the wind*

are negligible for ordinary surface ships. Sailing before the wind, a yacht is propelled by:

(*a*) the vector change of momentum of the wind, deflected by the sails;
(*b*) lift generated by the sails acting as aerofoils; because an aerofoil requires an angle of incidence, yachtsmen prefer sailing with wind on the quarter rather than dead astern, particularly when flying a spinnaker, to give them more thrust.

When sailing into the wind, the yacht is propelled only by a component of the lift due to the sails, acting as aerofoils. Lift and associated drag depend upon the set of the sails, their sizes, shapes, stretch and material, the angle of heel of the boat, the relative wind velocity and the presentation of the sails to the wind and to each other. Because the yacht does not quite point in the direction in which it is going and is also heeled, the hull too acts as an aerofoil, experiencing hydrodynamic lift and drag which exactly balance out the air forces when the boat is in steady motion.

The transverse couple produced by air and water forces is reacted to by the hydrostatic righting moment to keep the boat in stable equilibrium. Large angle stability and dynamical stability are clearly of great importance.

Longitudinally, the relative position of hydrodynamic lift, the centre of lateral resistance and the lateral component of air lift determine whether the rudder carries weather helm, as shown in Fig. 16.24, or lee helm. Lee helm is dangerous because, if the tiller is dropped or the rudder goes free by accident, the boat will not come up into the wind but veer away and increase heel. Ideally, to minimize rudder drag, a yacht should carry slight weather helm in all attitudes and it is towards this 'balance' that yacht designers aim.

The resistance of a yacht calculated or measured in the manner conventional for surface ships, is not a very helpful guide to the yacht designer. Minimum resistance is required at small angles of yaw and small angles of heel, and these are different from the conventional figures. Resistance in waves is also of considerable importance and varies, of course, with the response of the boat to a particular sea—because of augment of resistance due to pitching, a yacht may well sail faster in Force 4 conditions than it does in Force 5, or better in the Portland Reaches than off Rhode Island.

A yacht designer must therefore achieve minimum resistance yawed, heeled and in waves, good longitudinal balance in all conditions and satisfactory stability. The rig must not upset the longitudinal balance and must give maximum performance for sail area permitted, in all conditions and direction of wind, particularly close to the wind. While some theory helps, the process overall remains much an art.

Annex (Related to Chapter 11)

The Froude 'constant' notation (Froude, 1888)

This form of presentation is truly non-dimensional, although appearing a little strange at first to those used to the more common forms of non-dimensional presentation used in general engineering. It is not used these days but there is a lot of earlier data in this form. For that reason it is necessary to give students an introduction to the notation and to retain the Imperial units in which the data was derived.

As the characteristic unit of length, Froude used the cube root of the volume of displacement and denoted this by U. To define the ship geometry he used the following

$$\text{\textcircled{M}} = \text{length constant} = \frac{\text{wetted length}}{U}$$

$$\text{\textcircled{B}} = \text{breadth constant} = \frac{\text{wetted breadth}}{U}$$

$$\text{\textcircled{D}} = \text{draught constant} = \frac{\text{draught at largest section}}{U}$$

$$\text{\textcircled{S}} = \text{wetted surface constant} = \frac{\text{wetted surface area}}{U^2}$$

$$\text{\textcircled{A}} = \text{section area constant} = \frac{\text{section area}}{U^2}$$

As performance parameters he used relationships between speed and displacement and speed and length

$$\text{\textcircled{K}} = \frac{\text{speed of ship}}{\text{speed of wave of length } U/2} = V\left(\frac{4\pi}{gU}\right)^{\frac{1}{2}}$$

$$\text{\textcircled{L}} = \frac{\text{speed of ship}}{\text{speed of wave of length } L/2} = V\left(\frac{4\pi}{gL}\right)^{\frac{1}{2}}$$

And as a resistance constant

$$\text{\textcircled{C}} = \frac{1000 \times \text{resistance}}{\Delta \text{\textcircled{K}}^2}$$

In verbal discussions, $\text{\textcircled{M}}$ is referred to as 'circular M', $\text{\textcircled{B}}$ as 'circular B', and so on.

The resistance used in calculating C can be the total, residuary or frictional resistance and these can be denoted by C_T, C_R and C_F respectively.

Standard expressions can be derived for each constant as in Table A.1. In this form they are dimensional.

For comparison of design data, the results from model tests were plotted in the form of C–K curves with the data corrected to represent a 16 ft model. Separate curves are drawn for each displacement value at which the model is run. (Fig. A.1.) Superimposed on the plots are curves showing the skin friction correction in passing from the 16 ft model to ships of various length.

This plotting is included in the *Elements of Form Diagram* which is the standard method of summarizing the important information on ship's form and resistance. The principal dimensions and form coefficients are given in tabular form. The curve of areas, waterline and midship section for the design displacement are plotted non-dimensionally by dividing length dimensions by U and areas by U^2. To facilitate comparison of forms, the curves are drawn on a common base length and the ordinates of the curve of areas adjusted to keep the area under the curve unchanged.

The data from a large selection of models is assembled as plots of C against M for a range of K values. Generally, displacement and length are fairly well established by other design considerations so that for a given speed M and K are known. The best form of those already tested can be selected and the data for this form provides the first approximation to the resistance of the new design.

Table A.1
Froude 'constant' notation (sea water).

Parameter	Symbol	Expression British units (tonf, ft, knots)	Metric units (tonne, m, knots)
Characteristic length	U	$3.271\, \Delta^{\frac{1}{3}}$	$0.992\, \Sigma^{\frac{1}{3}}$
Length	M	$0.3057\, L/\Delta^{\frac{1}{3}}$	$1.0083\, L/\Sigma^{\frac{1}{3}}$
Breadth	B	$0.3057\, B/\Delta^{\frac{1}{3}}$	$1.0083\, B/\Sigma^{\frac{1}{3}}$
Draught	D	$0.3057\, T/\Delta^{\frac{1}{3}}$	$1.0083\, T/\Sigma^{\frac{1}{3}}$
Wetted surface	S	$0.0935\, S/\Delta^{\frac{2}{3}}$	$1.0167\, S/\Sigma^{\frac{2}{3}}$
Section area	A	$0.0935\, A/\Delta^{\frac{2}{3}}$	$1.0167\, A/\Sigma^{\frac{2}{3}}$
Speed	K	$0.5834\, V/\Delta^{\frac{1}{6}}$	$0.5848\, V/\Sigma^{\frac{1}{6}}$
	L	$1.055\, V/\sqrt{L}$	$0.5824\, V/\sqrt{L}$
Resistance	C	$\dfrac{427.1\,\text{e.h.p.}}{\Delta^{\frac{2}{3}} V^3}$	$\dfrac{579.7\, P_E}{\Sigma^{\frac{2}{3}} V^3}$

Note: The U.K. nautical mile, 6080 ft (1853.18 m) is greater than the international nautical mile which is 1852 m. P_E is the power in kW and Σ is the mass displacement.

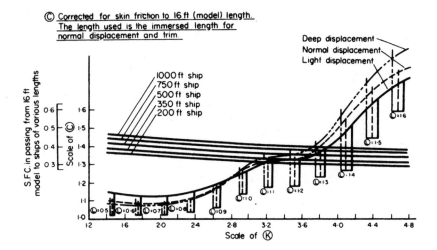

Ⓒ Corrected for skin friction to 16 ft (model) length.
The length used is the immersed length for
normal displacement and trim.

<table>
<thead>
<tr><th colspan="4">Unit model</th></tr>
<tr><th></th><th>Light</th><th>Normal</th><th>Deep</th></tr>
</thead>
<tbody>
<tr><td>Length constant</td><td></td><td></td><td></td></tr>
<tr><td>Draught   "</td><td></td><td></td><td></td></tr>
<tr><td>Breadth   "</td><td></td><td></td><td></td></tr>
<tr><td>Skin   "</td><td></td><td></td><td></td></tr>
<tr><td>Largest sec:area</td><td></td><td></td><td></td></tr>
<tr><td> "   " coeff^t</td><td></td><td></td><td></td></tr>
<tr><td>Prismatic coeff^t</td><td></td><td></td><td></td></tr>
</tbody>
</table>

<table>
<thead>
<tr><th colspan="4">Ship Dimensions</th></tr>
<tr><th></th><th>Light</th><th>Normal</th><th>Deep</th></tr>
</thead>
<tbody>
<tr><td>Length</td><td></td><td></td><td></td></tr>
<tr><td>Beam (max)</td><td></td><td></td><td></td></tr>
<tr><td>Draught (mean)</td><td></td><td></td><td></td></tr>
<tr><td>Displacement</td><td></td><td></td><td></td></tr>
<tr><td>Largest sec area</td><td></td><td></td><td></td></tr>
<tr><td>Prismatic coeff^t</td><td></td><td></td><td></td></tr>
<tr><td>Skin</td><td></td><td></td><td></td></tr>
<tr><td>L.C.B. Abaft F.P.</td><td></td><td></td><td></td></tr>
</tbody>
</table>

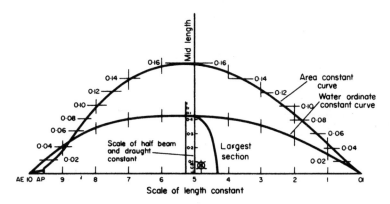

Fig. A.1 Elements of form diagram

Table A.2
O and f values. R. E. Froude's frictional data. Values at standard temperature
$= 15°C = 59°F$ *(British Units), f related to S in ft^2, V in knots, R in lbf.*

Length (ft)	O	f	Length (ft)	O	f
5	0.15485	0.012585	40	0.10043	0.009791
6	0.1493	0.012345	45	0.09839	0.009691
7	0.1448	0.012128	50	0.09664	0.009607
8	0.1409	0.011932	60	0.0938	0.009475
9	0.13734	0.011751	70	0.09164	0.009382
10	0.13409	0.011579	80	0.08987	0.009309
11	0.1312	0.011425	90	0.0884	0.009252
12	0.12858	0.011282	100	0.08716	0.009207
13	0.1262	0.011151	120	0.08511	0.009135
14	0.12406	0.011033	140	0.08351	0.009085
15	0.1221	0.010925	160	0.08219	0.009046
16	0.12035	0.010829	180	0.08108	0.009016
17	0.11875	0.010742	200	0.08012	0.008992
18	0.11727	0.010661	250	0.07814	0.008943
19	0.1160	0.010596	300	0.07655	0.008902
20	0.1147	0.010524	350	0.07523	0.008867
21	0.1136	0.010468	400	0.07406	0.008832
22	0.11255	0.010413	450	0.07305	0.008802
23	0.11155	0.010361	500	0.07217	0.008776
24	0.1106	0.010311	550	0.07136	0.008750
25	0.10976	0.010269	600	0.07062	0.008726
26	0.1089	0.010224	700	0.06931	0.008680
27	0.1081	0.010182	800	0.06818	0.008639
28	0.1073	0.010139	900	0.06724	0.008608
29	0.1066	0.010103	1000	0.06636	0.008574
30	0.1059	0.010068	1100	0.06561	0.008548
35	0.10282	0.009908	1200	0.06493	0.008524

Froude method

Using the 'constant' notation:

$$\textcircled{C}_F = \frac{1000 \text{ (frictional resistance)}}{\Delta \textcircled{K}^2}$$

$$= \frac{1000}{\rho g U^3} \times \frac{fSV^{1.825}}{4\pi V^2/gU}$$

$$= O . \textcircled{S} \textcircled{L}^{-0.175}$$

where

$$O = \frac{1000f}{4\pi\rho(gL/4\pi)^{0.0875}} = \text{'Circular } O\text{'}$$

Since for both model and ship

$$\textcircled{C}_T = \textcircled{C}_R + \textcircled{C}_F$$

$$[\textcircled{C}_T]_{\text{ship}} = [\textcircled{C}_T]_{\text{model}} - [O_m - O_s]\textcircled{S}\textcircled{L}^{-0.175}$$

Table A.3
f values. R. E. Froude's skin friction constants (Metric units)

Length (m)	f	Length (m)	f	Length (m)	f
2	1.966	11	1.589	40	1.464
2.5	1.913	12	1.577	45	1.459
3	1.867	13	1.566	50	1.454
3.5	1.826	14	1.556	60	1.447
4	1.791	15	1.547	70	1.441
4.5	1.761	16	1.539	80	1.437
5	1.736	17	1.532	90	1.432
5.5	1.715	18	1.526	100	1.428
6	1.696	19	1.520	120	1.421
6.5	1.681	20	1.515	140	1.415
7	1.667	22	1.506	160	1.410
7.5	1.654	24	1.499	180	1.404
8	1.643	26	1.492	200	1.399
8.5	1.632	28	1.487	250	1.389
9	1.622	30	1.482	300	1.380
9.5	1.613	35	1.472	350	1.373
10	1.604				

f in metric units $= f$ (Imperial units) $\times 160.9$.

$$R_F = fSV^{1.825}$$

$R_F =$ Frictional resistance, N
$S =$ wetted surface, m^2
$V =$ speed, m/s
$L =$ waterline length, m

f in metric units $= f$ in Imperial units $\times 47.87$ when V is in knots.

where O_m and O_s are the 'circular O' values for model and ship respectively.

In other words, the total resistance of the ship expressed non-dimensionally can be obtained from that of the model by making a correction which is dependent on the skin friction. For this reason, the term $(O_m - O_s)(S)(L)^{-0.175}$ is known as the *skin friction correction*.

To assist in applying the above method O and f values are tabulated in Tables A.2 and A.3 where:

Frictional resistance $= fSV^{1.825}$

and

$$\textcircled{C}_F = O \cdot (S)(L)^{-0.175}$$

The values are those agreed by the International Conference of Ship Tank Superintendents held in Paris in 1935 Values of $L^{-0.175}$ are tabulated in Table A.4.

If the model tests are carried out at other than $59\,°F(15\,°C)$, then the data have to be corrected for this by increasing or decreasing the $\textcircled{C}_F$ value by 2.4 per cent for every $10\,°F$ the temperature is below or above this value. Thus, if the experiments are conducted at a temperature of $t_1\,°F$

Table A.4
Value of $(L)^{-0.175}$

(L)	$(L)^{-0.175}$	(L)	$(L)^{-0.175}$	(L)	$(L)^{-0.175}$
0.00	∞	1.00	1.0000	2.00	0.8858
0.05	1.6892	1.05	0.9915	2.05	0.8819
0.10	1.4962	1.10	0.9835	2.10	0.8782
0.15	1.3937	1.15	0.9758	2.15	0.8746
0.20	1.3253	1.20	0.9686	2.20	0.8711
0.25	1.2746	1.25	0.9617	2.25	0.8677
0.30	1.2345	1.30	0.9551	2.30	0.8644
0.35	1.2017	1.35	0.9488	2.35	0.8611
0.40	1.1739	1.40	0.9428	2.40	0.8580
0.45	1.1500	1.45	0.9370	2.45	0.8549
0.50	1.1290	1.50	0.9315	2.50	0.8518
0.55	1.1103	1.55	0.9262	2.55	0.8489
0.60	1.0935	1.60	0.9210	2.60	0.8460
0.65	1.0783	1.65	0.9161	2.65	0.8432
0.70	1.0644	1.70	0.9113	2.70	0.8404
0.75	1.0516	1.75	0.9067	2.75	0.8378
0.80	1.0398	1.80	0.9023	2.80	0.8351
0.85	1.0288	1.85	0.8979	2.85	0.8325
0.90	1.0186	1.90	0.8938	2.90	0.8300
0.95	1.0090	1.95	0.8897	2.95	0.8275
				3.00	0.8251

$$[(C)_F]_{model} = [1 + 0.0024(59 - t_1)]O_m(S)(L)^{-0.175}$$

The wetted surface area used in determining (S) is taken as the mean wetted girth of sections multiplied by the length on the waterline. It is therefore less than the true wetted surface area as the inclination of the surface to the middleline plane of the ship is ignored.

Some authorities calculate the appendage resistance. This can introduce an error and this is allowed for in the QPC factor which is deduced from ship trials. The National Physical Laboratory introduced a scaling factor β by which the model appendage resistance can be multiplied, although they recommended that the value of β should generally be taken as unity.

In the absence of firmer data for a ship, the value of (S) can be obtained by applying the Haslar formula:

$$(S) = 3.4 + \frac{(M)}{2.06}$$

or the Taylor formula given in the main text.

Worked example

The (C)-(K) curve for a 16 ft model of a ship 570 ft long and 11,500 tonf displacement, corrected to standard temperature, is defined by the following table:

(K)	2.2	2.3	2.4	2.5	2.6	2.7	2.8
(C)	1.130	1.130	1.132	1.138	1.140	1.141	1.150

(K)	2.9	3.0	3.1	3.2	3.3	3.4
(C)	1.153	1.156	1.172	1.193	1.236	1.283

Deduce a plot of e.h.p. against speed for the clean condition assuming that (S) is given by the Haslar formula with a multiplying factor of 1.03. Also, calculate the e.h.p. for the ship 6 months out of dock assuming that the skin frictional resistance increases by $\frac{1}{4}$ per cent per day out of dock.

Solution: From the formulae quoted in the text the following relationships can be deduced.

$$(K) = 0.5834 \frac{V}{\Delta^{\frac{1}{6}}}$$

Hence, in this case, $V = 8.144\,(K)$ where V is in knots

$$(L) = 1.055 \frac{V}{\sqrt{L}} = \frac{1.055}{\sqrt{570}} V$$
$$U^3 = 11{,}500 \times 35 = 402{,}500\,\text{ft}^3$$

Hence $U = 73.83\,\text{ft}$.

Table A.5
Froude analysis

(K)	V (knots)	(C)	SFC	Clean ship		(L)	$(L)^{-0.175}$	$\delta(C)_F$	Dirty ship	
				(C)	e.h.p.				(C)	e.h.p.
2.2	17.9	1.130	0.371	0.759	5200	0.79	1.042	0.243	1.002	6770
2.3	18.7	1.130	0.368	0.762	5950	0.83	1.033	0.241	1.003	7810
2.4	19.5	1.132	0.366	0.766	6800	0.86	1.027	0.239	1.005	8890
2.5	20.4	1.138	0.363	0.775	7850	0.90	1.019	0.237	1.012	10,200
2.6	21.2	1.140	0.361	0.779	8800	0.94	1.011	0.236	1.015	11,500
2.7	22.0	1.141	0.357	0.784	10,040	0.97	1.005	0.234	1.018	13,000
2.8	22.9	1.150	0.354	0.796	11,400	1.01	0.998	0.232	1.028	14,700
2.9	23.7	1.153	0.351	0.802	12,800	1.05	0.992	0.231	1.033	16,500
3.0	24.4	1.156	0.349	0.807	14,050	1.08	0.987	0.230	1.037	18,000
3.1	25.2	1.172	0.347	0.825	15,750	1.11	0.982	0.229	1.054	20,150
3.2	26.1	1.193	0.345	0.848	18,000	1.15	0.976	0.227	1.075	22,800
3.3	26.9	1.235	0.343	0.892	20,800	1.19	0.970	0.226	1.118	26,000
3.4	27.7	1.283	0.341	0.942	24,000	1.22	0.966	0.225	1.167	29,600

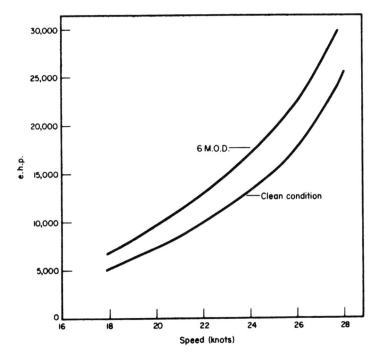

Fig. A.2 Plot of results

$$\textcircled{M} = \frac{L}{U} = \frac{570}{73.83} = 7.72$$

$$\textcircled{S} = 1.03\left[3.4 + \frac{\textcircled{S}}{2.06}\right] = 7.205$$

$$\textcircled{C}_F = O\,\textcircled{S}\,\textcircled{L}^{-0.175}$$

where, for $L = 570$, $O = 0.0711$

$$\therefore\quad \textcircled{C}_F = 0.0711 \times 7.205 \times \textcircled{L}^{-0.175}$$

$$= 0.512\,\textcircled{L}^{-0.175}$$

For the 16 ft model, $O = 0.1204$

and

$$\textcircled{C}_F = 0.867\textcircled{L}^{-0.175}$$

The reduction to be applied to the model $\textcircled{C}$ to obtain the ship $\textcircled{C}$ is 0.355 $\textcircled{L}^{-0.175}$

This is the skin friction correction, SFC.

$\delta(C)_F$ for the ship in the dirty condition $= 0.456(C)_F$

$$= 0.233(L)^{-0.175}$$

e.h.p. $= \dfrac{\Delta^{\frac{2}{3}}}{427.1}(C)_T V^3 = 1.193(C)_T V^3$ in this case

Table A.5 can now be constructed and the e.h.p./speed curves plotted as in Fig. A.2.

Bibliography

References are grouped by chapter, or group of chapters, with a final listing for general reading. Those references quoted will lead on to other useful references for the student to pursue.

The transactions, and conference proceedings, of the learned societies are among the main sources. Others are technical journals, research organisations and the universities. Much information on these sources can be gleaned from the Internet.

The following abbreviations have been used, with the letter T standing for Transactions:

RINA   The Royal Institution of Naval Architects (*INA* prior to 1960).
SNAME   The Society of Naval Architects and Marine Engineers.
DTMB   David Taylor Model Basin, USA.
IMO   International Maritime Organisation.
HMSO   Her Majesty's Stationery Office, now the Government Bookshop.

Chapter 10, Powering of ships: general principles;
Chapter 11, Powering of ships: application

Bertram, V. (2000) *Practical Ship Hydrodynamics*, Butterworth-Heinemann.
Breslin, J. P. and Andersen, P. (1994) *Hydrodynamics of Ship Propellers*, Cambridge University Press.
Carlton, J. S. (1994) Marine propellers and propulsion,
Froude, R. E. (1888) On the 'constant' system of notation, *TINA*.
Gawn, R. W. L. (1953) Effect of pitch and blade width on propeller performance, *TINA*.
Hadler, J. B. (1958) Coefficients for International Towing Tank Conference 1957
Lerbs, H. (1952) Moderately loaded propellers with a finite number of blades and an arbitrary distribution of circulation, *TSNAME*.
Model–ship correlation line, DTMB Report 1185.
Standard procedure for resistance and propulsion experiments with ship models, National Physical Laboratory Ship Division Report No 10.
Taylor, D. W. (1943) *The Speed and Power of Ships*. US Government Printing Office.
Re-analysed by Gertler, M. in David Taylor Model Basin Report 806 (1954).
Van Lammeren, W. P. A. et al. (1969) The Wageningen B-screw series, *TSNAME*.

Chapter 12, Seakeeping

Havelock, T. H. (1956) The damping of heave and pitch: a comparison of two-dimensional and three-dimensional calculations, *TINA*.
Lewis, F. M. (1929) The inertia of the water surrounding a vibrating ship, *TSNAME*.
Lloyd, A. R. J. M. and Andrew, R. N. (1977) Criteria for ship speed in rough weather, 18th American Towing Tank Conference.
Lloyd, A. R. J. M. (1998) Seakeeping. Ship behaviour in rough weather.
Newman, J. N. (1978) The theory of ship motions, *Advanced Applied Mechanics*.
Salvensen, N., Tuck, E. O. and Faltinsen, O. (1970) Ship motions and sea loads, *TSNAME*.

St. Denis, M. and Pierson, W. J. (1953) On the motions of ships in confused seas, *TSNAME*.
Schmitke, R. T. (1978) Ship sway, roll and yaw motions in oblique seas, *TSNAME*.

Chapter 13, Manoeuvrability

Burcher, R. K. (1991) The prediction of the manoeuvring characteristics of vessels, *The dynamics of ships, Proceedings of the Royal Society, London*.
Dand, I. W. (1981) On ship-bank interaction, *TRINA*.

Chapter 14, Major ship design features

Ware, H. D. (1988) Habitability in surface warships, *TRINA*.

Chapter 15, Ship design

Carreyette, J. (1977) Preliminary ship design cost estimates, *TRINA*.
Friedman, N. (1979) *Modern Warship Design and Development*, Conway.
Goss, R. O. (1982) *Advances in Maritime Economics*, UWIST Press.
Rawson, K. J. (1989) Ethics and fashion in design, *TRINA*.
Reliability of system equipment, *British Standard 5760, BSI*.
Van Griethuysen, W. J. (1993) On the choice of monohull warship geometry, *TRINA*.

Chapter 16, Particular ship types

Brown, D. K. and Tupper, E. C. (1989) The naval architecture of surface warships, *TRINA*.
Claughton, A. R., Wellicome, A. J. F. and Shenoi, R. A. (1998) *Sailing Yacht Design: Theory* (Vol. 1), *Practice* (Vol. 2), Addison, Wesley and Longman.
Dawson, P. (2000) *Cruise Ships*, Conway Maritime Press.
Dorey, A. L. (1989) High speed small craft, *TRINA*.
Kaplan, P. et al. (1981) Hydrodynamics of SES, *TSNAME*.
Patel, M. H. (1989) *Dynamics of Offshore Structures*, Butterworths.
Pattison, D. R. and Zhang, J. W. (1994) The trimaran ships, *TRINA*.

General

Merchant Shipping (Crew accommodation) Regulations, HMSO.
Merchant Shipping (Dangerous Goods) Regulations, HMSO.
Merchant Shipping (Fire Appliances) Regulations, HMSO.
Merchant Shipping (Grain) Regulations, HMSO.
Merchant Shipping (Life Saving Appliances) Regulations, HMSO.
Merchant Shipping (Passenger Ship Construction and Survey) Regulations, HMSO.
Merchant Shipping (Tonnage) Regulations, HMSO.
International Maritime Dangerous Goods Code, IMO.
MARPOL. Regulations and Guidelines, IMO.
SOLAS, Regulations and Guidelines, IMO.

Bishop, R. E. D. and Price, W. G. (1979) *Hydroelasticity of Ships*, Cambridge University Press.
Friedman, N. (1979) *Modern Warship Design and Development*, Conway.

Kuo, C. (1998) *Managing Ship Safety*, LLP Ltd.

Nishida, S. (1992) *Failure Analysis in Engineering Applications*, Butterworth-Heinemann.

Schneekluth, H. and Bertram, V. (1998) *Ship Design for Efficiency and Economy*, Butterworth-Heinemann.

Taylor, D. A. (1996) *Introduction to Marine Engineering Revised*, Butterworth-Heinemann.

Tupper, E. C. (2000) *An Introduction to Naval Architecture*, Butterworth-Heinemann.

Watson, D. G. M. (1998) *Practical Ship Design*, Elsevier Ocean Engineering Book Series.

Significant Ships. Annual publication of the RINA reviewing some of the ships entering service.

Some useful web sites

Much useful data can be gleaned from the Internet. As an example, the RINA makes available all its technical papers which are issued for discussion. Other sites give details of the facilities at various research establishments. The Lloyds Register site gives information on the software they have available.

The following are some sites the student may find helpful. Many others relating top shipbuilders and equipment manufacturers, are regular contained in the advertisements in technical publications, such as *The Naval Architect* which is the journal of the RINA.

Learned societies

Royal Institution of Naval Architects	*www.rina.org*
Society of Naval Architects and Marine Engineers, USA	*www.sname.org*
Institute of Marine Engineers	*www.imare.org.uk*
The Nautical Institute	*www.nautinst.org*

International and government organizations

International Maritime Organisation	*www.imo.org*
Dept. of the Environment, Transport and the Regions	*www.detr.gov.uk*
UK Maritime & Coastguard Agency	*www.mcagency.org.uk*
US Coast Guard	*www.uscg.mil*
Defence Evaluation and Research Agency (UK)	*www.dera.gov.uk*
MARIN (Netherlands)	*www.marin.nl*
David Taylor Model Basin (USA)	*www50.dt.navy*

Classification Societies

International Association of Classification Societies	*www.iacs.org.uk*
American Bureau of Shipping	*www.eagle.org*
Bureau Veritas	*www.veristar.com*
China Classification Society	*www.ccs.org.cn*
Det Norske Veritas	*www.dnv.com*
Germanischer Lloyd	*www.GermanLloyd.org*
Korean Register of Shipping	*www.krs.co.kr*
Lloyds Register of Shipping	*www.lr.org*
Nippon Kaiji Kyokai	*www.classnk.or.jp*
Registro Italiano Navale	*www.rina.it*
Russian Maritime Register of Shipping	*www.rs-head.spb.ru*

Answers to problems

Chapter 10

1. 781 N, 1.63 m/s, 4.05×10^5.
2. 24.25, 27.64, 15.96, 12.37, 10.45 knots.
3. 34.2 knots, 19.7 knots.
4. 36 knots.
5. 0.524, 0.582, 13.4 MW.
6. 10.29 MW, 26.8 MW.
7. 28.14 knots.
8. 10.00, 12.11, 14.07 knots.
9. 28.257, 28.381, 28.370, 28.418, 28.319 knots.
10. 28.381 knots; 0.789, −0.453, 0.098, 0.034, −0.008.
11. 9.68 tonnef (94910 N).
12. 2.58 MW, 0.97, 0.14.

Chapter 11

1. 12.54 knots.
2. $L = 233$ m, $B = 21.7$ m, $T = 7.2$ m, $\Delta = 200$ MN, 22.9 MW at 30 knots.
3. —.
4. 14 knots, 3.16 MW.
5. 35 per cent, 32.4 per cent.
6. 7.78 MW, 9.42 MW, $L = 150$ m, $B = 24$ m, $T = 9$ m.
7. 28 knots.
8. 14.9 MW, 44 MW
9. 4.36 m, 17.9 MW, 0.56, 4.88 m, 229 r.p.m.
10. 2.77 m, 0.716, 3.60 m.
11. 21700 N/m^2, 769 r.p.m.
12. 550 r.p.m., 33.7 N, 103 Ncm
13. 0.60, 0.961.
14. 12.14 tonnef.
15. 95.4 newtons.
16. 448 r.p.m., 0.12, 0.110, 1.009.
17. 0.70, 2.45 m/sec.
18. 0.04, 0.0458, 0.66, 28.8 MW.
19. 27.8 knots, 26.5 knots.

Chapter 12

1. 3.39 m, 76.5 m, 10.93 m/s.
2. 8.69, 3.61, 3.47 secs.
3. $91\frac{1}{2}$ degrees, $120\frac{1}{2}$ degrees.
4. $a{:}b{:}c$ as $1{:}\ \pi/2 : 4/\pi$.

5. 10.07 m.
6. 13.16 tonnef.
7. 3.71°, 5.00°, 0.93°, 16.5 s.
8. Heave = 0.41 (wave height) at $\omega_E = 0.60$.
9. $a = 0.083$, $b = 0.0079$; 3.62°.
10. —.
11. 8 degrees approx., 33.6 kW.
12. 5.56.
13. $a = 0.12$, $b = 0.015$.
14. 2.03 m^2 s.
15. —.

Chapter 13

1. 0.222 MN m.
2. $3\frac{1}{2}$ degrees.
3. 1.96 m.
4. 1.25 MN, 1.4 m.
5. 0.67 MN, 0.184 MN m.
6. 1.19 MN, -0.194 MN m.
7. —.
8. 16 tonnef·m, 237 tonnef/m.
9. 2.540 m^2, 2.127 m, 3.403 m^4, 0.833 m^4.
10. $\pm 29°20'$.
11. 55.2, 47.3, 41.4 m^2.
12. 14.8 MN m.
13. A stable, B unstable; 19.4 m, 38.5 m.
14. 40.2 m, 70.7 m.
15. 40 m, stable, 2.38 knots.

Chapter 14

11. 328 kN/m^2, 196 kN/m^2.
12. 112.7 MN/m^2 and 470 kW, 1.97 MN/m^2 and 8.2 kW.
13. 1046 litres/min at 0.99 MPa; 23.2 kW for smooth pipes, 1046 litres/min at 1.06 MPa; 25.6 kW for rough pipes.
14. 2110 J/s, 940 J/s.
15. 10,700 J/s.
16. Approx. 7 °C, about 25 cc.
17. About 4.65×10^3 J/kg from 16.6/15.6 to 21/17.
18. Slope 0.49, off Coil 14/13, mix 30/23.6; 1.54 m^3/sec. (including fresh air) is one solution.
19. Approx. 1020 m^2.
20. (a) 64 per cent, (b) 57 per cent approx.
21. 30000 GT, 23232 NT, (a) the same, (b) 23307.

Index